Future Scenarios of Land Use and Land Cover Change

Future Scenarios of Land Use and Land Cover Change

Guest Editors

Rob Marchant
Jessica Thorn
Rebecca Kariuki

MDPI

Basel • Beijing • Wuhan • Barcelona • Belgrade • Novi Sad • Cluj • Manchester

Guest Editors
Rob Marchant
York Institute for Tropical Ecosystems
University of York
York
UK

Jessica Thorn
Centre for Environmental Policy
Imperial College London
London
UK

Rebecca Kariuki
School of Sustainability
Arizona State University
Tempe, AZ
USA

Editorial Office
MDPI AG
Grosspeteranlage 5
4052 Basel, Switzerland

This is a reprint of the Special Issue, published open access by the journal *Land* (ISSN 2073-445X), freely accessible at: https://www.mdpi.com/journal/land/special_issues/3N3796V6ZR.

For citation purposes, cite each article independently as indicated on the article page online and as indicated below:

Lastname, A.A.; Lastname, B.B. Article Title. *Journal Name* **Year**, *Volume Number*, Page Range.

ISBN 978-3-7258-6404-1 (Hbk)
ISBN 978-3-7258-6405-8 (PDF)
https://doi.org/10.3390/books978-3-7258-6405-8

Cover image courtesy of Rob Marchant
Fields of wheat with nitrogen fixing Acacia trees on the road to Narok, Kenya

Contents

Preface

This Special Issue collates recent advances in the application of land use scenario tools to explore future land use and land cover change trajectories across a range of spatial scales and thematic foci. The emphasis is on innovative and transformative adaptation pathways. The nine manuscripts examine (i) the drivers and impacts of land use and land cover change alongside other drivers of environmental change; (ii) interactions between development and environment processes; (iii) participatory approaches to land use and land cover change analysis; (iv) the interplay between institutions, technologies, and cultural practices; and (v) the barriers and contingency factors that enable or hinder the acceptance and adoption of scenario outcomes. Collectively, these studies demonstrate how land use scenarios go beyond linear projections, accounting for deep uncertainty, enabling decision-making to critically explore diverse and plausible futures in a broad possibility space. Engaging with complexity, uncertainty, and contested values will help anticipate risks, adapt strategic assumptions, and guide more effective responses to navigate the future in the coming years, decades, and beyond.

Rob Marchant, Jessica Thorn, and Rebecca Kariuki
Guest Editors

land

MDPI

Article

Past and Future Land Use and Land Cover Trends across the Mara Landscape and the Wider Mau River Basin, Kenya

Evans Napwora Sitati [1,2,*], Siro Abdallah [3], Daniel Olago [1,2] and Robert Marchant [4]

1 Institute for Climate Change and Adaptation, University of Nairobi, Nairobi P.O. Box 30197-00100, Kenya; dolago@uonbi.ac.ke
2 Department of Earth and Climate Sciences, University of Nairobi, Nairobi P.O. Box 30197-00100, Kenya
3 Conservation Department, Spatial Planning Unit, WWF-Kenya, Nairobi P.O. Box 62440-00200, Kenya; sabdalla@wwfkenya.org
4 York Institute for Tropical Ecosystems, Department of Environment and Geography, University of York, York YO10 5DD, UK; robert.marchant@york.ac.uk
* Correspondence: sitatievans@students.uonbi.ac.ke or ens514@york.ac.uk or esitatinapwora@gmail.com

Abstract: The Maasai Mara and the wider Mau River Basin in East Africa provide fundamental ecosystem services that support people, wildlife, livestock and agriculture. The historical indigenous land use of the Mara and wider Mau basin was wildlife conservation and pastoralism with highland agriculture. However, land policy changes, the rise of community conservancies and the increase in human populations have mediated unprecedented land use shifts over time. We analyze land use and land cover change (LULCC) trends from 1990 to 2040 in the Mara and the wider Mau River Basin landscape. The study examines land use and land cover change trends, establishes factors driving the trends, and assesses the implications of these trends on biodiversity. Multi-temporal satellite images, together with physical and social economic data, were collated to generate future scenarios for transitions for forest, shrubland, grassland, cropland, wetlands and built-up areas between 1990 and 2040. Agricultural expansion is the chief driver of LULCC in the Mara and the wider Mau River Basin, particularly since 2015. There was insignificant change to the forest cover after 2015, which was in part due to government intervention on forest encroachment and boundaries. The anthropogenic choice of tilling the land in the basin caused a decline in grasslands, forests and expanded shrublands, particularly where there was clear tree cutting in the Mau forest. Land use and land cover trends have generated undesirable impacts on ecosystem services that support wildlife conservation.

Keywords: land cover change; forests; grassland; shrubland; agriculture conservation

1. Introduction

Land is a repository of natural capital where forests, wetlands, grasslands and mountains support life and development [1], and it is an interface where human activities, ecological processes and the environment interact [2,3]. Land use creates habitats and shapes land cover, which subsequently modifies habitat types and species behavior [4]. The conversion of land from one specific cover type to another, land use and land cover change (LULCC), is driven by people who want to meet their social, cultural and spiritual needs [5]. Anthropogenic activities have influenced and modified land cover for thousands of years with change accelerating in recent centuries, particularly due to social economic developments [6], population growth [7] and changing climates [8]. This has accelerated biodiversity loss and made it challenging to support people, livelihoods and ecosystems [9]. The African Elephant (*Loxidonta africana*) population, for instance, has declined [10,11], while the Mountain Bongo (*Tragelaphus eurycerus* ssp. *Isaaci*) in the Mau forest is massively impacted and highly threatened [10].

People interacting with environmental processes [11], market trends operating at regional and international scales [12], global historical trends [13], technological advancements, institutions, local and global policies [14] have all driven and shaped LULCC [15]. These are mediated by local policies and acts; for example, Kenya's Sessional Paper, No. 1 of 2017 on National Land Use Policy [16], grants authority and liberty for use of land for any purpose. The Kenya Land Act 2012 (under review) permits the conversion of land from one category to another within the precepts of the law [17]. However, in most cases, these legal frameworks are often more of a guidance than providing a detailed planning framework. In circumstances where the law is followed, politics and local power relationships usually influence the process because of individual interests. A particular example is the Mau forest, where politicians and influential citizens allocated themselves huge chunks of land and incited illegal settlers to resist eviction from the forest [18] at the expense of the forest ecosystem. The forest excision from 1989 to 2009 and blatant human encroachments resulted in the forest ecosystem undergoing the conversion of an estimated 25% (107,000 ha) to agriculture and other land uses [19].

Research has provided insights into the relationships between human activities, rainfall, soil fertility, wildlife, livestock grazing, and deforestation across Eastern Africa [20] over decadal and centennial scales [21,22], particularly with respect to pastoral communities and their links to the conservation of rangelands and wildlife [23]. Anthropogenic pressures are affecting the diverse composition of fauna and flora, thus influencing species interactions, and in extreme cases driving multiple species extinction [24]. LULCC dynamics across eastern Africa have particularly been shaped by rapid population growth that has amplified demands for food, shelter and incomes, leading to an increased ecological isolation of protected areas [25]. These trends have been exacerbated by large-scale land use changes influenced by foreign and local interests, particularly agricultural expansion and wood extraction for local and international consumption [26,27].

Although protected areas are vital for biodiversity conservation, they do not provide complete protection to biodiversity nor completely resolve conservation development-related conflicts [28]. Often, the effects of landscape transformations permeate into the protected areas [28], hence affecting species behavior, abundance and composition as well as ecosystem processes and habitat integrity. More often, nature suffers collateral damage from economic and development activities. LULCC resulting from direct human activities such as settlement [29], livestock grazing [30–32], the harvesting of wild plants and animals and logging impacts have driven species declines, particularly in areas where inhabitants depend on natural resources to support their livelihoods [29].

The Mau-Mara ecosystem in southwestern Kenya comprises the Masai Mara National Reserve (MMNR), neighboring conservancies (Enoonkishu, Lemeki, Ol Chorroua, Oloisukut, Mara North and Mara Conservancy) and the Mau forest (North) and is characterized by diverse land uses. The vast rangeland supports 30% of Kenya's wildlife population [33]. The national reserve, owned and controlled by the County Government of Narok, is exclusively demarcated for wildlife tourism and conservation. The adjacent group ranches and land extending to the northern forested highlands are owned privately or communally and have multiple land uses, ranging from pastoralism, small-scale farming, mechanized farming and wildlife tourism. This landscape also lies on the borders with the Serengeti National Park (SNP) in Tanzania, thus allowing wildlife from Tanzania to migrate and occupy the MMNR and the adjoining conservancies. The resident wildlife species also migrate between the reserve and the connecting dispersal areas. The free ranging migrations by wildlife demonstrate the viability of migratory space for wild species in the landscape and that what happens in the contiguous conservancies has a direct influence on wildlife in the protected areas. A significant number of wildlife (84%) are found in rangelands outside the MMNR [34]. However, the unprotected dispersal ranges are being increasingly transformed by agro-pastoral human communities. Expanding commercial farming [31,35], tourism and other human activities on land adjacent to the MNNR are threatening the sustainable coexistence of pastoral communities with wildlife [36].

Most studies in the Mara landscape have focused on tourism and on specific species declines (e.g., elephants, cheetahs, lions, vultures). Specifically, little research has been conducted to quantify and provide insights into the spatial and temporal dynamics of LULCC at the landscape level. We used satellite imagery software to model and construct the future of the Mau Mara basin in temporality for forests, grasslands, shrubland, agriculture and built-up areas. This study analyzes long-term trends of LULCC in the wider Mau-Mara landscape, extending from the Mau highlands through the Mara catchment to the MMNR. The study establishes where the changes are occurring, examines the potential drivers of these changes and recommends mitigation strategies applicable to reduce the impacts of the changes on the wildlife, people and livestock.

2. Materials and Methods

2.1. Study Area

The study area (Figure 1) was restricted to the Mau-Mara landscape in Narok County. The Mara River Basin covers 8992 km^2 extending 160.8 km north to the upper Mau forest blocks (Transmara, Olposimoru, eastern Mau and south western Mau) and 124.2 km eastward across the MMNR and adjacent rangelands. The basin intersects three counties: Narok, Bomet and Nakuru with over a 60% stretch located in Narok County. The Narok County population in 2019 was 1,157,873; Nakuru 2,162,202 while Bomet County was 875,689 [37]. The Mara landscape (from Transmara to Loita Forest area) is situated in the Narok West subcounty that covers 5452 km^2 with an estimated population of 195,287 [37].

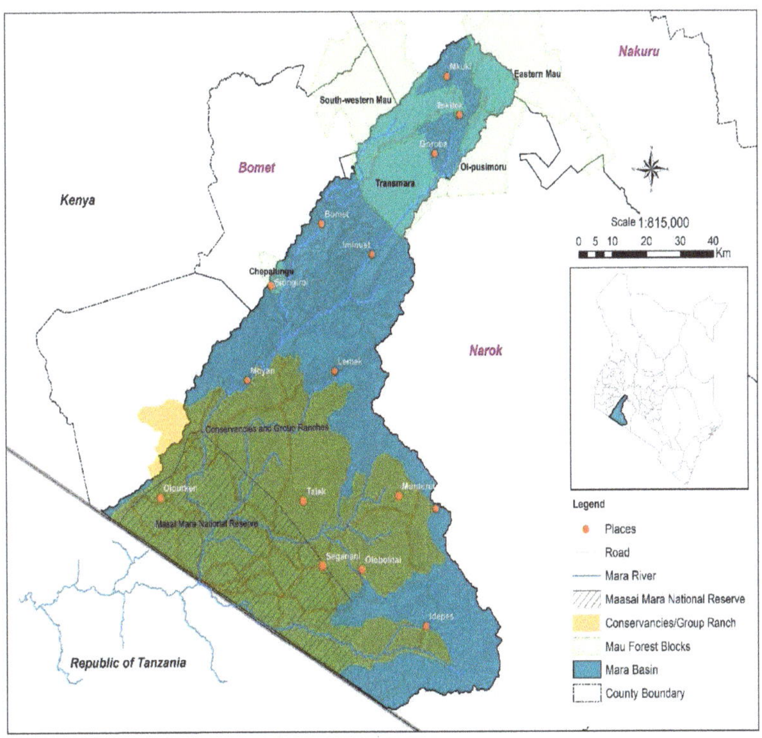

Figure 1. Map of Mara River Basin, Masai Mara National Reserve and the surrounding conservancies, group ranches and the Mau forest blocks and built-up areas.

Rainfall is bimodal (the wet season beginning from June to November and dry season from July to October) with the short rains falling during November–December and the long rains from March to June [38]. Annual rainfall ranges from 800 to 1200 mm with a northwest to southeast declining gradient [36]. The vegetation is predominantly open grass plains and savanna woodlands [38] with undulating ranges characterized by savanna/woodlands and forest patches southward and northward, respectively. The Mara River flows through the MMNR, which is characterized by the spectacular annual migration of wildebeest (*Connochaetes taurine*), zebras (*Equus burchelli*), and Thomson's gazelle (*Gazella thomsoni*) from the SNP (Tanzania) to MMNR (Kenya) and dispersing to the nearby conservancies.

The Ewaso Ny'iro, Sondu, Mara and Njoro are the main rivers that discharge their waters from the Mau complex to Lakes Natron, Victoria and Nakuru, respectively [39]. Most of the land in the Mau Mara landscape forms part of the unprotected area that surrounds the MMNR and is characterized by a mixture of private and communally owned land. Historically, land was held in a trust for the semi-nomadic Maasai community by the government, and some areas in the east of the landscape still retain this arrangement. Since the 1970s, trust lands were converted into group ranches under local administration [40]. Subdivisions of these group ranches into parcels of privately owned land has embedded significant land transformations including mechanized farming and tourism businesses through conservancy establishments in the north [40].

The Narok County is home to 30% of Kenyas wildlife concentrations [28]. MMNR (1510 km^2) forms a small part of a much larger protected area complex (including the SNP, Ngorongoro Conservation Area, and a number of game reserves), which together cover over 25,000 km^2 [41] (Figure 2). Most of the wildlife is in the Greater Mara Ecosystem that has an estimated 6600 km^2 stretch, consisting of the MMNR and the surrounding conservancies. Voluminous wildlife concentrations sighted in this ecosystem depend on the Mara River flows, its tributary rivers (Talek and Sand Rivers) and springs to supply it with waters.

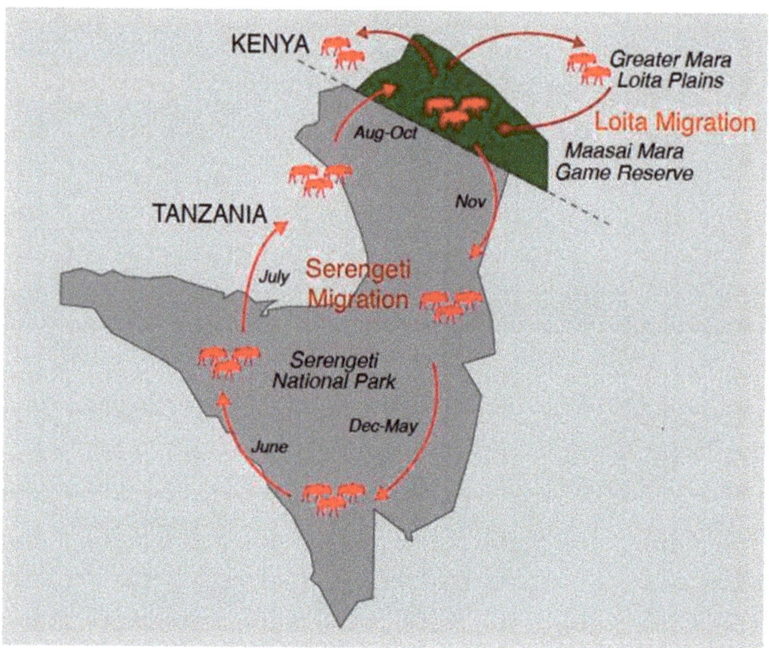

Figure 2. Wildebeest migration in and outside the Serengeti in the Greater Mara Ecosystem.

2.2. Land Use and Land Cover Mapping

This study integrates information from multispectral remote sensing, modeling, socioeconomic datasets and experience on implementing integrated management programs in the Mara rangelands. The Systems for Land-based Emissions Estimation in Kenya (SLEEK) program has been producing annual land cover products from 1990 to 2015 based on the multispectral remote sensing of a moderate resolution (30 m) Landsat imagery series

(Landsat 5, 6, 7 and 8). The land cover class defines six natural vegetation classes, three human-developed classes and one non-vegetated land use and land cover class based on IPCC guidelines for UNFCCC reporting with Kenya's local definition [42]. To focus on cropland, forest, grassland, urban/rural settlements, shrubland and wetland LULCC, we modified SLEEK's 1990, 2000 and 2015 classified land use and reduced the number of classes from 10 to 7 (Table 1), collapsing the dense, moderate and open forests into one forest class, annual and perennial croplands to one cropland class and introduced urban/rural settlements. The study applied remote sensing to model socioeconomic consequences of LULCCs focusing on cropland, forest, grassland, urban/rural settlements, shrubland and wetlands from 1990 to 2040.

Table 1. List of land use/cover classes.

Code	Land Cover Class	Definition
C	Cropland	Land purposely managed for agricultural activities. It is land characterized by the presence of crops with evidence of tillage.
F	Forestland	Areas occupied by forests are characterized by tree grown cover $\geq$ 15%, an area $\geq$0.5 ha and a tree height $\geq$ 2 m. Includes areas managed for forestry where trees have not attained 2 m height but with potential to do so, and areas that are temporarily destocked.
OG	Grassland	Open grasslands, herbaceous cover, woody vegetation < 10%. Rangelands and pasture land which do not qualify as forestland and cropland. The category includes grassland in wildlands, shrub savannah, in managed and unmanaged systems.
S	Shrubland	Vegetated grasslands are lands with woody vegetation with a height less than 2 m. The total percent shrub cover exceeds 30%. The shrub foliage can be either evergreen or deciduous.
UR	Urban/Rural Settlement	This category includes a mosaic of scattered rural settlements and defined structured built setup in urban areas.
VW	Wetland	Vegetated wetland, permanent mixture of water and herbaceous or woody vegetation.
W	Waterbody	Open waters (oceans, seas, lakes, reservoirs, and rivers).

2.3. Accuracy Assessment

Accuracy validation was assessed with an error matrix and kappa coefficient [43]. This approach produced the overall accuracy of the classified image by comparing how each of the pixels classified versus the definite land cover conditions obtained from their corresponding ground truth data.

2.4. Modeling Projected Land Use/Cover for 2040

The Methods of Land Use Change Evaluation (MOLUSCE) plugin in QGIS was used to simulate future land use land cover scenarios for 2040. The selected plugin was used because it provides a set of algorithms for land use change simulations such as Artificial Neural Networks (ANNs), Linear Regression (LR), and Multi-Criteria Evaluation (MCE). The plugin uses a classic realization of multilayer perceptron. The Artificial Neural Network (ANN), a biologically inspired computational network that uses multilayer perceptrons with backpropagation learning algorithms based on a supervised procedure comprising of three layers: input, hidden, and output [44], was applied. This algorithm was selected since it is deemed effective in predicting the quantitative changes in land use types and related transitions while providing capabilities to simulate the spatial and temporal evolution of land use at greater accuracies [45]. The aggregated neurons were instructed to perform particular transformations with the signals passing through the multiple layers to generate predictive modeling. The input data are the LULC 1990 and LULC 2015 layer with the target output LULC 2040 created using the following transformation.

$$(C - 1)\,(2N + 1)^2 + B\,(2N + 1)^2 \text{ input neurons;}$$

$$w(n+1) = r \times dw(n) + m \times dw(n-1)$$

$$E = \frac{t_i - o_i}{d}$$

M output neurons (it depends on sampling mode); in this case, the sampling mode was 25 years hence 2040 forward projection;
C is the count of land use categories
N is the neighborhood size specified by the user,
B is the summary band count of factor rasters (for example, if two factors are one band and three band, then $B = 4$).
M depicted a count of unique categories in the change map (C2).

The module uses a classic backpropagation algorithm with momentum for the learning procedure. Weights correction are performed as

where w is a vector of neuron weights
dw is a vector of weight changes
n is an iteration number
r is learning rate
m is momentum.

The training set is divided in two parts: the learning set (80% of samples by default) and a validation set (20% of samples). The module uses online learning with stochastic integration: where a random sample is selected from the learning set, the weights of the net are updated during forward/backward propagation. An error of fitting (for a sample) is the average square error of partial outputs of the net:

E is a sample error
t_i is the target value of an output neuron for a given sample
o_i is the real output value of the neuron
d is the count of output neurons.

The 2015 simulation was validated against the 2015 land cover to test the accuracy and correctness of the prediction, which also implied 2040's prediction possibility. An overall percentage of correctness of 93.15 and overall Kappa of 0.89 were reached for the 2015 simulation.

2.5. Land Use/Cover Change Detection and Analysis

A post-classification change detection method was employed in performing LULCC detection. Pixel-based comparison was used to generate per-pixel change information and thus relate the changes more efficiently, taking advantage of the "-from, -to" information. Classified image pairs were compared based on cross-tabulation in order to derive qualitative and quantitative aspects of the changes for the periods 1990 to 2000, 2000 to 2015, 2015 to 2040 and most significantly long-term changes from 1990 to 2015 and 1990 to 2040. LULCC gains and losses per class/category were then compiled using a land change modeler extension for ArcGIS version 10.6 Arcmap. Mapping and spatial extension tool boxes were used to run analysis and calculations for respective areas.

Predictive data analysis was then applied to comprehend the contemporary trends based on the available historical facts to deduce conclusions that generate predictions about the future trends through information technology and landscape modeling [38] of LULCC in the Mau-Mara landscape. This predictive modeling used the QGIS land use modeler add-on to establish the likely future drivers that might instigate LULCC as elaborated in Section 2.4. Exploratory analysis was applied to understand the connectedness of factors driving LULCC in the Mau-Mara ecosystem in order to generate an understanding of future developments [46].

3. Results

3.1. Current Baseline and Projection Timeline

Our current baseline year is 2015, while the projected timeline is 2040. It falls within an era when the forest cover change was evident because of deforestation when analyzing classified images from the year 1989 to 2020 [47,48]. The results were obtained after analyzing multi-temporal satellite imageries and classifying LULCCs in different categories thus illustrating major vegetation changes. A pre-analysis of pixel-based comparison classification reports between the years showed that the classifier hit remarkable consistency registering overall land use/cover classification levels for the three epochs ranging from 75% to 89% (Table 2), satisfying the minimum accuracy for satellite-derived land use/cover maps for this study [29].

Table 2. Summary of land use/cover classification accuracies for 1990, 2000 and 2015. The producer accuracy is based on the classification point of view, while the user accuracy shows the reality on the ground based on field validation.

Land Use/Cover	1990		2000		2015	
	Producer's %	User's %	Producer's %	User's %	Producer' %	User's %
Cropland	88.7	92.8	78.0	72.7	74.2	78.3
Forestland	86.5	86.1	73	61.3	77.0	71.3
Grassland	78.8	89.7	72.9	81.6	73.8	69.8
Shrubland	94.7	87.3	90.5	89.4	76.2	78.2
Urban/Rural Settlement	92.1	92.1	83.5	90.1	82.8	80.8
Wetland	100	50.0	75	100	77.5	72.5
Waterbody	97.7	97.7	100	100	80	83.7
Overall accuracy	89.5%	–	84.6%	–	75.6%	–
Kappa statistic	**0.8510**	–	**0.7828**	–	**0.7025**	–

Agriculture in particular increased slightly (Table 2) particularly in the Mau area, which has a forest and is surrounded by small and large-scale agricultural farms. In the Mara region, the conservancies act as buffer areas against the expansion of agriculture from Mulot (Bomet County), Olulung'a and Transmara (County) urban areas. The conservancies also prohibit cropland expansion in the Mara landscape as they shield the Mara River from water abstraction, thus denying irrigated agriculture. Culturally, the livelihood of the Maasai community is also more oriented to pastoralism than crop farming, which is perhaps the reason why there is a low expansion of agriculture in the Mara.

3.2. Overall Land Use/Cover Trends in Mara Basin

Dominant categories as of 1990 were open grasslands (54.1%), shrubland (19.8%), forestland (14.4%) and cropland (11.7%), respectively. However, a significant shift in the landscape was registered by the year 2015 as open grasslands declined to 49.4%, shrubland expanded to 27.4%, and croplands increased to 15%. Forests sharply declined to 8.4% (Table 3) because of clear logging on private lands for agriculture and commercial purposes. The projected LULCC scenario for 2040 (as illustrated in Section 2.4 modeling projected land use/cover for 2040) assuming business as usual (no efforts are employed to address LULCC and their impacts) predicts a continuous decline in grasslands (47.1%), forest (7.1%), and a minimal increase in croplands (15.1%). Shrubland is expected to grow to 30.5% by 2040, while settlements are expected to register an increase to 0.1% from 0.002% in 1990.

Urban and rural settlement areas expanded from 13 ha in 1990 to 74.49 ha in 2000, and 505 ha in 2015 with a projection to attain 1009 ha by 2040 (Table 3). This is similar to the shrubland ecosystem that increased from 163,240 ha (19.8%) in 1990 to a predicted 251,298 ha (31%) by 2040. Wetlands are predicted to increase from 43 ha (0.01%) in 1990 to 52 ha in 2040.

An estimated 370 inhabitants were removed from the forest by the government (Figure 3) between 2010 and 2015. Data for this study were conducted in 2019 when people had been removed from the forest habitat, which is the reason why there is a negative. The decision to evict illegal settlers was to reverse the lost 10,000 ha of forest land through logging and 2000 ha in grasslands that had been converted to agriculture. On the contrary, shrubland and crop land gained an estimated acreage of 60,000 ha and 50,000 ha, respectively. The temporal trends experienced in the basin over the three periods point to significant changes in three vegetative categories: shrubland; forestland; grassland and one land use category: cropland. The shrubland category registered the biggest expansion gaining over 88,058 ha, followed by croplands, which gained an estimated 28,091 ha over the same period (Table 4). Alarmingly, 59,840 ha of forests was lost, followed closely by grasslands, which lost over 57,260 ha. Table 4 provides the rates of change for each of the four categories:

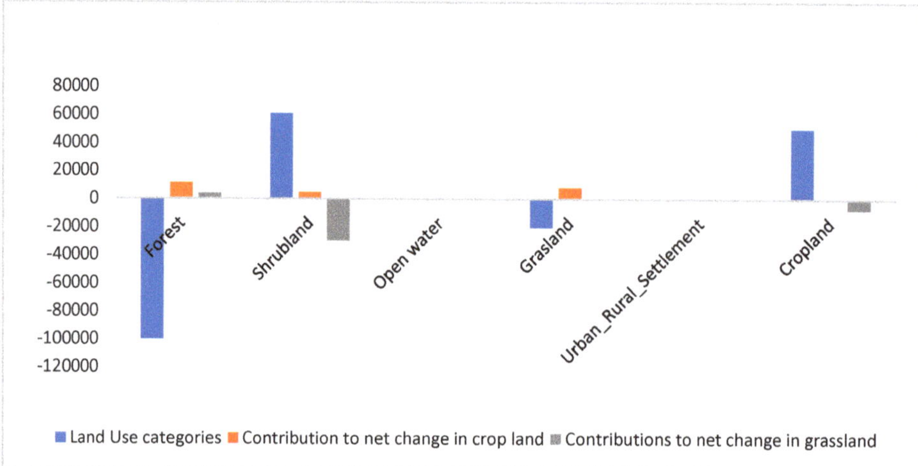

Figure 3. Overall land use/cover change in the Mara basin (1990 to 2015).

Table 3. Area and amount of change in different land use/cover categories in the Mau-Mara River Basin.

Years	1990		2000		2015		2040	
Land Use Cover Categories	**Ha**	**%**	**Ha**	**%**	**Ha**	**%**	**Ha**	**%**
Cropland	96,543	11.7	110,272.1	13.4	123,322	15	124,635	15
Forest	118,587	14.4	79,716.19	9.7	69,104	8.4	58,747	7
Grassland	445,534	54.1	428,306.5	52	404,985	49.2	388,273	47
Open Water	54	0.01	51.8	0.01	56	0.01	0	0
Shrubland	163,240	19.8	205,592.7	25	226,048	27.4	251,298	31
Urban/Rural	13	0	74.5	0.01	505	0.1	1009	0.2
Wetland	43	0.01	0.060729	0	0	0.01	52	0.01
Total	**824,014**	**100**	**824,013.9**	**100**	**824,020**	**100**	**824,014**	**100**

Table 4. Major vegetation changes in Mara basin (1990–2040).

Land Use/ Cover Category	1990–2000		2000–2015		1990–2040		Annual Rate Gain/Loss Current Projected		
	Ha	**%**	**Ha**	**%**	**Ha**	**%**	**Ha**	**%**	**%**
Cropland	+13,728.6	+14	+13,050.3	+14	+28,091.46	29	+1121.4	+1.16	+0.79
Forest	−38,870.8	−33	−10,619.8	−9	−59,840.16	−50	−233.9	−1.94	+1.41
Open Grassland	−17,227.6	−4	−23,321.8	−5	−57,260.9	−13	−1638.8	−0.37	+0.3
Shrubland	+42,353.1	+26	+20,455.7	+13	+88,058.3	54	+2799.5	+1.7	+1.5

i. Trends in Forest land—where does change occur?

It is projected that a total of 50% of forests in the basin, representing approximately 59,840 ha, will be lost by the year 2040. The annual rate of loss of forests in the basin currently stands at 1.94% or 2333 ha. Most of the forest (around 38,870 ha) was lost from 1990 to 2000; this rate is expected to decline to 1.41% beyond the year 2015, yet it is still alarming to forest cover and links to water productivity and management in the basin. As of 2015, the main contributors to forest loss (Figure 4) were shrubland conversion, which accounted for the loss of over 28,518.51 ha of forest land, followed by conversion to cropland and grassland, which cumulatively accounted for over 14,814 ha loss of forest land in the basin. Significant loss in the forest habitat will have a tremendous impact on species behavior, distribution and numbers. More specific is the potential in the rise of the bush meat trade, because part of the forest is shared by communities that are not wildlife conservators.

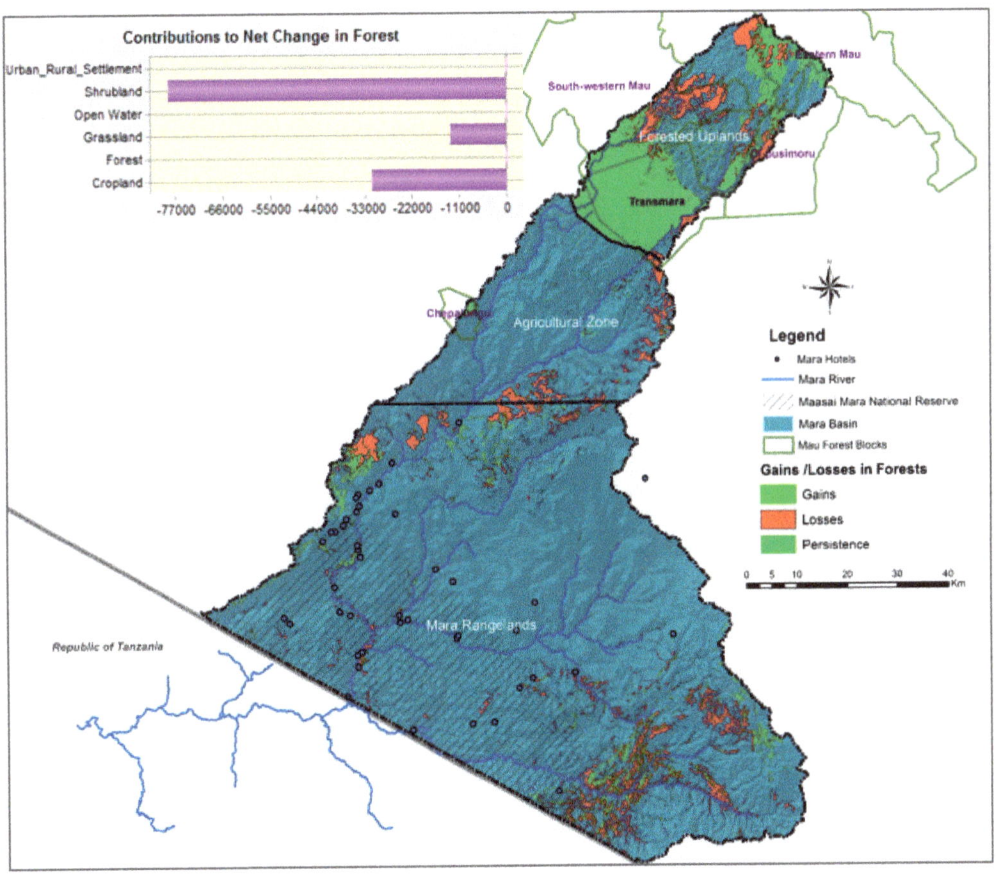

Figure 4. Gains and losses in forest cover in Mara River Basin (1990–2015).

While the conversion of forests to cropland and grassland is attributed to anthropogenic disturbance, mainly agriculture, other factors include overgrazing, encroachment by human settlement, and historical natural resource conflicts in the area, among other non-sustainable land use practices that commonly occurred in the middle agricultural and forested upland zones of the basin, e.g., around Chepalungu, southwestern Mau and

Eastern Mau forest blocks (Figure 4). Forest to shrubland conversion predominantly in the southeastern parts of the Mara rangelands in the basin was not evident.

ii. Trends in Grasslands an Shrubland—where does change occur?

The majority of the Mara River Basin is predominantly a grassland savannah ecosystem dominated by open grasslands at 49% and shrubland at 27.4% as of 2015 (Table 3). However, the two categories have been registering contrasting trends over the 1990–2015 period. While grassland gained over 36,000 ha at an annual rate of 1.7% (2561.03 ha) (Figure 5), shrublands lost over 37,037 ha at an estimated annual rate of 0.37% (1499 ha). This study noted that the biggest threat to shrinking grasslands in the basin is its conversion to shrubland (Figure 3), accounting for the loss of over 29,629 ha of grassland followed by encroachment from agriculture accounting for around 7407 ha of the grassland area. Most of the grassland conversion occurs in the southeastern parts of the basin, including areas south of Montorben toward the Siana and Olderkesi conservancies regions. Grasslands lost through agriculture expansion are common in the border areas of the rangelands and the middle agricultural section of the basin around Olchoroua and Enonkishu conservancies that borders Mulot town in Bomet County. Bomet town is mostly inhabited by the Kipsigis community that practices crop farming. Their proximity to the Mara rangelands signals a likely expansion of crop agriculture in the future that will trigger LULCC particularly, thus exerting pressure on the Enonkishu and Olchoroua conservancies.

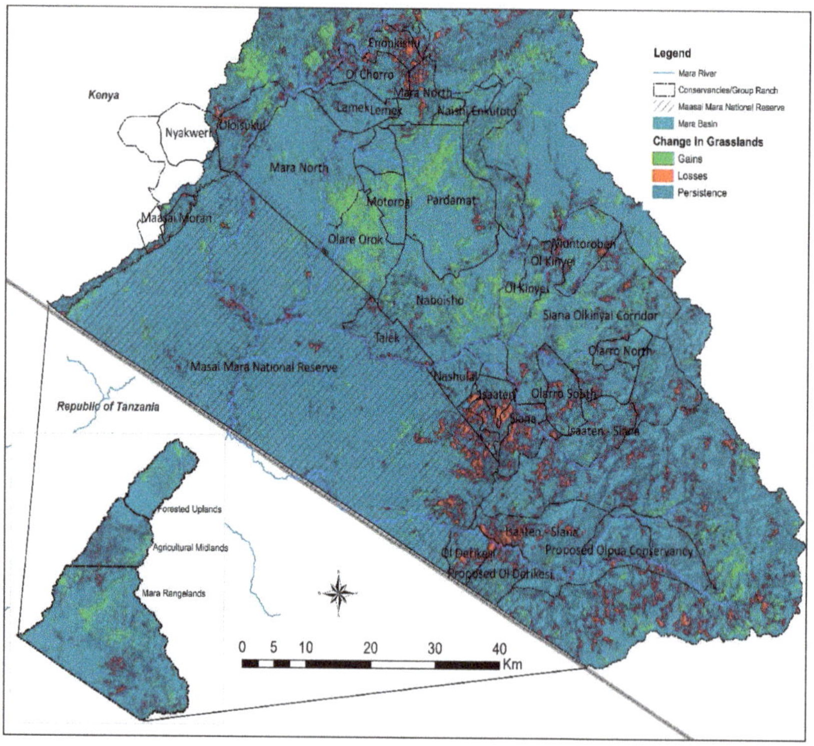

Figure 5. Grassland changes (gains and losses) in Mara basin in the period 1990–2015.

iii. Trends in Cropland—where does change occur?

Croplands predominantly occupy the densely populated areas, i.e., the agricultural and forest uplands zones of the Mau Mara landscape. Most croplands are a result of encroachment into the forest (11,000 ha), grassland (8000 ha), and shrubland (4500 ha) (Figure 6).

New and existing agriculture fields had drastically fragmented the forest into blocks. The agricultural gains were mostly at the edge of the remaining forest blocks. Agriculture areas that had continuously been cultivated had little or no live tree canopies.

At 2040, cropland will have shrunk part the forest with the grasslands and build up area also expanding (Figure 7). The most encroached forest blocks as a result are southwestern Mau and eastern Mau.

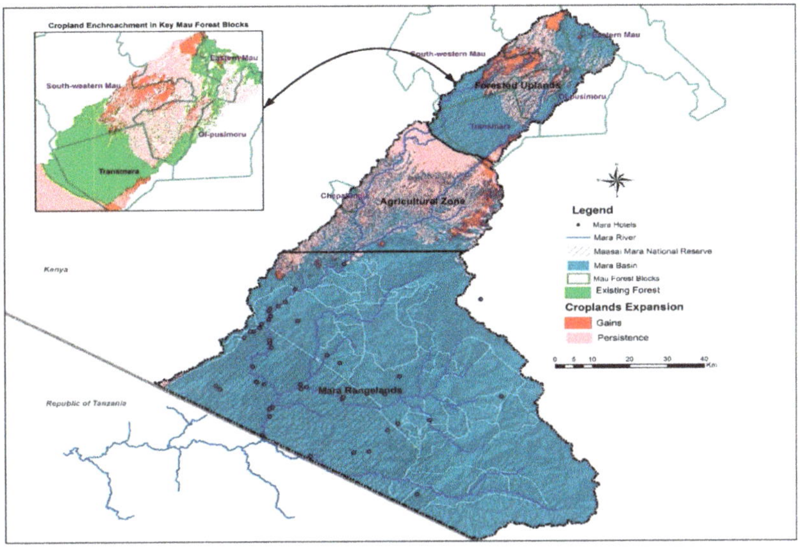

Figure 6. Cropland expansion and related forest encroachment in Mara basin (1990–2015).

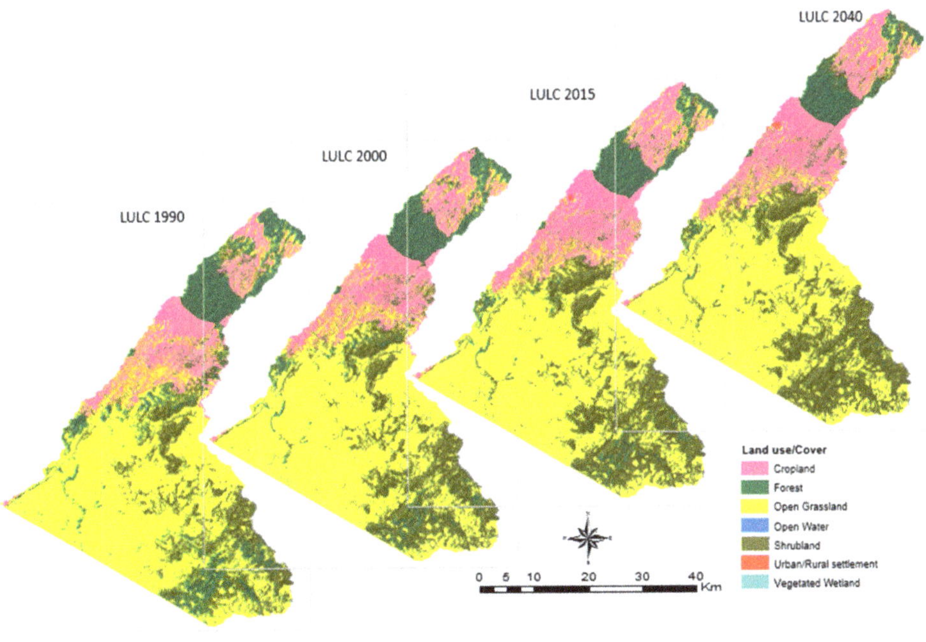

Figure 7. Land use/cover maps of Mara basin derived from satellite data for 1990, 2000, and 2015 and projected for 2040.

4. Discussion

4.1. Land Use Transition in Rangelands

The severe environmental and biodiversity degradation (gullies, emergency of invasive species, loss of habitats forage and other plant species) resulting from the need for more land to increase food production will worsen by 2040, which is when food demand will increase by 70% [49]. The primary drivers of this outward aggregate shift in food demand are population growth, improved wages per person, and urbanization [50]. In the Maasai Mara rangelands, the use of fires and livestock to manage grass, emerging farms and the acceleration of built-up areas are defining and giving the landscape a new shape. There has been widespread mechanized farming on selected mono-crops such as maize and wheat varieties [51] in areas like Lemek, Nkorinkori, Ololunga, Nairage Enkare, Suswa, etc. [52], which might have generated shifts in the preferred locations for the ungulates and in their numbers due to human influence. For example, noticeable and dramatic LULCC in Tanzania rangelands from 1975 to 2015 [3] have destabilized hotpots for free-ranging ungulate populations (wildebeest) in the SNP [53,54] as land is being converted to small-scale agriculture on the protected area peripheries with routine livestock encroachment in search of pasture [55]. This trend is similar in the northwest of Tanzania around the Loliondo Greater Conservation Area, the northern parts of Lolkisale, and the wider Arusha District, where the construction of infrastructure and expanding agriculture will exert pressure on wetlands, settlements and forests [15].

In southern Kenya (Amboseli, Loita and Maasai Mara), increased rainfall variability (23%), an increase in formal school enrollment (15%), the need for financial stability (13%), land subdivision (12%), socioeconomic development (10%), and population growth (8%) are all key and interacting drivers behind land use change [15]. Diverse park management regime approaches and the use of 13,635 [56] fires as a management tool to burn c. 8001 km^2 (24.1%) of Mara–Serengeti protected areas and 116 km^2 (3.2%) of the buffer zones between 2001 and 2014 have been key drivers in the rangeland change [56]. Perhaps that is why grasslands dominate in the confines of Mara–Serengeti while more woodlands are evident in the contiguous surrounding lands (conservancies and community areas). This variation has impacts on species distribution, as some species prefer grasslands (i.e., gazelles and cheetahs), while other dominate in the forest (i.e., giraffes).

4.1.1. Impacts of Deforestation on Mau Mara Landscape

Water resource availability and increased demand mean that the water resource is highly stressed [57], which is a situation exacerbated by LULCC [58]. Agricultural expansion is the principal driving force of deforestation, land degradation and the associated loss of ecosystem biodiversity [59] across the Mau-Mara landscape. Large-scale commercial farming, primarily tea farming, has reduced the size of the forest from 15% in the year 1989 to 13% in 2010 [39]. There was a gradual 0.6% yearly increase in small-scale farmlands from forested areas for a period of 40 years from 1973 with the greatest change being evident from 1994 to 2008 [46]. The 40,000 occupants of the forest from Enoosokon, Enkaroni, Enoosaen, Enekishomi and Sisiyian group ranches had asked the environmental court to seek redress for forceful eviction in 2018 [60], which is a factor that the environmental court did not consider in its 2022 ruling.

Although the sustainable management of agricultural operations can preserve and restore critical habitats, help protect watersheds, and improve soil health and water quality, they can also yield substantial negative impacts on people and the environment if managed unsustainably [61]. Agriculture activities accounted for 40% of tropical deforestation between 2000 and 2010 with the local expansion of subsistence agriculture accounting for 33% [58]. From 2000 to 2010 is the period when the Mau forest was undergoing immense anthropogenic influence, particularly illegal logging and the unlawful settlement of people, which also led to a decrease in water levels of the Mara River [39,62]. Degrading a forest ecosystem like the Mau that supports multiple environmental niches usually causes multiple cascading effects on people, livelihoods and biodiversity in adjacent ecosystems.

The reduced forest cover might affect the local hydrological cycle that supports social economic activities like agriculture, grass for livestock and wildlife (like in the grasslands of the Mara landscape) [23]. A reduction in water levels because of less precipitation endangers the aquatic and other terrestrial lives either directly or indirectly and would have a big impact on the nation's economy. The tourism sector that contributes 10% to Kenya's GDP would be impacted, as a significant portion of this socioeconomic activity is wildlife oriented, particularly in the Maasai-Mara region, which depends on the Mara River Basin ecosystem for its survival [63].

The rivers that flow through communally owned lands like the Ewaso Nyiro might not discharge adequate water to support livestock and wildlife during the dry season periods. Rivers that flow through agricultural zones like Amala and Nyangores (that source their waters from the Mau forest and discharge them into the Mara River) might trigger a change from rain-fed agriculture to irrigated food production. This might cause an over-abstraction of water, which can change a river's volume flow status from permanent to seasonal. Furthermore, as the farms are located on steep land, when it rains, the fertile topsoil is washed into the rivers, which is a tragic loss of an important resource. Hydrological modeling indicates that the depleting forest cover is making flows less stable with higher peak flows and reduced dry season flows [63] resulting in decreased resilience across the landscape. Intensified agriculture has resulted in deforestation, a reduction in shrubland, water abstraction, water access and biomass competition with wildlife, agrochemical pollution and excess nutrient discharge [57]. Domestic and agriculture (large-scale) water demands have skyrocketed to 20,967 m^3 as of 2010 with a projected increase of up to 33,918 m^3 in 2030 [61]. The total five daily water demand in the Mara River Basin was estimated as 20,967 m^3 in 2010 and projected to be up to 33,918 m^3 in 2030 [64]. Livestock water demand in the Mara catchment was estimated to increase by 69 m^3 yr^{-1} from 1990 to 86 m^3 yr^{-1} in 2010 [65]. A lack of capacity to enforce water abstraction permits has increased water usage by farmers [66]. In addition to flow, the use of inorganic fertilizers might cause eutrophication that increases algal and/or epiphyte biomass growth related to anthropogenic nutrient loading; this can be followed by more severe impacts such as a loss of submerged aquatic vegetation, oxygen depletion, imbalanced food webs, reduced biodiversity, altered biogeochemical cycling, fish kills and the formation of "dead zones" [67,68].

There were 32% and 14% decreases in dense and open woodland forest cover. This is attributed to clear logging to pave the way for the expansion of tea plantations between 1973 and 2000. The rangelands (shrubland, grassland and savannah) have decreased by 34% as a consequence of the expansion of cropland, which has increased by 20% [64]. Meanwhile, the wetland area in Tanzania has increased by 131% because of sediment build-up downstream as a consequence of soil erosion in the catchment [64]. This was perhaps caused by poor agricultural methods on the slopes of the Mau forest hills and amplified by tree logging upstream, thus necessitating easy soil removal during the rainy season. High sediment load estimates of between 113 and 432 tons/day were swept into wetlands and water bodies as a result of human actions [69]. This scenario might well amplify in the future, as unsustainable land utilization persists and as inorganic farming supplements are embraced to meet the food demand threshold. The riparian vegetation might also be jeopardized particularly when land is tilled to river bank buffers, thus limiting water sinks and absorption capability, which is a factor that will necessitate more sediment and chemical disposal.

4.1.2. Impact of the Forest on Wildlife

The current 75% decline for Mara wildlife populations [26], resulting from large-scale cultivation [32,35,70], in-migration [71], sedentarization [70,72], and settlement expansion [70], would be massively impacted by change in the Mara River flows. All the poaching cases recorded during the elephant dung estimate survey exercise in the Mau forest in 2020 was evident particularly in the southwestern Mau, Transmara and northern

Tinderet forests blocks. Three carcasses of wild animals were recorded: one for elephant (old), colobus monkey, and bush buck [73]. This is substantially less than in 2016 when 19 carcasses were recorded: 4 for elephants, 9 bushbuck, 3 monkeys, 1 red duiker, 1 rat and 1 mongoose. The drop in poaching incidences is a result of law enforcement in the forest that has minimized human activities. Poaching incidence cases were mostly intended for bush meat.

Critically endangered species of the Mountain Bongo found in the forest would likely become extinct. The inferential figure for this critically endangered antelope [74] that lives in forest ecosystems was less than 100 individuals [10], which is significantly below the critically endangered threshold of 250 mature individuals [72]. In May 2013, new discoveries of this antelope species were made with the preliminary results indicating that the Maasai Mau Forest Complex hosts some 20 individuals, while the southwest of the Mau Forest supports less than 10 individuals [72]. An elephant dung survey in 1995 recorded 1003 individuals compared to 652 in 2016, representing a deplorable 35% decline [11]. The 1990s and early 2000s was a period when Kenya lost a significant number of elephants and Rhinos to poaching. The situation will further be aggravated by prolonged droughts [31,38] that can be a trigger for poaching [73] as a fallback for survival.

As the forest habitat is depleted, human wildlife conflict incidence cases and threshold levels also become rampant because of natural resource competition. The areas adjacent to the MMNR, for example, have high rates of Human–Wildlife Conflicts (HWCs) per year. In 2015, Narok County, for example, reported 258 HWC cases through the KWS community department. The cases entailed 26 fatalities caused by retaliation attacks, 112 cases of property damage caused by leopards, snakes, and lions, 58 cases of crop destruction caused by zebras, buffaloes, and waterbucks, 57 incidents of livestock predation whereby 90 animals were predated upon, and 5 cases of injuries caused by buffaloes, snakes, and hippopotamus [74].

4.1.3. Impact of Changing Land Use Policies

Land subdivision of the entire Mara landscape is almost complete with the Loita Plains and Olderkesi being the only remaining areas currently undergoing the demarcation process. It is presumed that owning a land title deed permits freedom of tilling, selling, and obtaining returns from wildlife [35]. Land use change and production study in the Masai-Mara landscape confirmed that over 50,000 ha of the subdivided land had already been sold to peasant non-Maasai farmers (Kisii, Kalenjin and Kikuyu) [66]. This figure might be higher now because of the earmarked infrastructure developments (an international airport in Transmara, upgrading of three roads, the Narok-Sekenani (complete), one from Narok via Mararianta to Lemek area and the second to Mulot via Nyekweri Forest Conservation Area, Oloisukut Conservancy, Emarti, Enoonkishu Conservancy). The Transmara landscape has recently also been a target for the more affluent in Kenya, who are strategically buying extensive tracts of land for future investments. As the land is subdivided, the likelihood of it falling to peasant farmers will persist especially in Siana and Olderkesi, where land allocations are as small as 2 acres per person. This fragmentation will hinder large-scale production that is usually embraced in large farms.

4.2. New Steps and Feeding Land Use and Land Cover Change into Policy for Better Management Decisions

The Mau-Mara landscape is transitioning from its traditional land uses as the contemporary needs and demands also change. Massive "land grabbing", infrastructure developments like roads and the development of an international airport in the landscape will spur sporadic developments that will need policy interjection. The need to review existing or enact new policies to mitigate future LULCC might guarantee the sustainability of the landscape. The completed spatial plans for Bomet County and Narok Counties plus the Maasai Mara Greater Ecosystem Management plan is a clear demonstration of the good intentions of the multiple stakeholders in the landscape in arresting future impacts of land uses on ecosystems, people and wildlife. However, political will (at the national and county level) and assimilating other stakeholders' views can mitigate some of the unprecedented LULCC outcomes, which is a crucial element of future sustainable land use planning.

Limitations of the Study

The length of Mara River is approximately 395 km from Kenya to Tanzania with 60% being in Kenya [75]. This study was limited to the Mara River Basin on the Kenyan side. It is anticipated that the findings of the study will guide the County Government of Narok in developing policies and frameworks that will assist address contemporary and future threats facing the Mara River basin. Similar studies in Tanzania are recommended to guide the government on interventions they can pursue to counter problems facing the river basin. The studies from both countries can aid the two countries avoid pitfalls and overlaps when implementing recommendations. Periodic satellite imagery updates on land use change driver categories like agriculture and built-up areas and their impacts on ecosystem (wildlife, forests, water, grasslands) should be undertaken to advise on prompt measures that governments can pursue to descale and mitigate their acceleration. Focus studies on the impacts of the river and the forest on specific wildlife species along the river basin and within the Mara and Serenget landscapes is encouraged to understand their demographics and behaviors for swift decision action.

5. Conclusions

The Mau Mara landscape remains important for the sustenance of people, livelihoods and biodiversity for contemporary and future generations. The diverse stakeholders in the landscape (tourism sector, farmers, pastoralists, conservancies, civil society organizations, researchers, national and county government) can play a fundamental role in mitigating myriad potential threats. The development of joined up landscape management plans to organize land use outside the protected area and the Narok county spatial plan are fundamental to instigate change in land uses across the landscape. This will entail the methodical distribution and organizing of land use activities earmarked for safeguarding the available natural resources in protected areas [66]. Understanding potential pathways of future land cover change can open debates as to what policy and management interventions are required to achieve conservation and development targets, or at least to avoid undesirable futures.

Author Contributions: Conceptualization, E.N.S., S.A., R.M. and D.O.; methodology, S.A. and E.N.S.; software, S.A.; validation, S.A. and E.N.S.; formal analysis, S.A. and E.N.S.; investigation, S.A. and E.N.S.; resources, E.N.S., R.M. and S.A.; data curation, R.M. and S.A.; writing—original draft preparation, E.N.S.; writing—review and editing, E.N.S., S.A., R.M. and D.O.; visualization, E.N.S. and S.A.; supervision, D.O. and R.M.; project administration, E.N.S.; funding acquisition, E.N.S. and R.M. All authors have read and agreed to the published version of the manuscript.

Funding: This study was supported through commonwealth funding scholarship awarded to E.N.S.; as part of his PhD studies at the University of Nairobi (Kenya) and University of York, (United Kingdom). S.A.; and E.N.S.; utilized WWF Kenya resources (time, vehicles, and equipments and previous data) to consolidate the information thus enriching the content of the study.

Data Availability Statement: Data are contained within the article.

Conflicts of Interest: The authors declare no conflict of interest. The funders had no role in the design of the study; in the collection, analyses, or interpretation of data; in the writing of the manuscript, or in the decision to publish the results.

References

1. Pflugmacher, D.; Rabe, A.; Peters, M.; Hostert, P. Mapping pan-European land cover using Landsat spectral-temporal metrics and the European LUCAS survey. *Remote Sens. Environ.* **2019**, *221*, 583–595. [CrossRef]
2. Cui, J.; Zhu, M.; Liang, Y.; Qin, G.; Li, J.; Liu, Y. Land Use/Land Cover Change and Their Driving Factors in the Yellow River Basin of Shandong Province Based on Google Earth Engine from 2000 to 2020. *ISPRS Int. J. Geo-Inf.* **2022**, *11*, 163. [CrossRef]

3. Kija, H.K.; Ogutu, J.O.; Mangewa, L.J.; Bukombe, J.; Verones, F.; Graae, B.J.; Kideghesho, J.R.; Said, M.Y.; Nzunda, E.F. Land Use and Land Cover Change Within and Around the Greater Serengeti Ecosystem, Tanzania. *Am. J. Remote Sens.* **2020**, *8*, 1–19. [CrossRef]
4. Phukan, P.; Thakuriah, G.; Saikia, R. Land Use Land Cover Change Detection Using Remote Sensing and GIS Techniques. A case study of Golaghat District of Assam, India. *Int. Res. J. Earth Sci.* **2013**, *1*, 11–15.
5. Sang, C.C.; Olago, D.O.; Nyumba, T.O.; Marchant, R.; Thorn, J.P.R. Assessing the Underlying Drivers of Change over Two Decades of Land Use and Land Cover Dynamics along the Standard Gauge Railway Corridor, Kenya. *Sustainability* **2022**, *14*, 6158. [CrossRef]
6. Molotoks, A.; Stehfest, E.; Doelman, J.; Albanito, F.; Fitton, N.; Dawson, T.P.; Smith, P. Global Projections of Future Cropland Expansion to 2050 and Direct Impacts on Biodiversity and Carbon Storage. *Glob. Chang. Biol.* **2018**, *24*, 5895–5908. [CrossRef] [PubMed]
7. Maja, M.M.; Ayano, S.F. The Impact of Population Growth on Natural Resources and Farmers' Capacity to Adapt to Climate Change in Low-Income Countries. *Earth Syst. Environ.* **2021**, *5*, 271–283. [CrossRef]
8. Bradshaw, C.J.A.; Ehrlich, P.R.; Beattie, A.; Ceballos, G.; Crist, E.; Diamond, J.; Dirzo, R.; Ehrlich, A.H.; Harte, J.; Harte, M.E.; et al. Undertra the Challenges of Avoiding a Ghastly Future. *Front. Conserv. Sci.* **2021**, *1*, 615419. [CrossRef]
9. Ren, Y.; Lu, Y.; Comber, A.; Fu, B.; Harris, P.; Wu, L. Spatially Explicit Simulation of Land Use/Land Cover Changes: Current Coverage and Future Prospects. *Earth-Sci. Rev.* **2019**, *190*, 398–415. [CrossRef]
10. Wildlife Research and Training Institute (WRTI) and Kenya Wildlife Service (KWS). *National Wildlife Census 2021 Report*; WRTI: Naivasha, Kenya, 2021; ISBN 987-9914-40-397-8.
11. Mwangi, G.; Kiwanuka, D. A Review of the Population Status of Forest Elephants in Mau Forest, Kenya. *Tech. Rep.* **2017**, 1–15. [CrossRef]
12. Magliocca, N.R.; Rudel, T.K.; Verburg, P.H.; McConnell, W.J.; Mertz, O.; Gerstner, K. Synthesis in Land Change Science: Methodological Patterns, Challenges, and Guidelines. *Reg. Environ. Chang.* **2015**, *15*, 211–226. [CrossRef] [PubMed]
13. Ellis, E.C.; Kaplan, J.O.; Fuller, D.Q.; Vavrus, S.; Goldewijk, K.K.; Verburg, P.H. Used Planet: A Global History. *Proc. Natl. Acad. Sci. USA* **2013**, *110*, 7978–7985. [CrossRef] [PubMed]
14. Meyfroidt, P.; Chowdhury, R.R.; de Bremond, A.; Ellis, E.C.; Erb, K.H.; Filatova, T. Middle-Range Theories Of Land System Change. *Glob. Environ. Chang.* **2018**, *53*, 52–67. [CrossRef]
15. Kariuki, R.W.; Munishi, L.K.; Courtney-Mustaphi, C.J.; Capitani, C.; Shoemaker, A.; Lane, P.J. Integrating Stakeholders' Perspectives and Spatial Modelling To Develop Scenarios of Future Land Use and Land Cover Change in Northern Tanzania. *PLoS ONE* **2021**, *16*, e0245516. [CrossRef]
16. Republic of Kenya. *Sessional Paper, No. 1 of 2017 on National Land Use Policy*; October, 2017; Ministry of Lands and Physical Planning: Nairobi, Kenya, 2017.
17. Republic of Kenya. *The Land Act, 2012 No. 6 Of 2012, Laws of Kenya*; National Council for Law Reporting: Nairobi, Kenya, 2012.
18. Mutugi, M.; Kiiru, W. Biodiversity, Local Resource, National Heritage, Regional Concern, and Global Impact: The Case of Mau Forest, Kenya. *Eur. Sci. J.* **2015**, *1*, 681–691.
19. Government of Kenya. Rehabilitation of the Mau Forest Ecosystem. In *A Project Concept Prepared by the Interim Coordinating Secretariat, Office of the Prime Minister, on behalf of the Government of Kenya*; GoK: Nairobi, Kenya, 2009.
20. Sitters, J.; Kimuyu, D.M.; Young, T.P.; Claeys, P.; Olde Venterink, H. Negative Effects of Cattle on Soil Carbon and Nutrient Pools Reversed by Mega Herbivores. *Nat. Sustain.* **2020**, *3*, 360–366. [CrossRef]
21. Phelps, L.; Chevalier, M.; Shanahan, T.; Aleman, J.C.; Courtney-Mustaphi, C.; Kiahtipes, C.A.; Broennimann, O.; Marchant, R. Asymmetric response of forest and Grassy biomes to climate variability across the African Humid Period: Influence by anthropogenic disturbance? *Ecography* **2020**, *43*, 1118–1142. [CrossRef]
22. Githumbi, E.; Kariuki, R.; Shoemaker, A.; Courtney, C.J.; Mustaphi, M.; Chuhila, R.S.; Lane, P.J.; Marchant, R. Pollen, People and Place: Paleo-environmental, Archaeological, and Ecological Perspectives on Vegetation change in the Amboseli Landscape, Kenya. *Front. Earth Sci.* **2018**, *5*, 1–26. [CrossRef]
23. DeFries, R.; Hansen, A.; Newton, A.C.; Hansen, M.C. Increasing Isolation of Protected Areas in Tropical Forests over the Past Twenty Years. *Ecol. Appl.* **2005**, *15*, 19–26. [CrossRef]
24. Gross, D.; Dubois, G.; Pekel, J.-F.; Mayaux, P.; Holmgern, M.; Prins, H.H.T.; Rondinini, C.; Boitani, L. Monitoring Land cover changes in African protected areas in the 21st Century. *Ecol. Inform.* **2013**, *14*, 31–37. [CrossRef]
25. Pfeifer, M.; Burgess, N.D.; Swetnam, R.D.; Platts, P.J.; Willcock, S.; Marchant, R. Protected Areas: Mixed Success in Conserving East Africa's Evergreen Forests. *PLoS ONE* **2012**, *7*, e39337. [CrossRef]
26. Gemeda, D.O.; Meles, S.K. Impacts of Human-Wildlife Conflict in Developing Countries. *J. Appl. Sci. Environ. Manag.* **2018**, *22*, 1233–1238. [CrossRef]
27. Schroeder, H.; Grab, H.; Kessler, A.; Poveda, K. Human-Mediated Land Use Change Drives Intraspecific Plant Trait Variation. *Front. Plant Sci.* **2021**, *11*, 592881. [CrossRef] [PubMed]
28. Ogutu, J.O.; Piepho, H.-P.; Said, M.Y.; Ojwang, G.O.; Njino, L.W.; Kifugo, S.C. Extreme Wildlife Declines and Concurrent Increase in Livestock Numbers in Kenya: What Are the Causes? *PLoS ONE* **2016**, *11*, e0163249. [CrossRef]
29. Ogutu, J.O.; Piepho, H.-P.; Dublin, H.T.; Bhola, N.; Reid, R.S. Dynamics of Mara-Serengeti Ungulates in Relation to Land Use Changes. *J. Zoöl.* **2009**, *278*, 1–14. [CrossRef]

30. Norton-Griffiths, M.; Said, M.Y.; Serneels, S.; Kaelo, D.S.; Coughenour, M.; Lamprey, R.H.; Thompson, D.M.; Reid, R.S. Land use economics in the Mara area of the Serengeti ecosystem. In *Serengeti III: Human Wildlife Interactions*; Packer, C., Sinclair, A.R.E., Eds.; University of Chicago Press: Chicago, IL, USA, 2009; pp. 379–416.
31. Ogutu, J.O.; Owen- Smith, N.; Piepho, H.-P.; Said, M.Y. Continuing Wildlife Population Declines and Range Contraction in the Mara Region of Kenya during 1977–2009. *J. Zool.* **2011**, *285*, 99–109. [CrossRef]
32. Kenya Wildlife Service [KWS]. Effects of Current Drought on Wildlife Dispersal. KWS News. 2016. Available online: https://www.kws.go.ke/content/effects-current-drought-wildlife-dispersal#:~:text=The%20ongoing%20drought%20situation%20in%20the%20country%20has,resulting%20to%20an%20increase%20in%20human%20wildlife%20conflicts (accessed on 26 September 2022).
33. World Wide Fund for Nature [WWF]. 84% of Wildlife Counted in the Mara Live in Community Conservancies, Our News. 2023. Available online: https://www.wwfkenya.org/?235151/Census-84-of-wildlife-counted-in-Maasai-Mara-live-in-community-conservancies (accessed on 17 February 2024).
34. Thompson, M.; Homewood, K. Entrepreneurs, Elites and Exclusion in Maasailand: Trends in Wildlife Conservation and Pastoralist Development. *Hum. Ecol.* **2002**, *30*, 107–138. [CrossRef]
35. Ndegwa, C.; Murayama, Y. Analysis of Land Use/Cover Changes and Animal Population Dynamics in a Wildlife Sanctuary in East Africa. *Remote Sens.* **2009**, *1*, 952–970. [CrossRef]
36. Kenya National Bureau of Statistics. *Kenya Population and Housing Census Report*; Kenya National Bureau of Statistics: Nairobi, Kenya, 2019.
37. Ogutu, J.O.; Piepho, H.-P.; Dublin, H.T.; Bhola, N.; Reid, R.S. Rainfall Influences on Ungulate Population Abundance in the Mara-Serengeti Ecosystem. *J. Anim. Ecol.* **2008**, *77*, 814–829. [CrossRef]
38. Maasai Mara Wildlife Conservancies Report [MMWCA]. *State of Mara Conservancies Report 2019*; MMWCA: Narok, Kenya, 2019.
39. Kenya Forest Service. *KFS Field Manual for Biophysical Forest Resources Assessment in Kenya May 2016*; Kenya Forest Service: Nairobi, Kenya, 2016. Available online: https://www.kefri.org/assets/publications/tech/Improving%20Capacity%20in%20Forest%20Resources%20Assessment%20in%20Kenya%20(IC-FRA)-%20Field%20Manual%20for%20Tree%20Volume%20and%20Biomass%20Modelling.pdf (accessed on 17 December 2023).
40. Sophia, S.; Ndambuki, J.M. Accuracy Assessment of Land Use/Land Cover Classification Using Remote Sensing and GIS. *Int. J. Geosci.* **2017**, *8*, 4.
41. The County Government of Narok. *The Maasai Mara National Reserve Management Plan 2023-32*; Narok County: Narok, Kenya, 2022.
42. Park, Y.S.; Lek, S. *Artificial Neural Networks: Multilayer Perceptron for Ecological Modeling*; Elsevier: Amsterdam, The Netherlands, 2016; Volume 28, pp. 123–140; ISBN 9780444636232. [CrossRef]
43. Kumar, V.; Garg, M.L. Predictive Analytics: A Review of Trends and Techniques. *Int. J. Comput. Appl.* **2018**, *182*, 31–37. [CrossRef]
44. Bhasin, H. 8 Types of Analysis in Research. Marketing. 2019. Available online: https://www.marketing91.com/types-of-analysis-in-research (accessed on 21 November 2022).
45. Lin, Z.; Peng, S. Comparison of multimodel simulations of land use and land cover change considering integrated constraints—A case study of the Fuxian Lake basin. *Ecol. Indic.* **2022**, *142*, 109254. [CrossRef]
46. Food and Agriculture Organization of the United Nations [FAO]. 'How to Feed the World in 2050', Food and Agriculture Organization of the United Nations, Paper Prepared for How to Feed the World in 2050: High-Level Expert Forum, 12–13 October 2009, Rome. 2009. Available online: www.fao.org/fileadmin/templates/wsfs/docs/expert_paper/How_to_Feed_the_World_in_2050.pdf (accessed on 9 September 2022).
47. Boitt, M.K. Impacts of Mau Forest Catchment on the Great Rift Valley Lakes in Kenya. *J. Geosci. Environ. Prot.* **2016**, *4*, 137–145. [CrossRef]
48. Jebiwott, A.; Ogendi, G.M.; Agbeja, B.O.; Alo, A.A.; Kibet, R. Mapping the Trends of Forest Cover Change and Associated Drivers in Mau Forest, Kenya. *Remote Sens. Appl. Soc. Environ.* **2021**, *23*, 100586. [CrossRef]
49. Kariuki, R.W.; Western, D.; Willcock, S.; Marchant, R. Assessing Interactions between Agriculture, Livestock Grazing and Wildlife Conservation Land Uses: A Historical Example from East Africa. *Land* **2021**, *10*, 46. [CrossRef]
50. Kideghesho, J.R.; Nyahongo, J.W.; Hassan, S.N.; Tarimo, T.C.; Mbije, N.E. Factors and Ecological Impacts of Wildlife Habitat Destruction in the Serengeti ecosystem in northern Tanzania. *Afr. J. Environ. Assess. Manag.* **2006**, *11*, 17–32.
51. Muchane, M.N.; Karanja, D.; Wambugu, G.M.; Mutahi, J.M.; Masiga, C.W.; Mugoya, C.; Muchai, M. Land use practices and their implications on soil macro-fauna in Maasai Mara ecosystem. *Int. J. Biodivers. Conserv.* **2012**, *4*, 500–514. [CrossRef]
52. Serneels, S.; Said, M.Y.; Lambin, E.F. Land Cover Changes around a Major East African Wildlife Reserve: The Mara Ecosystem (Kenya). *Int. J. Remote Sens.* **2001**, *22*, 3397–3420. [CrossRef]
53. Sinclair, A.R.E.; Arcese, P. *Serengeti II: Dynamics, Management, and Conservation of an Ecosystem*; University of Chicago Press: Chicago, IL, USA, 1995; Volume 2.
54. Veldhuis, M.P.; Ritchie, M.E.; Ogutu, J.O.; Morrison, T.A.; Beale, C.M.; Estes, A.B.; Mwakilema, W.; Ojwang, G.O.; Parr, C.L.; Probert, J.; et al. Cross-boundary Human Impacts Compromise the Serengeti-Mara Ecosystem. *Science* **2019**, *363*, 1424–1428. [CrossRef]
55. Dessu, S.B.; Melesse, A.M.; Bhat, M.G.; McClain, M.E. Assessment of water resources availability and demand in the Mara River Basin. *CATENA* **2014**, *115*, 104–114. [CrossRef]

56. Muiruri, V.; Marchant, R.; Rucina, M.S.; Scott, L.; Lane, P.J. Late Holocene Environmental Change and Anthropogenic: Ecosystem Interaction on the Laikipia Plateau, Kenya. *AMBIO* **2022**, *51*, 785–798. [CrossRef] [PubMed]
57. Tran, N.B.; Dhunna, S.; Bhatta, N. A Case Study on Water Quality and Ecosystem Management in Mara River Basin-Kenya and Tanzania, Research Gate. 2017. Available online: https://www.researchgate.net/publication/335826938 (accessed on 9 January 2024).
58. Food and Agriculture Organization of the United Nations [FAO]. Status of the World's Forest 2020. 2022. Available online: https://www.fao.org/state-of-forests/en/ (accessed on 5 September 2022).
59. Nyambura, M. Over 40,000 Mau Forest Evictees to Appeal Court, Citizen Digital. 2022. Available online: https://citizen.digital/news/over-40000-mau-forest-evictees-to-appeal-court-ruling-n307466 (accessed on 21 October 2022).
60. World Wide Fund for Nature (WWF). Sustainable Agriculture. 2022. Available online: https://www.worldwildlife.org/industries/sustainable-agriculture (accessed on 5 September 2022).
61. Githumbi, E.N.; Courtney, C.J.; Marchant, R. Late Pleistocene and Holocene Afromontane Vegetation and Headwater Wetland Dynamics within the Eastern Mau Forest, Kenya. *J. Quat. Sci.* **2021**, *36*, 239–254. [CrossRef]
62. World Wide for Nature (WWF). Better Land Management in the Mara. In *How Hundreds of Farmers Reduced Soil Erosion Fresh Water Case Study*; WWF: Nairobi, Kenya, 2017.
63. Conley, D.J.; Paerl, H.W.; Howarth, R.W.; Boesch, D.F.; Seitzinger, S.P.; Havens, K.E.; Lancelot, C.; Likens, G.E. Controlling Eutrophication: Nitrogen and Phosphorus. *Science* **2009**, *323*, 1014–1015. [CrossRef] [PubMed]
64. Lake Victoria Basin Commission. *WWF-ESARPO. Assessing Reserve Flows for the Mara River*; WWF: Nairobi, Kenya, 2010.
65. Ferreira, J.G.; Andersen, J.H.; Borja, A.; Bricker, S.B.; Camp, J.; da Silva, M.C.; Garcés, E.; Heiskanen, A.; Humborg, C.; Ignatiades, L.; et al. Overview of eutrophication indicators to assess environmental status within the European Marine Strategy Framework Directive. *Estuar. Coast. Shelf Sci.* **2011**, *93*, 117–131. [CrossRef]
66. Nyariki, D.M.; Mwang'ombe, A.W.; Thompson, D.M. Land Use Change and Livestock Production Challenges in an Integrated System: The Masai-Mara Ecosystem, Kenya. *J. Hum. Ecol.* **2009**, *26*, 163–173. [CrossRef]
67. World Wide Fund for Nature [WWF]. Spatial Planning. 2021. Available online: https://www.wwfkenya.org/spatial_planning/ (accessed on 5 November 2022).
68. Probert, J.R.; Parr, C.L.; Holdo, R.M.; Anderson, T.M.; Archibald, S.; Courtney Mustaphi, C.J.; Dobson, A.P.; Donaldson, J.E.; Hopcraft, G.C.; Hempson, G.P.; et al. Anthropogenic Modifications to Fire Regimes in the wider Serengeti-Mara Ecosystem. *Glob. Change Biol.* **2019**, *25*, 3406–3423. [CrossRef]
69. Homewood, K.M.; Trench, C.P.; Brockington, D. Pastoralist livelihoods and wildlife revenues in East Africa: A case for coexistence? Pastoralism: Research, Policy and Practice. SpringerOpen Journal. 2012. Available online: http://www.pastoralismjournal.com/content/2/1/19 (accessed on 6 March 2023).
70. Waithaka, J. Maasai Mara: An ecosystem under siege: An African Case Study On the Societal Dimension of Rangeland Conservation. *Afr. J. Range Forage Sci.* **2009**, *21*, 79–88. [CrossRef]
71. Mati, B.; Khaka, E.; Malesu, M.M.; Oduor, A.; De Bock, T.; Nyabenge, M. Mapping the Potential of Rainwater Harvesting Technologies in Africa. In *A GIS Overview on Development Domains for the Continent and Nine Selected Countries*; Technical Manual No. 7; World Agroforestry Centre [ICRAF], Netherlands Ministry of Foreign Affairs: Nairobi, Kenya, 2006; pp. 120–12; ISBN 929059 2117. Available online: https://apps.worldagroforestry.org/downloads/Publications/PDFS/MN15297.pdf (accessed on 6 January 2024).
72. IUCN SSC Antelope Specialist Group. *Tragelaphus eurycerus* ssp. isaaci. *IUCN Red List. Threat. Species* **2017**, e.T22057A50197212. [CrossRef]
73. Mwiu, S.S.; Kiambi, J.; Mukeka, D.; Leto, J.; Edebe, R.; Chepkwony, P.; Omondi, L.K.; Vanleeuwe, H. *Estimating the Population Status of Savanna Elephant (Loxodonta Africana) in the Mau Forest Complex: Wildlife Research and Training Institute*; WRTI: Naivasha, Kenya, 2021.
74. Machoka, L.N. Factors Influencing Human Wildlife Conflict in Communities Surrounding Protected Areas: A Case of Kenya Wildlife Service Focusing on Maasai Mara National Reserve, Narok County, Kenya. Ph.D. Dissertation, University of Nairobi, Nairobi, Kenya, 2017.
75. The County Government of Narok. *The Greater Maasai Mara Ecosystem Management Plan, 2023–2033*; Narok County: Narok, Kenya, 2022.

land

Article

Scenario-Based Land Use and Land Cover Change Detection and Prediction Using the Cellular Automata–Markov Model in the Gumara Watershed, Upper Blue Nile Basin, Ethiopia

Haile Belay [1,2,*], Assefa M. Melesse [3,*] and Getachew Tegegne [3,4]

1 Africa Center of Excellence for Water Management, Addis Ababa University, Addis Ababa P.O. Box 1176, Ethiopia
2 Department of Hydraulic and Water Resources Engineering, Dilla University, Dilla P.O. Box 419, Ethiopia
3 Department of Earth and Environment, Florida International University, Miami, FL 33199, USA; getachewtegegne21@gmail.com
4 Department of Civil Engineering, Sustainable Energy Center of Excellence, Addis Ababa Science and Technology University, Addis Ababa P.O. Box 16417, Ethiopia
* Correspondence: haile.belay@aau.edu.et (H.B.); melessea@fiu.edu (A.M.M.)

Abstract: Land use and land cover (LULC) change detection and prediction studies are crucial for supporting sustainable watershed planning and management. Hence, this study aimed to detect historical LULC changes from 1985 to 2019 and predict future changes for 2035 (near future) and 2065 (far future) in the Gumara watershed, Upper Blue Nile (UBN) Basin, Ethiopia. LULC classification for the years 1985, 2000, 2010, and 2019 was performed using Landsat images along with vegetation indices and topographic factors. The random forest (RF) machine learning algorithm built into the cloud-based platform Google Earth Engine (GEE) was used for classification. The results of the classification accuracy assessment indicated perfect agreement between the classified maps and the validation dataset, with kappa coefficients (K) of 0.92, 0.94, 0.90, and 0.88 for the LULC maps of 1985, 2000, 2010, and 2019, respectively. Based on the classified maps, cultivated land and settlement increased from 58.60 to 83.08% and 0.06 to 0.18%, respectively, from 1985 to 2019 at the expense of decreasing forest, shrubland and grassland. Future LULC prediction was performed using the cellular automata–Markov (CA–Markov) model under (1) the business-as-usual (BAU) scenario, which is based on the current trend of socioeconomic development, and (2) the governance (GOV) scenario, which is based on the Green Legacy Initiative (GLI) program of Ethiopia. Under the BAU scenario, significant expansions of cultivated land and settlement were predicted from 83.08 to 89.01% and 0.18 to 0.83%, respectively, from 2019 to 2065. Conversely, under the GOV scenario, the increase in forest area was predicted to increase from 2.59% (2019) to 4.71% (2065). For this reason, this study recommends following the GOV scenario to prevent flooding and soil degradation in the Gumara watershed. Finally, the results of this study provide information for government policymakers, land use planners, and watershed managers to develop sustainable land use management plans and policies.

Keywords: land use and land cover prediction; Google Earth Engine; random forest; CA–Markov; scenario-based prediction

1. Introduction

Recently, land use and land cover (LULC) change detection and prediction have become popular research topics in the area of remote sensing; thus, these issues have attracted the attention of several researchers and land use planners [1,2]. This is due to the significant effect of LULC changes on altering the hydrological characteristics of basins and watersheds [3,4], climate change [5], and other environmental issues [6]. Globally, different LULC classes of basins and watersheds have undergone considerable changes from one category to another due to increasing anthropogenic activities such as deforestation, agricultural expansion, urbanization, mining, and other related activities [7–9]. For this reason, regularly updating LULC maps is an important task for capturing the dynamic nature of land use [10].

For spatiotemporal LULC mapping, remotely sensed multispectral images available at various spatiotemporal resolutions provide vital information. Some of these images include Moderate Resolution Imaging Spectroradiometer (MODIS), Pourl'Observation de la Terre (SPOT), Synthetic Aperture Radar (SAR), Landsat series, Sentinel-2 missions, and Rapid Eye Earth Imaging System (REIS) images [1]. From these images, the Landsat series has been used in a number of LULC mapping studies [5,11].

In LULC classification studies, the use of cloud-free composites of time-series images is more strongly recommended than the use of single images [10,12]. However, the traditional methods of searching, filtering, cloud masking, compositing, downloading, and classifying time series images require high computing power and a large volume of data storage. For this purpose, Google Earth Engine (GEE), which is a cloud-based platform, can solve the problems associated with big data processing [7,13]. This platform has recently been considered a powerful tool for processing and analyzing large volumes of remote sensing data. In the GEE environment, there are various supervised machine learning (SML) classification algorithms. Some of these include decision tree (DT), random forest (RF), naive Bayes (NB), minimum distance (MD), and classification and regression tree (CART) methods [6,14,15]. In this regard, several studies comparing SML classifiers have recommended the RF algorithm because it has the highest classification accuracy and the potential to handle high-dimensional data with a minimum number of parameters [16–18]. Moreover, the use of topographic factors and spectral indices can improve classification accuracy [19].

In addition to the LULC classification discussed above, LULC prediction is important for simulating future landscape conditions and determining the potential driving factors that cause land use change. According to the comprehensive review of [1], there are various temporal prediction models (for example, the Markov chain and system dynamics (SD)) and spatial prediction models (for example, cellular automata (CA) and conversion of land use and its effects (CLUE)) and hybrids of the two model categories. Recently, LULC prediction using hybrid models, such as the cellular automata–Markov (CA–Markov) model, has become an effective approach because it integrates both spatial and temporal simulation powers of the CA and Markov models. For this reason, a number of researchers have applied the CA–Markov model for predicting future LULC dynamics at the basin, watershed, and even city levels [20–22].

The prediction of future LULC dynamics based on historical trends is a critical issue [23]. However, the assumption of the continuity of the current land use trend is mostly susceptible to uncertainties due to the dynamic nature of land use change driver variables [24,25]. For this purpose, scenario-based LULC change analysis is mostly recommended as a tool to explore such uncertainties [26]. Recently, five shared socioeconomic pathways (from SSP1to SSP5) have been proposed for possible future socioeconomic development, climate change, and land use policies [27,28]. These SSP-based pathways provide information for research and policies related to land use and climate change to reduce the adverse impacts of rapid population growth, intensive agricultural cultivation, and greenhouse gas emissions [28]. Therefore, several scenario-based LULC change analysis studies have also been conducted at different spatiotemporal scales. For example, Wang, et al. [29] employed the CA–Markov model for simulating future LULC under environmental protection (EPS), crop protection (CP), and spontaneous (SP) scenarios in Tianjin city. The authors found a dramatic increase in built-up areas under the SP scenario relative to those under the EP and CP scenarios. Similarly, Gebresellase, et al. [30] simulated future LULC dynamics in the Upper Awash Basin of Ethiopia using the CA–Markov model for the years 2030 and 2060 under the BAU and governance (GOV) scenarios considering various spatial driving variables. The authors found a significant increase in cultivated land and settlement under the BAU scenario and an increase in forest area under the GOV scenario.

In Ethiopia, several studies have been conducted on historical LULC change detection at the basin and watershed scales [31–33]. These historical LULC change detection studies are vital for better understanding the interactions between human activities and their surroundings [34,35]. In addition to historical LULC change detection, information on future LULC change trends over a long period of time is also critical. As LULC change rapidly varies in Ethiopia due to human and natural factors [36], modelling and documenting future LULC dynamics is important for providing information for land use planners and policymakers to develop sustainable land use management plans [37]. However, in Ethiopia, less attention has been given to the prediction of future LULC dynamics, except for a few recent studies conducted in the Omo-Gibe River Basin [38], Upper Awash River Basin [30], Nashe watershed [21], Goang watershed [22], Majang Forest biosphere Reserve [2], and Addis Ababa and surrounding areas [39].

The Gumara watershed is a subwatershed of the Lake Tana subbasin in the Upper Blue Nile Basin (UBNB) of Ethiopia. Previous studies conducted in watersheds frequently used the maximum likelihood (ML) algorithm for LULC classification [40,41], which requires normally distributed data for classification [42]. However, less consideration has been given to robust machine learning classifiers such as the RF algorithm, which does not require the assumption that the data need to be normally distributed. The Gumara watershed was selected as the study area since it is highly prone to soil erosion in the uplands and flooding in the lowlands resulting from LULC change [36,43]. For these reasons, the present study aims (1) to map the historical LULC dynamics using the RF algorithm built into the GEE environment for the years 1985, 2000, 2010, and 2019; (2) to detect LULC changes in the historical period (1985–2019); and (3) to predict future LULC dynamics under the BAU and GOV for the years 2035 and 2065 using the CA–Markov model, which is available from the IDRISI Selva17.0 software package. In this regard, the BAU scenario is based on the current trends of socioeconomic development, urbanization, intensive crop production, rapid population growth, the development of large industrial parks, and the construction of infrastructure [39]. On the other hand, the GOV scenario is based on the Green Legacy Initiative (GLI) program, which was launched in 2019 by the Ethiopian government to regreen the country by initiating massive tree seedling plantations [44,45]. These two scenarios were selected after identifying the land use characteristics of the study watershed, available data for land use drivers, and reviewing studies related to scenario-based simulations conducted in Ethiopia [30,39,46]. This study differs from previous studies conducted in the Gumara watershed: (1) it employed the RF algorithm, which is a robust machine learning LULC classifier, and (2) it provided scenario-based future LULC predictions for the near future and far future periods.

2. Materials and Methods

2.1. Study Area

This study was conducted in the Gumara watershed, which is a subwatershed of the Lake Tana subbasin in the Upper Blue Nile Basin (UBNB) of Ethiopia (Figure 1a,b). The watershed boundary lies between 11°35′ N–11°55′ north latitude and 37°30′ E–38°15′ east longitude (Figure 1c). The main river of the watershed is the Gumara River, which originates from the Guna Mountains and drains into Lake Tana. The total drainage area of the watershed delineated using the 30 m × 30 m digital elevation model (DEM) is approximately 1425 square kilometers (km^2) (Figure 1c). The elevation of the watershed generally ranges between 1782 and 3712 m above mean sea level (amsl) (Figure 1c). The watershed is characterized by a mountainous terrain with a steep gradient in the upstream part and an undulating terrain with a gentle gradient in the downstream part of the watershed [41]. Climatologically, the watershed experiences a long rainy season from June to September, with an average annual areal rainfall of 1445 mm. The average annual temperature of the watershed is approximately 20.5 °C. The lowland part of the watershed is prone to flooding in the wet season (June to September), and it also deposits eroded soil from upland subwatersheds [47]. In the watershed, cultivated land is the major LULC class and accounts for more than 75% of the watershed area; the remaining watershed area is covered with grassland, shrublands, forests, and settlements [32]. The lower part of the watershed adjacent to Lake Tana is known for rice cultivation [43]. In watersheds, rapid transitions of grassland, shrubland, and forest to cropland have been observed in recent decades. The majority of the watershed is dominated by rural kebeles with scattered settlement and their livelihood is based on mixed farming system by growing crops (teff, rice, wheat, barely, maize beans, oats, potato, etc.) and raising of livestock (goat, sheep, cattle, hen, etc.) [41]. In addition, watersheds have the potential for growth in different sectors, such as agroindustry, ecotourism, agroforestry, livestock, and energy [48]. The watershed under study was selected because it is prone to soil erosion and flooding as a result of intensive agricultural cultivation, as shown in previous studies [49,50].

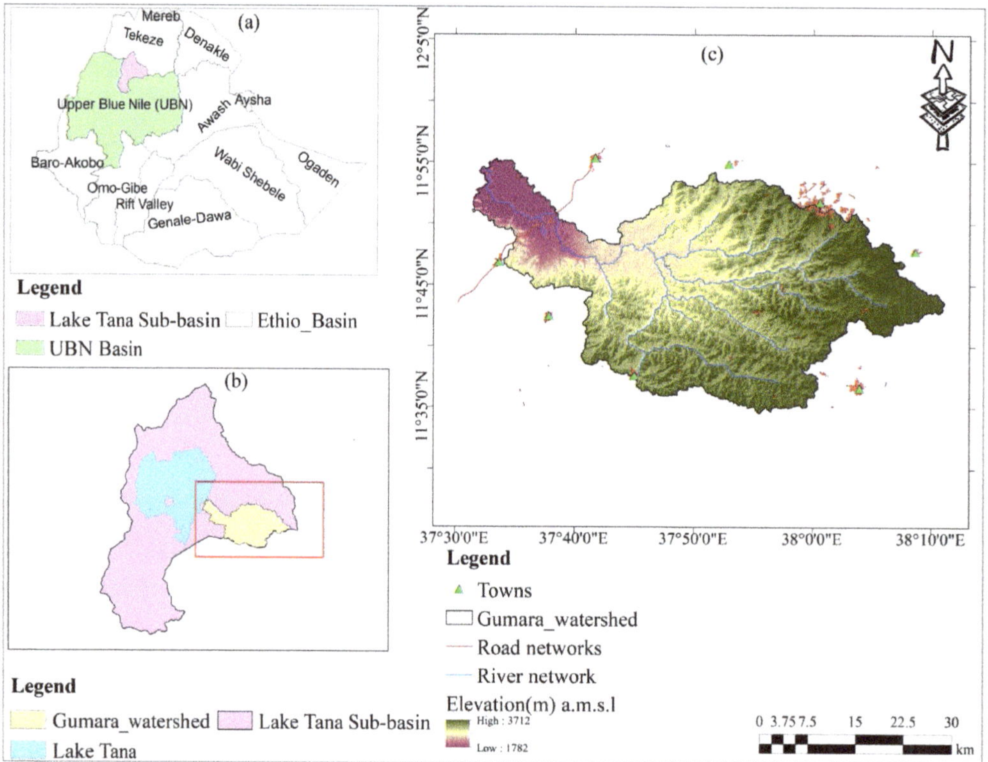

Figure 1. Location map of the study area: (**a**) River basins of Ethiopia, Upper Blue Nile Basin, and Lake Tana subbasin; (**b**) Lake Tana subbasin, Lake Tana, and Gumara watershed; and (**c**) Gumara watershed boundary, location of towns, road networks, river networks, and elevation map of the Gumara watershed.

2.2. Data Used

LULC maps of the study watershed were prepared using surface reflectance (SR) image collections from Landsat-5/Thematic Mapper (TM) for 1985, 2000, and 2010 and from Landsat-8/Operational Land Imager (OLI) for 2019. The images were accessed from the U.S. Geological Survey website via the GEE data catalog. Table 1 presents a description of the Landsat images used for LULC classification of the Gumara watershed. In the GEE environment, the images were filtered by the area of interest (AOI), the date of acquisition, and the percentage of cloud cover (less than 1%). Figure 2 shows the final median composite of the Landsat images for the years 1985, 2000, 2010, and 2019.

Table 1. Description of Landsat surface reflectance images used for LULC classification of the Gumara watershed.

Landsat Image	Date of Acquisition	Path/Row	No. of Image	Resolution (m)
Landsat-5/TM	1 January–30 April 1985	169/052	6	30
Landsat-5/TM	1 January–30 March 2000	169/052	6	30
Landsat-5/TM	1 January–30 March 2010	170/052	3	30
Landsat-8/OLI	1 January–30 March 2019	169/052	6	30

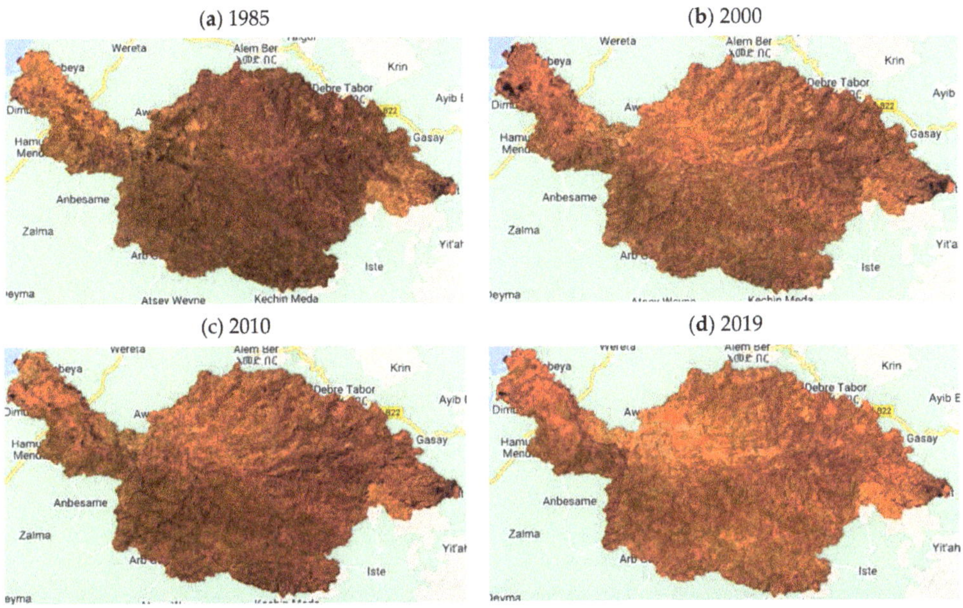

Figure 2. False color composites (NIR, red and green bands) of Landsat-5/TM (**a**–**c**) and Landsat-8/OLI (**d**) images used for LULC classification for the years (**a**) 1985, (**b**) 2000, (**c**) 2010, and (**d**) 2019. The deep red areas represent areas covered with scattered plants; the darker red areas represent densely vegetated areas.

In LULC classification using the SML algorithm, training and testing stages are important. For this study, training and testing data were collected through a field survey using a global positioning system (GPS) from January to February in 2022. Additionally, high-resolution historical satellite images from Google Earth Pro and Collect Earth Online (CEO) [51] were used to collect training and testing points. On this basis, points and polygons were randomly collected for each LULC class identified in Table 2. Accordingly, a total of 1784, 2226, 2910, and 2729 points were collected for the years 1985, 2000, 2010, and 2019, respectively. A table showing the number of training and testing/validation points collected for each LULC class can be found in the Supplementary Materials (S1). For classification, the shape files of the collected training/testing data were imported as assets into the GEE environment. Furthermore, other LULC change driving variables, such as elevation, slope, distance from roads, distance from streams, and distance from towns, were collected from different sources and processed using ArcGIS 10.4. The distance from roads was derived from the shape file of road network data acquired from the OpenStreetMap database (https://www.openstreetmap.org/#map=10/11.9090/38.2297, accessed on 2 February 2022). The distance from streams and the distance from towns were derived from the shape files of the stream network and town location, respectively. The elevation and slope were derived from the 30 m resolution DEM of the Shuttle Radar Topography Mission (SRTM) [52].

Table 2. Description of identified LULC classes in the Gumara watershed.

LULC Class	Description
Forest	Areas covered with open forest, dense forest, and woodland. This class mainly includes Eucalyptus tree and other woody plantations of the watershed.
Shrubland	Area of land covered with open and closed bushes and shrubs mainly found along the banks of rivers and streams.
Grassland	Areas covered with grasslands mainly used for grazing.
Cultivated land	Areas of agricultural land mainly used for crop cultivation. It also includes rice cultivation which concentrates at wetland part of the watershed.
Settlement	Areas of urban and rural settlements and other developments like roads.

2.3. Overview of the Methodology

This study mainly comprises two main sections: (1) LULC classification and change detection and (2) future LULC prediction. The first section was mainly focused on LULC classification for the years 1985, 2000, 2010, and 2019 and change detection (1985–2000, 2000–2010, 2010–2019, and 1985–2019), as presented in Figure 3. In this section, satellite image collection (Landsat-5/8), filtering image collection (by area of interest, date of acquisition, and metadata), computation of the median composite of filtered image collection, preparation of input variables (vegetation indices and topographic variables), collection and merging of training points, sampling region by integrating the median composite and merged training points, classification using the RF algorithm, accuracy assessment, and exporting of classified images were performed in the GEE environment. As GEE is a petabyte cloud-based platform, it makes image processing and classification easier without downloading the input satellite images [12]. Mapping of exported LULC images and change detection were performed using ARC-GIS 10.4. In this study, a change detection starting year (1985) was selected since substantial land use change was observed in most parts of Ethiopia during this year following the land reform proclamation of 1975 [35,53].

The second section of this study focused mainly on future LULC predictions, as presented in Figure 4, for the years 2035 and 2065. For this section, classified images from the first section and LULC change driver variables processed in ArcGIS 10.4 were used as inputs for the LULC prediction model. For the prediction, the CA–Markov model was employed because it combines the temporal prediction power of the Markov model with the spatial prediction power of the CA model. The procedures of this section were as follows: (1) preparation of input data (historical classified images and LULC change driver variables); (2) computation of transition probability matrices (TPMs) using the Markov model; (3) computation of transition suitability maps (TSMs); (4) prediction of LULC maps using the CA–Markov model for the reference year (2019); (5) validation of the CA–Markov model using the predicted and reference map of the year (2019); and (6) prediction of future LULC under the BAU and GOV scenarios for the years 2035 and 2065. In this study, LULC prediction years (2035 and 2065) were selected to examine the land use dynamics of the study watershed in the near future and in far future years.

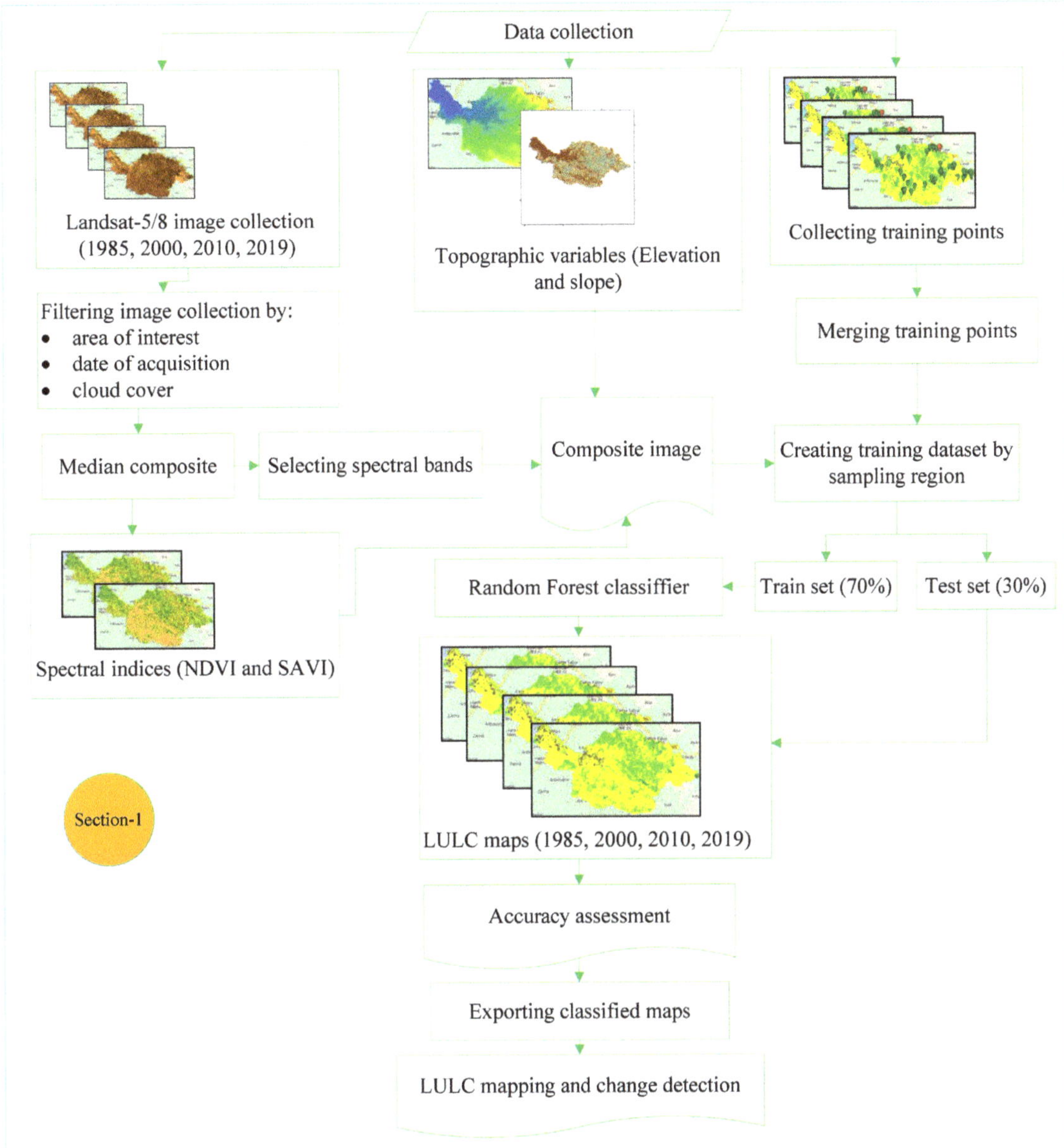

Figure 3. Methodological framework of LULC classification and change detection.

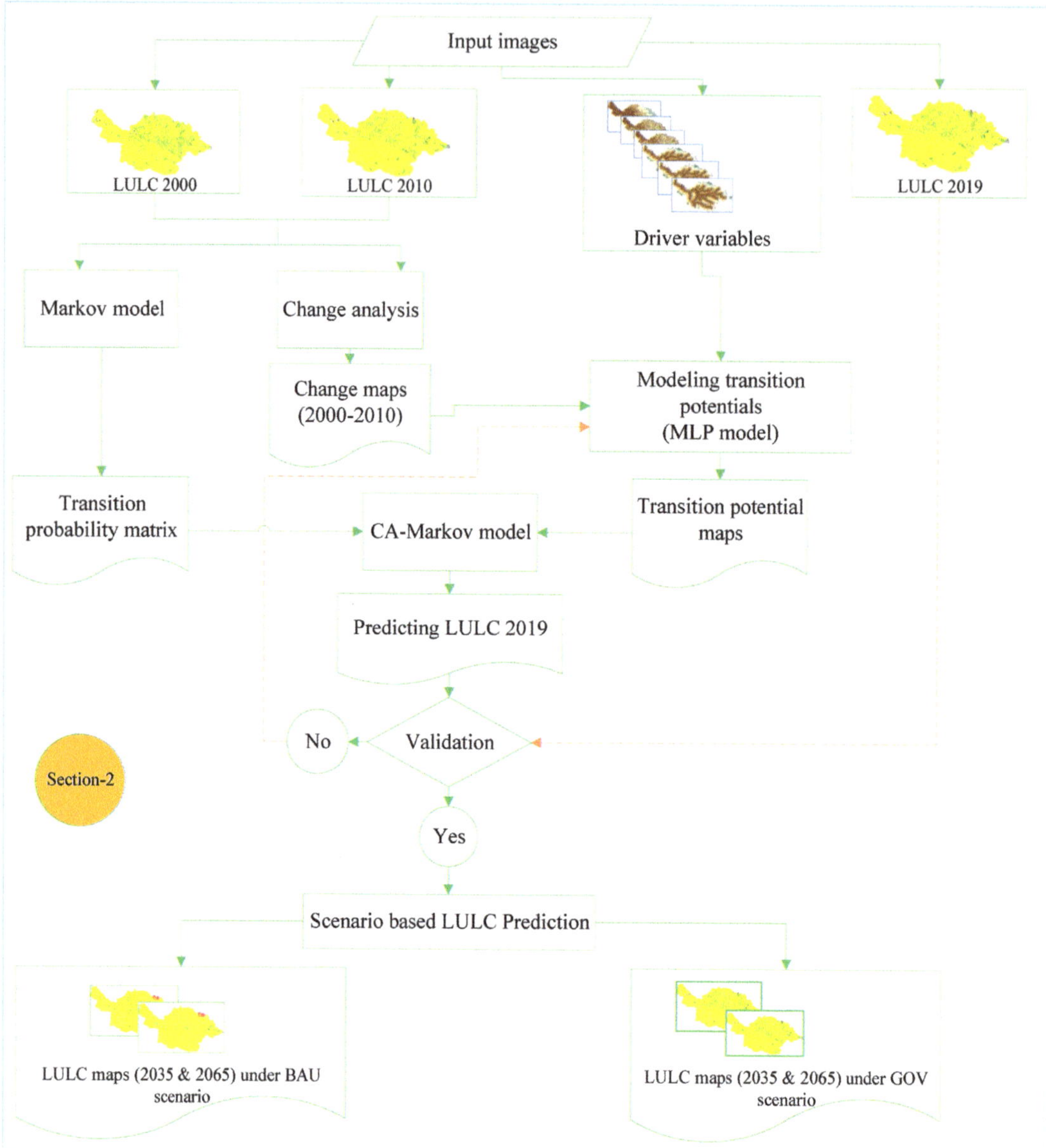

Figure 4. Methodological framework for future LULC prediction.

2.4. *LULC Classification and Change Detection*

The following subsections (Sections 2.4.1–2.4.5) discuss the details of LULC classification using the RF algorithm in the GEE environment, including variable selection, variable importance, classification accuracy assessment, and change detection methods.

2.4.1. LULC Classification

Among several classification algorithms available in the GEE environment, the RF algorithm, which was proposed by Breiman [17], was used for this study. As stated in previous studies [11,38,54], RF has gained popularity for image classification for a number of reasons: (a) it is capable of handling high-dimensional datasets; (b) it speeds up classification processes by selecting the most important variables; (c) it is an ensemble method; and (d) recent studies conducted on comparative analysis of classification algorithms have reported the superiority of RF over other SML algorithms for areas where agricultural cultivation is most dominant [6,16]. In addition, compared to other ML classification algorithms, RF requires a minimum number of parameters to be tuned. In LULC classification using SML algorithms, it is mostly recommended to use a larger percentage of the data for training than for testing to improve the performance of the classifier and to enhance the classifier to learn more about the spectral signatures of different LULC classes [10,18,19]. For this reason, the RF algorithm was trained using 70% of the ground truth data, and the accuracy of the classification was tested using 30% of the ground truth data. To split the data into training and testing sets, the "randomColumn()" function of GEE was used. For classification, the "smileRandomForest()" function of GEE was used. In the GEE environment, the "smileRandomForest()" function requires two main input parameters (N_{tree} and V_{split}). For this study, optimized values of these two parameters were estimated considering the classification accuracy and values of the parameters used in previous studies [55,56]. Accordingly, N_{tree} was found to be 200, and V_{split} was found to be 3, which is approximately the square root of the number of input variables. Using these optimized parameters of the RF algorithm, LULC classification was performed for the four historical years (1985, 2000, 2010, and 2019). Based on this, five LULC classes were identified for the Gumara watersheds, as shown in Table 2.

2.4.2. Input Variables

The spectral bands SR_B1 to SR_B5 and SR-B7 for Landsat-5/TM and SR_B2 to SR_B7 for Landsat-8/OLI were used as the input variables for the RF classifier. In addition to the spectral bands, vegetation indices and topographic factors (elevation and slope) were used to improve the classification accuracy. For the vegetation indices, the normalized difference vegetation index (NDVI) and soil adjusted vegetation index (SAVI) were used for this study. The NDVI is the most widely used indicator of vegetation greenness [57]. The NDVI values ranged from −1 to 1; low NDVI values (less than or equal to 0.1) corresponded to areas of bare land, rock, water, sand, or snow; moderate NDVI values (0.2 to 0.5) corresponded to sparse vegetation cover; and high NDVI values (greater than or equal to 0.6) corresponded to dense vegetation cover. The SAVI is a modified version of the NDVI that considers the effect of soil reflectance [58]. In the SAVI expression (Equation (1)), L takes different values depending on vegetation cover, where $L = 1$ in areas with no green vegetation cover, $L = 0.5$ in areas with moderate green vegetation cover, and $L = 0$ in areas with very high vegetation cover (which is equivalent to the NDVI equation). For this study, $L = 0.5$ was used considering the vegetation cover of the study watershed. Here, all the indices were computed in the GEE environment using the final median composite of the Landsat-5/8 images by applying Equations (1) and (2). Figures 5 and 6 show the computed spectral indices of the Gumara watershed for 1985, 2000, 2010, and 2019.

$$NDVI = \frac{NIR - RED}{NIR + RED} \quad (1)$$

$$SAVI = \left(\frac{NIR - RED}{NIR + RED + L}\right) * (1 + L) \quad (2)$$

where NIR is the near-infrared band (770–895 nm) and RED is the red band (630–690 nm).

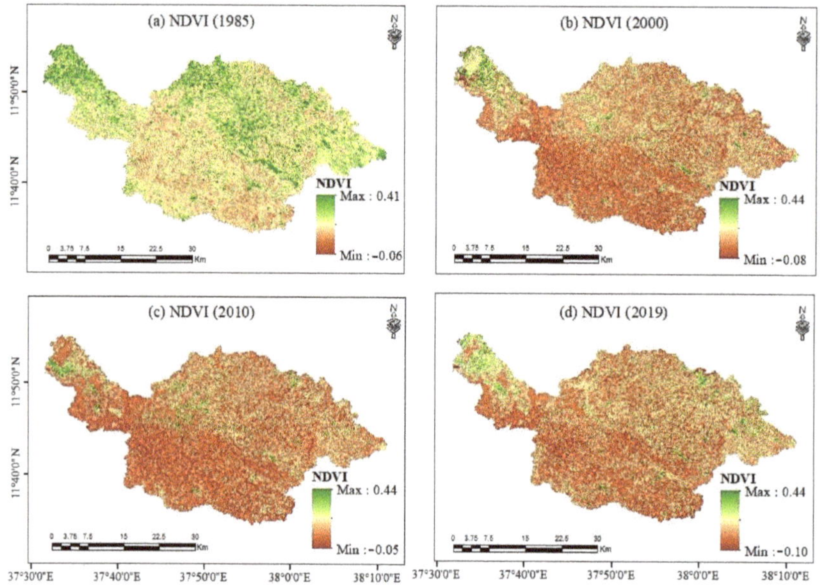

Figure 5. Computed NDVI images of the Gumara watershed for the years (**a**) 1985, (**b**) 2000, (**c**) 2010, and (**d**) 2019. In the figures, dark greens (maximum NDVI values example, NDVI = 0.4) represent vegetated areas, while dark reds (minimum NDVI values) represent bare soils or agricultural lands.

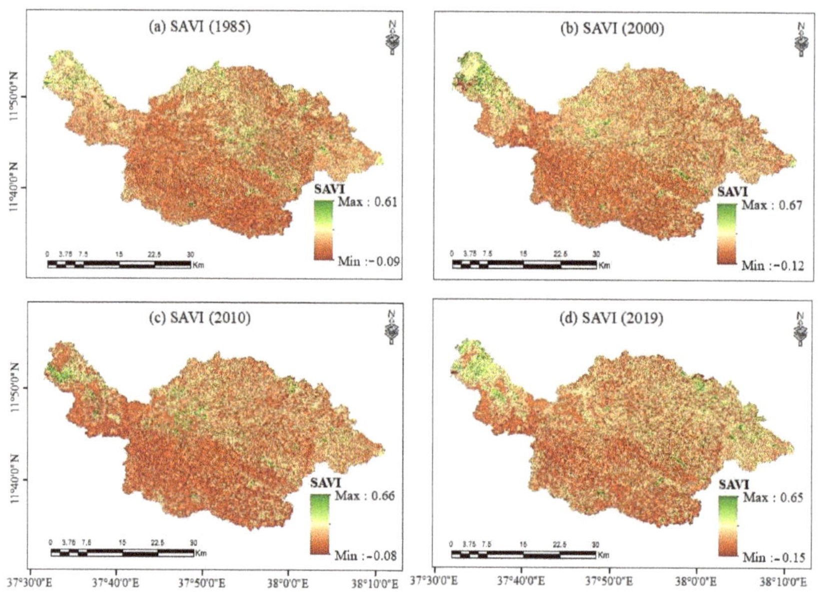

Figure 6. Computed SAVI images of the Gumara watershed for the years (**a**) 1985, (**b**) 2000, (**c**) 2010, and (**d**) 2019. In the figures, dark greens (maximum SAVI values, for example, SAVI $\geq$ 0.6) represent highly vegetated areas, while dark reds (minimum SAVIvalues) represent bare soils or agricultural lands.

2.4.3. Variable Importance

In this study, a total of 10 variables, which included spectral bands, vegetation indices and topographic factors, were used to train the RF algorithm as mentioned above (Section 2.4). In this case, all the input variables are not equally important when training the RF algorithm. In other words, each input variable has a different importance in separating one LULC class from another. Therefore, the impact of variables on LULC classification must be checked by computing variable importance (VI) [6,59]. For this purpose, the "explain()" function of GEE was used to generate the VI for each variable.

2.4.4. Accuracy Assessment

In image classification, it is vital to assess the performance of classifiers using test/validation datasets to determine whether the classification result is poor, satisfactory or strong [60,61]. For this study, a confusion matrix was produced in the GEE environment and exported as a comma separated value (csv) to evaluate the accuracy of the RF classifier. The accuracy was evaluated using the most widely used accuracy metrics, such as producer accuracy (PA), user accuracy (UA), overall accuracy (OA), and the kappa coefficient (K). Following Foody [61] PA, UA, OA, and K can be determined using Equations (3)–(6).

$$PA(\%) = \frac{X_{ij}}{x_{i+}} * 100 \tag{3}$$

$$UA(\%) = \frac{X_{ij}}{x_{+i}} * 100 \tag{4}$$

$$OA(\%) = \frac{\sum_{i=1}^{r} X_{ii}}{N} * 100 \tag{5}$$

$$K = \frac{N\sum_{i=1}^{r} X_{ii} - \sum_{i=1}^{r} x_{i+} * x_{+i}}{N^2 - \sum_{i=1}^{r} x_{i+} * X_{+i}} \tag{6}$$

where r is the number of rows in the confusion matrix, X_{ij} is the data in row i column j, X_{ii} is the number of correctly classified data in row i and column i (diagonal entries), x_{i+} is the marginal total of row i, x_{+i} is the marginal total of column i, and N is the total number of data points.

2.4.5. LULC Change Detection

LULC change detection analysis is important for determining the amount and type of land use changes that occurred over certain study periods. Among several change detection approaches, post classification change detection (PCCD), or map-to-map comparison approaches were employed for this study. This approach is most widely applied for comparing classified maps of different years [62,63]. In this study, for each of the LULC class percentage changes and rates of change for the four periods of 1985–2000, 2000–2010, 2010–2019, and 1985–2019 were computed to detect LULC changes in the study watershed using Equations (7) and (8).

$$P\Delta(\%) = \frac{A_2 - A_1}{A_1} * 100 \tag{7}$$

$$R\Delta\left(\text{km}^2/\text{year}\right) = \frac{A_2 - A_1}{t} \tag{8}$$

where $P\Delta$ is the percentage change, $R\Delta$ is the rate of change, A_1 is the area of the LULC class in km^2 in the initial year, A_2 is the area of the LULC class in km^2 in the final year, and t is the time interval of the period. Additionally, the amount of gain, loss, and net change were computed for each class in the four periods following the methodology provided in [64].

$$Gain\left(\text{km}^2\right) = CT - D \tag{9}$$

$$Loss\left(\text{km}^2\right) = RT - D \tag{10}$$

$$Net\ change\left(\text{km}^2\right) = Gain - Loss \tag{11}$$

where CT is the column total, RT is the row total, and D is the diagonal value of each LULC class that showed persistence.

2.5. LULC Prediction

Among the available future LULC prediction models, the combined cellular automata (CA) and Markov chain (CA–Markov) model has been frequently used in several studies [65–68]. This is because the CA–Markov model integrates the CA model for simulating the spatial dynamics of LULC changes and the Markov chain model for predicting the transition probabilities of LULC classes from one category to another over a certain period [2,23]. For this study, the main procedures followed to construct LULC maps of the Gumara watershed for the years 2035 and 2065 using the CA–Markov model available in the land change modeler (LCM) module of IDRISI Selva 17.0 are discussed in the following subsections.

2.5.1. LULC Change Driver Variables

There are many variables that drive LULC changes, which are related to land use policy, demographic distribution, socioeconomic development, climate change, and variability and topography [1,31,69]. However, these variables should be selected considering the specific conditions/characteristics of the study area. Based on these data, the elevation, slope, distance from streams, distance from roads, distance from towns, and evidence of likelihood of change were selected for this study, as illustrated in Figure 7. Elevation and slope were selected because these variables are important for determining suitable areas for cultivated land. The distance-to-socioeconomic development variables, such as distance from towns, distance from roads and distance from streams, were selected because these variables mainly influence the location of settlements. Here, the smaller the distance to towns, roads, or streams the greater the possibility for the LULC class to be converted to a settlement area. Evidence of likelihood of change was also selected to determine the probability of disturbance or to explain the change that occurred during the study period. In addition to the variables discussed here, population density is also a significant factor driving land use change. However, due to a lack of reliable census data, this factor was not considered in the present study. For this study, the variables presented in Figure 7a–e were processed in ArcGIS 10.4 using the Euclidian distance and surface function of the spatial analyst tool, while the evidence likelihood (Figure 7e) was processed in the land change modeler (LCM) module of IDRISI Selva 17.0. In the LCM module, driver variables can be modeled as dynamic or static. In this study, distance to roads, towns and streams and evidence of likelihood of change were modelled as dynamic variables to account for new developments in infrastructure in future years, whereas elevation and slope were modelled as static variables [70]. To facilitate future land use prediction, all the variables presented in Figure 7 were converted to the same pixel size (30 × 30 m), number of columns and rows (2398, 1279), and projection (WGS 1984 UTM Zone 37 N).

The quantitative association between the LULC changes and the driver variables was checked using Cramer's V test [71]. According to [72], the value of Cramer's V ranges between 0 and 1, with a high Cramer's V indicating good explanatory power of the variable. In this test, the p value also indicates the probability that Cramer's V value does not significantly differ from 0. Thus, a low p value is a good indicator that Cramer's V value should not be rejected, while a high p value is also a sure that the test is rejected.

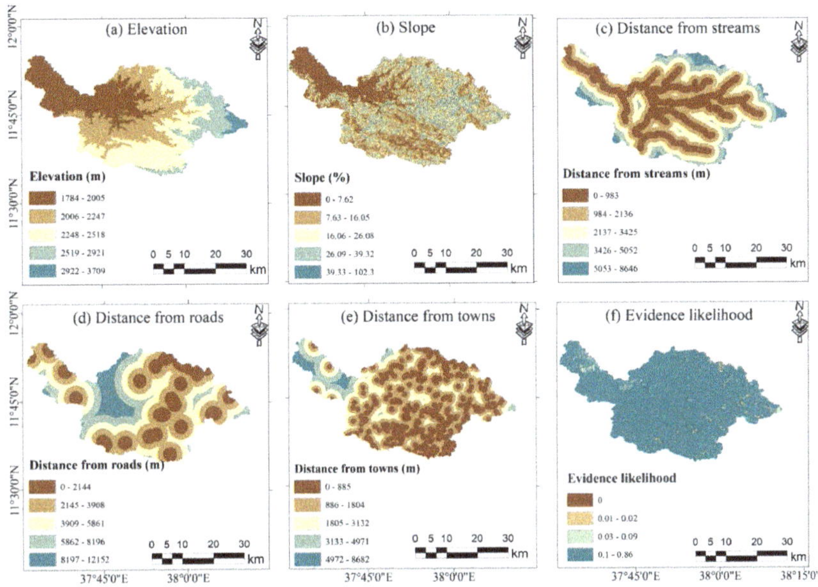

Figure 7. Map of driver variables: (**a**) elevation, (**b**) slope, (**c**) distance from streams, (**d**) distance from roads, (**e**) distance from towns, and (**f**) evidence likelihood.

2.5.2. Transition Probability Matrix (TPM)

Markov chain analysis is mostly applied for simulating complex processes by considering the initial and final states from an available set of states [8,68]. In the Markov chain model, the recent state of LULC (S_y) transforms to another state (S_{y+1}) at the next time, with the possibility indicated by the transition probability matrix (P_{jk}) (Equations (12)–(14)).

$$S_{y+1} = P_{jk} * S_y \tag{12}$$

$$P_{jk} = \begin{bmatrix} P_{11} & P_{12} & \ldots & P_{1n} \\ \vdots & \vdots & & \vdots \\ P_{n1} & P_{n2} & & P_{nn} \end{bmatrix} \tag{13}$$

$$0 \leq P_{jk} \leq 1 \; and \; \sum_{i=1}^{n} P_{jk} = 1, \;\; i, \, j = 1, \, 2, \, 3 \ldots ..n. \tag{14}$$

where P_{jk} is the transition probability matrix, which describes the likelihood of changing LULC from the (j) category to the (k) category in the LULC class at time S_y LULC; (y); S_{y+1} is the predicted LULC map at time $(y+1)$; and n is the number of LULC classes. A low value of the transition probability (approach to 0) indicates a decreased possibility of transition, while a high transition probability (approach to 1) indicates a high possibility of transition of a particular class to another [37]. For this study, the transition probability matrices was computed for the periods from 1985 to 2000, 2000 to 2010, 2010 to 2019, and 1985 to 2019. Afterwards, the generated TPMs were used to construct LULC maps for future years (for 2035 and 2065).

2.5.3. Transition Suitability Map (TSM)

In LULC prediction, the TSM provides information on the suitability of LULC for converting one class of cells to another [21,73]. In the LCM module of IDRISI Selva 17.0, the three most popular supervised machine learning algorithms, logistic regression (LR), similarity-weighted instance-based machine learning (SimWeight), and MLP-ANN, were used for generating TSMs. Among these algorithms, the MLP was used for this study since it is capable of modelling nonlinear relationships between dependent and independent variables [70]. In addition, multiple transition maps can be generated at once without the intervention of the modeler. The algorithm works as a feed forward ANN with one directional flow of data between the input, hidden, and output layers [14,74]. To generate the TSMs, historical LULC change patterns were used as the dependent variable, and topographic data (elevation and slope) and other proximity/distance-related driving factors were used as the independent variables (Section 2.5.1). For this study, the LULC TSMs were generated for the same periods as the TPM considering the most dominant LULC transition by setting a transition threshold of 2 km^2 in the LCM module. Based on these data, six dominant transitions, which included (1) transition from forest to shrubland, (2) forest to cultivated land, (3) shrubland to forest, (4) shrubland to cultivated land, (5) grassland to cultivated land, and (6) cultivated land to settlement, were identified. The reliability of the generated transition potential maps was evaluated by checking the accuracy of the MLP algorithm, where an accuracy greater than 80% was considered acceptable [70].

2.5.4. CA–Markov Model

As mentioned in Section 2.5.2, Markov chain analysis is important for estimating the TPM between earlier and later LULC maps. However, the model neglects spatial explicitness, and it does not provide information on the likely spatial distribution of LULC transitions [20]. Hence, by integrating CA with the Markov chain, this limitation can be overcome since the CA model uses spatial contiguity elements in its model structure [37,68]. CA is a bottom-up dynamic model that is mainly applied for spatiotemporal computations [68]. The model contains the cell space, cell, neighbors, set of rules, and time (Beroho et al., 2023). In the model, the neighbors can be identified by providing a contiguity filter to the CA model. It can simulate LULC changes over a certain time by using a set of rules that govern how a cell can change according to its initial state and its neighborhood, as expressed in Equation (15).

$$S_{y+1} = f(S_y,\ N) \tag{15}$$

where S_{y+1} is the later state that is predicted from the LULC map at time $(y+1)$; S_y is the initial state or the base LULC map at time (y); N is the neighborhood of each cell; and f is the governing transformation rule for the local space. In this study, future LULC maps of the Gumara watershed were generated for the years 2035 and 2065 using the CA–Markov model, which is available from IDRISI Selva 17.0. For prediction, earlier and latter LULC maps, Markov chain TPMs, and TSMs were used as inputs to the CA–Markov model.

2.5.5. Validation of the CA–Markov Model

In geospatial simulation, model validation is used to evaluate the accuracy, functionality and reliability of prediction performance [24]. For this study, model validation was performed by comparing the predicted LULC map from 2019 with the reference LULC map from 2019. For evaluation, commonly used kappa coefficients, namely, kappa for no information (K_{no}), kappa for location ($K_{location}$), and kappa for standard ($K_{standard}$), were computed using the "VALIDATE" module in IDRISI Selva 17.0. According to Congalton [75], a kappa coefficient less than 0 indicates no agreement, 0 to 0.2 indicates slight agreement, 0.2 to 0.41 indicates fair agreement, 0.41 to 0.60 indicates moderate agreement, 0.60 to 0.80 indicates substantial agreement, and 0.81–1.0 indicates perfect agreement.

2.5.6. Scenario-Based LULC Prediction

Recently, scenario-based ("what-if") prediction of future LULC has become a hot topic of research in the geospatial scientific community for identifying future land use dynamics under different socioeconomic conditions [9,20,30,66]. This scenario-based LULC prediction can serve as a robust tool for making and implementing decisions related to land use policies. For this study, two scenarios, (1) BAU and (2) GOV, were considered to predict the future LULC conditions of the Gumara watershed. These two scenarios were selected considering (1) the aim of the study, (2) the current land use conditions of the study watershed, (3) the availability of data for setting the scenarios, and (4) the future plans of the country Ethiopia (for example, the GLI program). The scenarios considered for this study were the same as the scenarios used in recent work of [30], which was conducted in the Upper Awash Basin of Ethiopia. However, the method used to determine the transition potential maps was different from this study.

The first scenario (BAU) assumes the continuity of the current trends in LULC changes and socioeconomic development (intensive agricultural cultivation, urbanization, and development infrastructures). The scenario was designed to answer the following question: "What will happen if the current trend of LULC changes and socioeconomic development continue in the future?" [9]. For this scenario, all the driving variables mentioned in the previous section (Section 2.5.1) and all dominant LULC "from–to" transitions were used to generate transition potential maps in the MLP algorithm. Under the BAU scenario, LULC drivers such as distance to roads, distance to towns, distance to streams, and evidence of likelihood of change were assumed to be dynamic variables to account for the future development of new infrastructure. Topographic factors (slope and elevation) were considered static variables. In this scenario, the expansion of agricultural lands and settlement areas is highly expected.

On the other hand, the second scenario (GOV) considers the GLI Program of Ethiopia. The GLI is a massive tree planting program that was first launched in 2019 with the objective of restoring degraded lands, increasing forest cover and reducing the impact of climate change [45,46]. Following the launch of the program, Ethiopia planted more than 20 billion tree seedlings from 2019 to 2022 and plans to plant 50 billion tree seedlings by 2026 [44]. In general, the objective of the program is to make Ethiopia a green and climate-resilient country [46]. In contrast to the BAU scenario, in the GOV scenario, increases in forests and other vegetation cover are expected due to controlled anthropogenic activities. For this scenario, only topographic/static variables (slope and elevation) were considered as driver variables. The other dynamic variables related to infrastructure development were neglected for this scenario. This scenario also assumes restricted transitions of different LULC classes to cultivated land and settlement, but it allows the transition of cultivated land to settlement.

3. Results

3.1. LULC Classification

Figure 8 shows the generated LULC maps of the Gumara watershed with five identified classes (forest, shrubland, grassland, cultivated land, and settlement). Based on the four LULC maps, cultivated land was the most dominant class, followed by shrubland, while settlements exhibited minimal areal coverage. For instance, in 1985, cultivated land accounted for 58.6%, followed by shrubland (31.72%), forest (5.22%), grassland (4.40%), and settlement (0.06%). In 2019, cultivated land accounted for more than 83.08%, followed by shrubland (12.79%), while forest, grassland, and settlement accounted for 2.59%, 1.36%, and 0.18%, respectively. This shows the significant expansion of cultivated land and settlement at the expense of reductions in shrubland, grassland and forest. As shown in Table 3 and Figure 9, between 1985 and 2019, the area of cultivated land and settlements significantly increased from 837.79 to 1187.48 km^2 and from 0.81 to 2.63 km^2, respectively. Consistently with these findings, Chakilu and Moges [32] reported significant expansion of cultivated lands in the Gumara watershed.

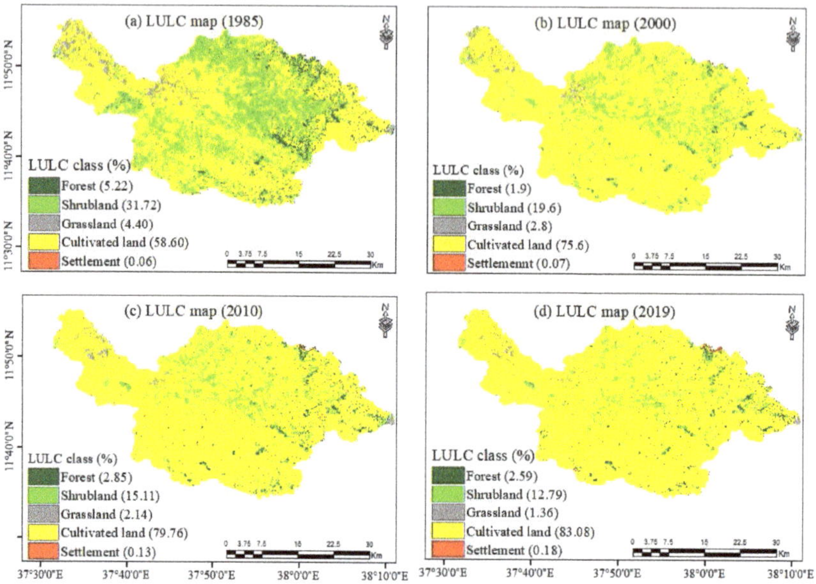

Figure 8. LULC maps of the Gumara watershed for (**a**) 1985, (**b**) 2000, (**c**) 2010, and (**d**) 2019. The values in the legend indicate the percentage of each LULC class.

Table 3. Area and percentage of each LULC class in the Gumara watershed for the four historical years (1985, 2000, 2010, and 2019).

LULC Class	Area (1985)		Area (2000)		Area (2010)		Area (2019)	
	(km^2)	(%)	(km^2)	(%)	(km^2)	(%)	(km^2)	(%)
Forest	74.60	5.22	27.62	1.93	40.77	2.85	37.01	2.59
Shrubland	449.49	31.72	280.00	19.58	216.00	15.11	182.81	12.79
Grassland	62.90	4.40	39.96	2.79	30.64	2.14	19.47	1.36
Cultivated land	837.79	58.60	1077.14	75.62	1136.11	79.76	1184.48	83.08
Settlement	0.81	0.06	0.99	0.07	1.88	0.13	2.63	0.18

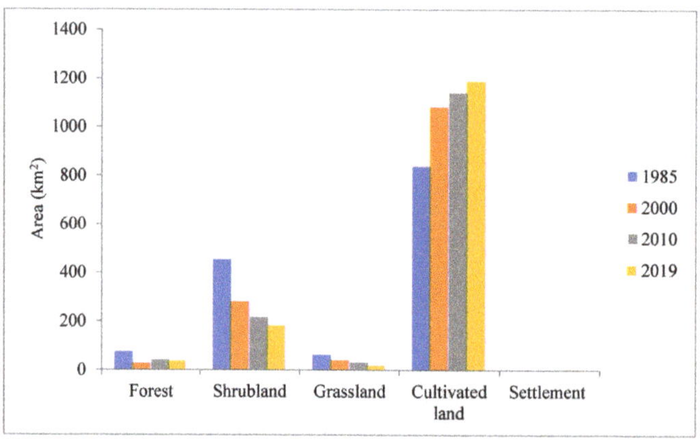

Figure 9. Area of each LULC class in the Gumara watershed for the four historical years (1985, 2000, 2010, and 2019).

3.2. Accuracy Assessment

Figure 10 and Table 4 present the results of the accuracy assessment in terms of UA, PA, OA, and K. Based on these results, the overall accuracies of the LULC maps of 1985, 2000, 2010, and 2019 were found to be 94.39%, 94.84%, 93.13%, and 91.13%, respectively (Table 4). The kappa coefficients were 0.92, 0.94, 0.90, and 0.88 for the corresponding years, respectively (Table 4). Based on the criteria provided in Congalton [75], all of the LULC maps showed perfect agreement with the validation dataset, with a kappa coefficient greater than 0.81. The confusion matrices used for accuracy assessment can be found in the Supplementary Materials (S2a–S2d).

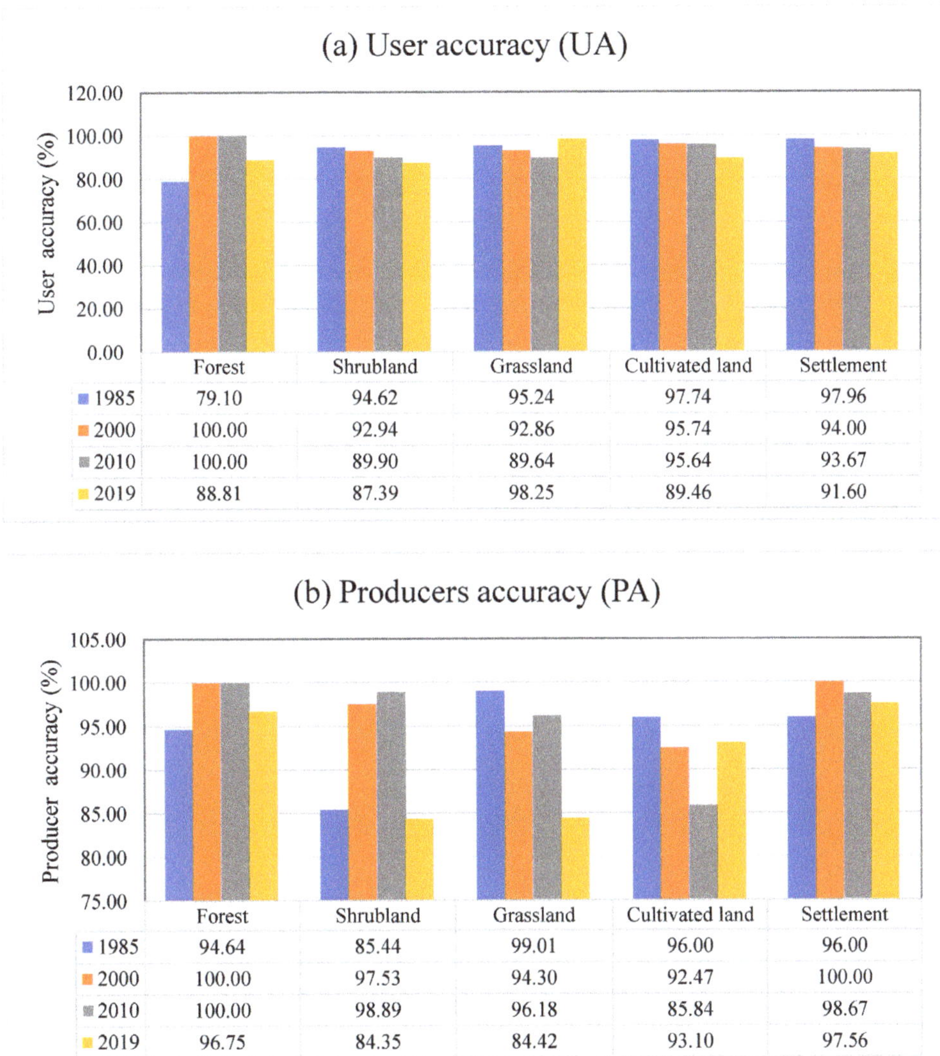

Figure 10. (**a**) UA and (**b**) PA assessment results for each class for the LULC maps for the years 1985, 2000, 2010, and 2019.

Table 4. Results of the classification accuracy assessment in terms of the overall accuracy (OA) and kappa coefficient (K).

LULC Map	Overall Accuracy	Kappa Coefficient	Status of Agreement
1985	94.39	0.92	Perfect agreement
2000	94.84	0.94	Perfect agreement
2010	93.13	0.90	Perfect agreement
2019	91.13	0.88	Perfect agreement

3.3. Variable Importance

In this study, a total of 10 variables (6 spectral band Landsat images, 2 vegetation indices and 2 topographic factors) were used to train the RF algorithm. Figure 11 shows the relative importance of these variables used for mapping LULC in the Gumara watershed. Among these variables, elevation was found to be the most important. Consistently with these findings, several studies have reported the importance of vegetation indices and topographic factors in improving LULC classification [55,76]. Among the vegetation indices, the NDVI was more important than the SAVI in most cases.

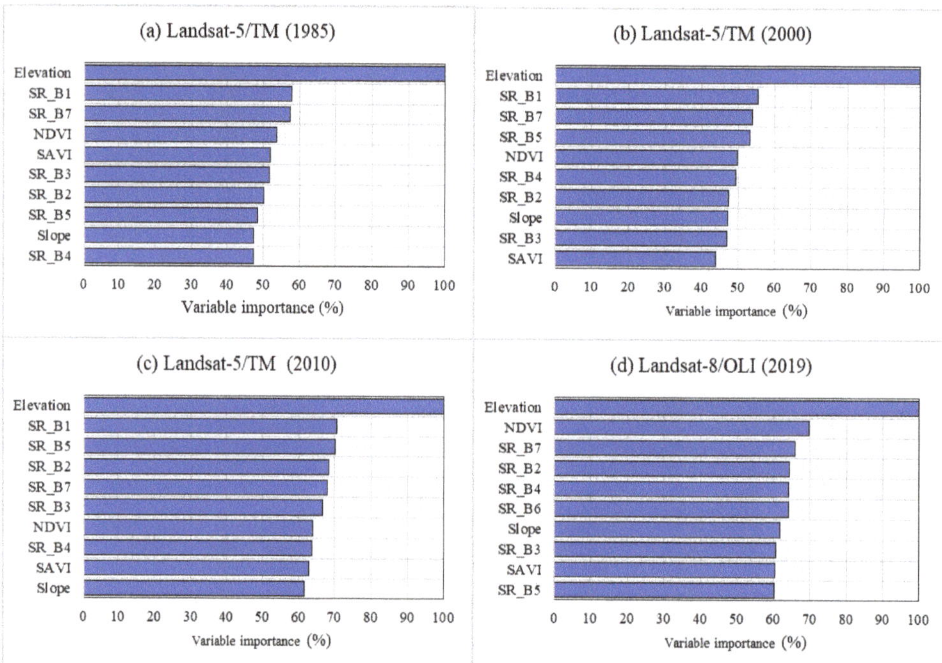

Figure 11. Relative variable importance (%) for the four datasets used for mapping LULC in the Gumara watershed: (**a**) Landsat-5/TM (1985), (**b**) Landsat-5/TM (2000), (**c**) Landsat 5/TM (2010), and (**d**) Landsat-8/OLI (2019).

3.4. Change Detection (1985–2019)

3.4.1. Percentage Change and Annual Rate of Change

Table 5 shows the percentage change and annual rate of change in each LULC class. According to the results, shrubland and grassland exhibited negative changes or reductions throughout the study period. Similarly, forestlands exhibited negative changes during all of the periods except for the period from 2000 to 2010. Considering all the percentage change values, the maximum positive change (41.74%) in cultivated land was detected for the period 1985 to 2019, while the maximum negative change (−69.05%) in grassland was detected for the same period. In terms of the rate of change, the maximum rate of increase (16.22 km^2/year) in cultivated land was detected for the period 1985 to 2000, while the maximum rate of decrease (−11.57 km^2/year) in shrubland was detected for the same period (1985–2000), as shown in Table 5.

Table 5. Percent change (PΔ) in (%) and annual rate of change (RΔ) in km^2/year for the four study periods.

LULC Class	1985–2000		2000–2010		2010–2019		1985–2019	
	PΔ	RΔ	PΔ	RΔ	PΔ	RΔ	PΔ	RΔ
Forest	−62.98	−3.13	37.64	1.32	−9.24	−0.42	−50.40	−1.11
Shrubland	−38.26	−11.57	−22.86	−6.40	−15.36	−3.69	−59.69	−7.96
Grassland	−36.48	−1.53	−23.33	−0.93	−36.46	−1.24	−69.05	−1.28
Cultivated land	29.05	16.22	5.45	5.90	4.15	5.26	41.74	10.28
Settlement	22.31	0.01	39.32	0.09	40.29	0.08	34.86	0.05

3.4.2. Gain, Loss, and Net Change

Table 6 shows the gains and losses of different LULC classes in the Gumara watershed for the four study periods computed using the LCM module of IDRISI Selva 17.0. The net change in each LULC class was also computed by taking into account the difference in the gain and the loss. For the first period (1985–2000), gains and losses of (322.28, −80.18), (91.32, −262.25), (23.87, −46.45), and (4, −52.45) km^2 were observed for cultivated land, shrubland, grassland, and forest, respectively (Table 6). For the second period (2000–2010), cultivated land gained 139.77 km^2, while shrubland and grassland lost 125.20 and 31.16 km^2, respectively. Similarly, in the other periods, larger values of gains than losses were found for cultivated land and settlement, while smaller values of gains than losses were found for the other classes. For this reason, the net changes were found to be positive for cultivated land and settlement and negative for the other classes (Figure 12). For example, for cultivated land, positive net changes of 242.10, 56.31, 49.59, and 348.01 km^2 were found for the corresponding periods. For shrubland, negative net changes of −170.93, −61, 03, −34.87, and −266.83 km^2 were found for the corresponding periods. For the forest class, negative net changes were found, with the exception of the second period (2000–2010).

Table 6. Gain and loss of each LULC class for the four study periods.

LULC Class	1985–2000		2000–2010		2010–2019		1985–2019	
	Gain (km^2)	Loss (km^2)	Gain (km^2)	Loss (km^2)	Gain (km^2)	Loss (km^2)	Gain (km^2)	Loss (km^2)
Forest	4.00	52.45	14.58	1.57	8.43	12.01	11.15	50.16
Shrubland	91.32	262.25	64.17	125.20	56.74	91.61	51.74	318.56
Grassland	23.87	46.45	21.29	31.16	9.86	21.72	9.15	53.45
Cultivated land	322.28	80.18	139.77	83.46	113.48	63.88	399.17	51.16
Settlement	0.71	0.57	2.00	0.57	2.29	1.72	2.86	0.71

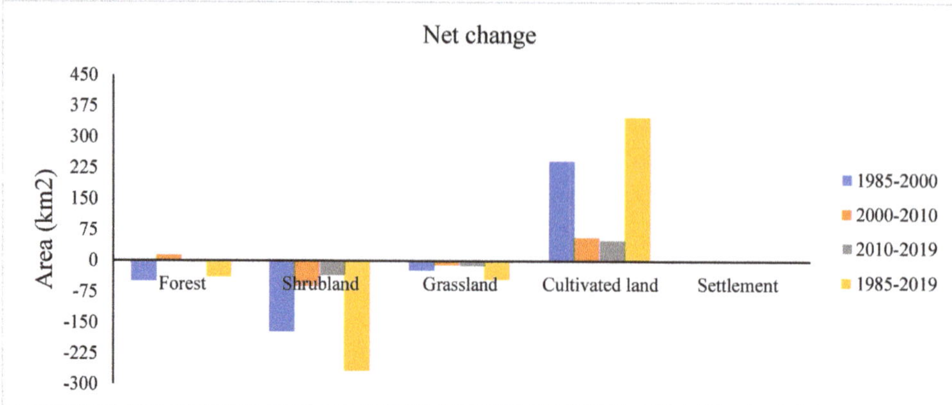

Figure 12. Net change (gain-loss) in each LULC class for the four study periods (1985–2000, 2000–2010, 2010–2019, and 1985–2019).

3.4.3. Contribution to the Net Change in Cultivated Land

To show the contribution of the four LULC classes to the net change in cultivated land over the study period, four graphs were generated, as shown in Figure 13a–d. In all the periods, shrubland was found to be the main contributor to the increase in the net change in cultivated land, followed by grassland. For example, in the first period (1985–2000), 200.54, 22.46, and 21.0 km^2 area of land were contributed by shrubland, grassland and forest, respectively, to increase the net change in cultivated land area (Figure 13a). Similarly, in the fourth period (1985–2019), areas of 280.67, 44.80, and 28.75 km^2 were contributed by shrubland, grassland, and forest, respectively, to increase the net change in cultivated land by 353.5 km^2 (Figure 13d). Overall, shrubland and grassland contributed to the increase in cultivated land in all four periods. Forests contributed to the increase in the net change in the first and fourth periods (Figure 13a,d) and contributed to the reduction in the net change in the second and third periods (Figure 13b,c); however, settlements contributed to the reduction in the net change in the second and third periods (Figure 13b,c), and they also contributed negligibly to the increase in the other periods.

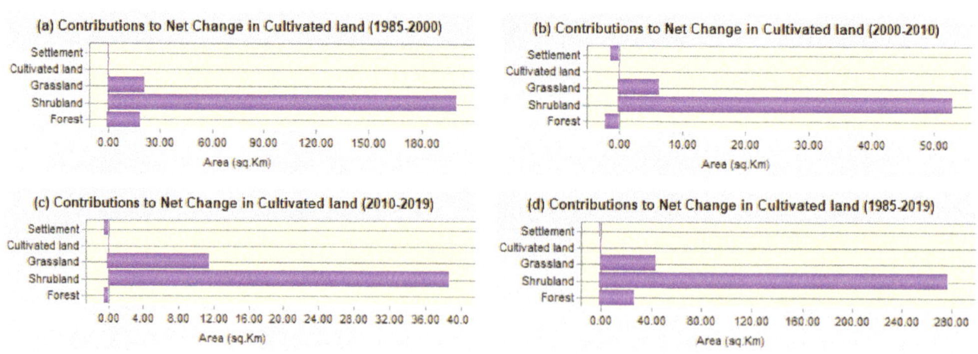

Figure 13. Contribution of each LULC class to the net change in cultivated land: (**a**) 1985–2000, (**b**) 2000–2010, (**c**) 2010–2019, and (**d**) 1985–2019.

3.5. LULC Change Driver Variables

For this study, topography and distance-/proximity-related driver variables that influence LULC changes were considered to generate the TSMs of the most dominant changes that occurred in the study watershed. These driver variables were modelled either as static or dynamic variables in the LCM module of IDRISI Selva 17.0. Before the drivers were used for predicting transition potential or suitability maps, the explanatory power of each variable was evaluated based on Cramer's V and p values (Table 7). Although Cramer's V value does not offer strong evidence for describing the explanatory power of the driving variable, it can be taken as a simplified approach to test the influence of the variables [8,21]. Based on the Cramer's V values presented in Table 7, the evidence of likelihood (probability disturbance) was found to be the most important variable, as it had the highest Cramer's V value (0.4885). Similarly to these findings, several studies have shown the importance of topographic factors and evidence of likelihood in influencing the spatial distribution of LULC changes [31,33].

Table 7. Cramer's V values of LULC driver variables.

Driver Variables	Cramer's V	p Value
Elevation	0.2105	0.0000
Slope	0.0437	0.0000
Distance from streams	0.0868	0.0000
Distance from roads	0.1004	0.0000
Distance from towns	0.0521	0.0000
Evidence likelihood	0.4885	0.0000

3.6. Transition Probability Matrix (TPM)

In LULC change studies, the TPM shows the probability of conversion of LULC classes from one category to another. For this study, four 5 × 5 Markovian TPMs were generated for the periods from 1985 to 2000, 2000 to 2010, 2010 to 2019, and 1985 to 2019 (Table 8a–d) using the Markov module of IDRISI Selva 17.0. In the matrices, the columns show the later or the newer year LULC classes (e.g., the LULC class of 2000), while the rows show the previous or earlier year LULC classes (e.g., the LULC class of 1985). The diagonal values shaded in grey show the proportion of the classes that showed persistence (LULC classes remaining unchanged), whereas the off-diagonal values show the proportion of the LULC classes that changed from one category to another between the earlier and the later years. High percentages of persistence were observed for the cultivated land in the four respective periods, with values of 91.41, 91.48, 93.81, and 96.99%. Similarly, settlements also exhibited persistent proportions of 87.44, 81.17, 82.49, and 81.80% for the corresponding periods. The other LULC classes (forest, shrubland, and grassland) exhibited minimum values of persistence percentages or probabilities. For instance, between 1985 and 2000 (Table 8a), persistent percentages of 18.48, 28.88, and 25.97% were observed for forest, shrubland, and grassland, respectively. Regarding the transition between LULC classes (off-diagonal), high percentages of conversion of shrubland, grassland, and forest to cultivated land were observed in the four periods. For example, high percentages of conversion of shrubland to cultivated land were observed, with values of 69.70, 66.78, 74.32, and 76.91% for the four respective periods, as presented in Table 8a–d. Similarly, percentages of 67.47, 76.50, 79.78, and 75.33% were observed for the conversion of grassland to cultivated land (Table 8a–d). These results indicate the expansion of agricultural cultivation and settlement areas in the Gumara watershed at the expense of forest, shrubland, and grassland. For this purpose, rapid population growth, demand for wood for charcoal and construction material, and extensive rice cultivation are the main factors influencing the expansion of cultivated land [41,43].

Table 8. (**a**) Transition probability matrix of LULC classes in the Gumara watershed from 1985 to 2000. (**b**) Transition probability matrix of LULC classes in the Gumara watershed from 2000 to 2010. (**c**) Transition probability matrix of LULC classes in the Gumara watershed from 2010 to 2019. (**d**) Transition probability matrix of LULC classes in the Gumara watershed from 1985 to 2019.

(a)

1985	2000				
	Forest	**Shrubland**	**Grassland**	**Cultivated Land**	**Settlement**
Forest	0.1848 [1]	0.4651	0.0043	0.3447	0.0010
Shrubland	0.0053	0.2888	0.0087	0.6970	0.0002
Grassland	0.0013	0.0643	0.2597	0.6747	0.0000
Cultivated land	0.0020	0.0614	0.0218	0.9141	0.0007
Settlement	0.0000	0.0048	0.0000	0.1208	0.8744

(b)

2000	2010				
	Forest	**Shrubland**	**Grassland**	**Cultivated Land**	**Settlement**
Forest	0.8073	0.0872	0.0017	0.0526	0.0012
Shrubland	0.0710	0.2508	0.0092	0.6678	0.0012
Grassland	0.0072	0.0997	0.1275	0.7650	0.0006
Cultivated land	0.0045	0.0598	0.0191	0.9148	0.0018
Settlement	0.0018	0.0248	0.0071	0.1546	0.8117

(c)

2010	2019				
	Forest	**Shrubland**	**Grassland**	**Cultivated Land**	**Settlement**
Forest	0.4571	0.2724	0.0030	0.2653	0.0022
Shrubland	0.0298	0.2198	0.0059	0.7432	0.0013
Grassland	0.0050	0.0305	0.1657	0.7978	0.0010
Cultivated land	0.0051	0.0461	0.0087	0.9381	0.0020
Settlement	0.0036	0.0246	0.0034	0.1434	0.8249

(d)

1985	2019				
	Forest	**Shrubland**	**Forest**	**Cultivated Land**	**Forest**
Forest	0.5800	0.1526	0.0020	0.2594	0.0059
Shrubland	0.0082	0.2225	0.0000	0.7691	0.0002
Grassland	0.0006	0.0126	0.2335	0.7533	0.0000
Cultivated land	0.0034	0.0204	0.0051	0.9699	0.0012
Settlement	0.0016	0.0020	0.0000	0.4784	0.8180

[1] The diagonal values shaded in grey colour show the proportion of the LULC classes that showed persistence (remaining unchanged).

3.7. Transition Suitability Maps (TSMs)

In this study, the six most dominant "from–to" transition potentials were considered, as discussed in section Section 2.5.3. To predict transition potentials, a feed-forward enhanced MLP neural network model was employed using the IDISI Selva 17.0 LCM module. In this regard, the main outputs of the MLP model are transition potential or suitability maps, which provide information on the suitability for change [70]. The TSMs that show potential for transition considered in this study can be found in the Supplementary Materials (S3). On the basis of these maps, the potentials for the transition from shrubland to cultivated land and from cultivated land to settlement are presented in Figure 14a,b. Hence, in the region of the watershed indicated by a circular shape, either the expansion of cultivated land or the expansion of settlement areas is highly expected in future years. This part of the watershed includes the town of Debre Tabor, which is the capital of the South Gonder Zone in the Amhara regional state.

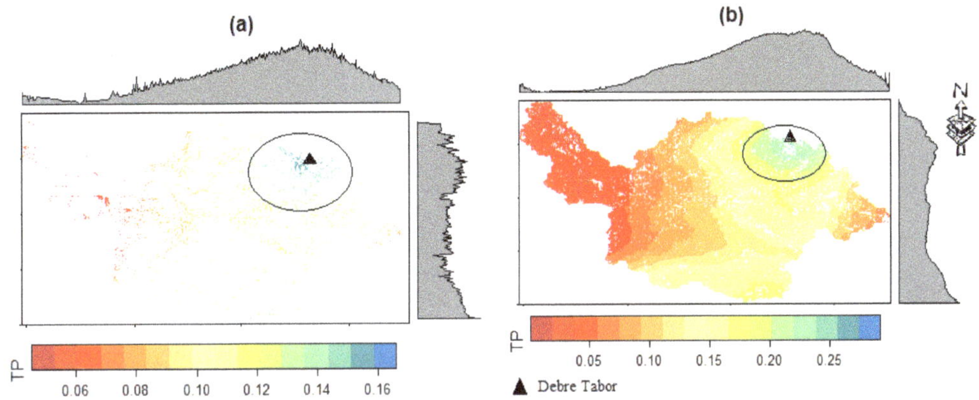

Figure 14. Potential for transition: (**a**) shrubland to cultivated land and (**b**) cultivated land to settlement. TP is the transition potential. The greater the TP is, the greater the possibility of a transition from one class to another. The gray shaded regions show the orientation gradients of the transition potential, wherein the maximum transitions are oriented along the northeastern part of the watershed for both transitions. The areas bordered by circles indicate the maximum values of transition suitability. The triangle symbol in both of the figures indicates the location of the town Debre Tabor.

3.8. Validation of the CA–Markov Model

In this study, the performance of the CA–Markov model in predicting future LULC dynamics was evaluated by comparing the reference and predicted (simulated) LULC maps from 2019 (Figure 15). For validation, the LULC map of 2019 under the BAU scenario (Figure 15b) was generated using the LULC map of 2000 as an earlier image and the LULC map of 2010 as a later image, along with the transition probability matrix and the transition potential maps. Kno, Klocation, and Kstandard were found to be 0.89, 0.86, and 0.94, respectively. These results confirm the high reliability of the CA–Markov model for predicting the future LULC dynamics of the Gumara watershed since all the kappa values are greater than 0.81 based on the criteria provided in Congalton [75]. After evaluating the performance of the CA–Markov model, future LULC maps were predicted under the BAU and GOV scenarios for 2035 and 2065. Figure 16 also shows a comparison of the areas of the LULC classes between the reference and predicted maps for 2019. Based on these findings, the areas of the LULC classes are nearly similar except for slight overprediction of the cultivated land and slight underprediction of all the other classes by the CA–Markov model compared to the areas of the LULC classes of the reference map (Figure 16). As mentioned above, LULC validation was performed to evaluate the prediction capability of the CA–Markov model under the BAU scenario only. As the GOV scenario was recently launched by the Ethiopian government in 2019, validation was not performed for this scenario.

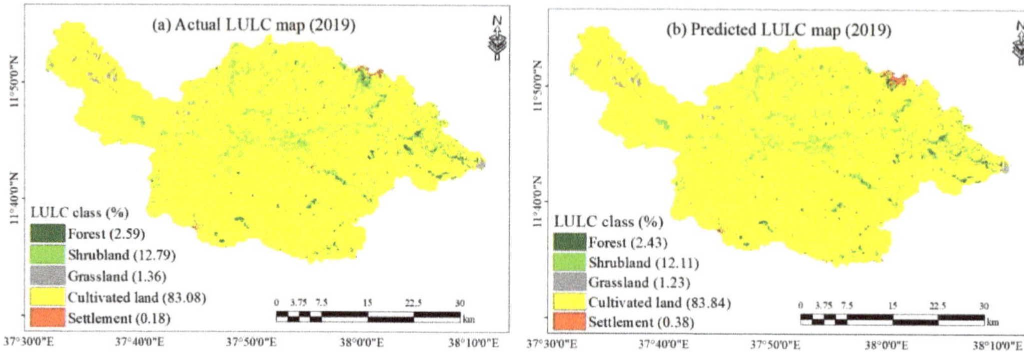

Figure 15. LULC maps (2019): (**a**) reference LULC map and (**b**) CA–Markov model-predicted LULC map under the BAU scenario.

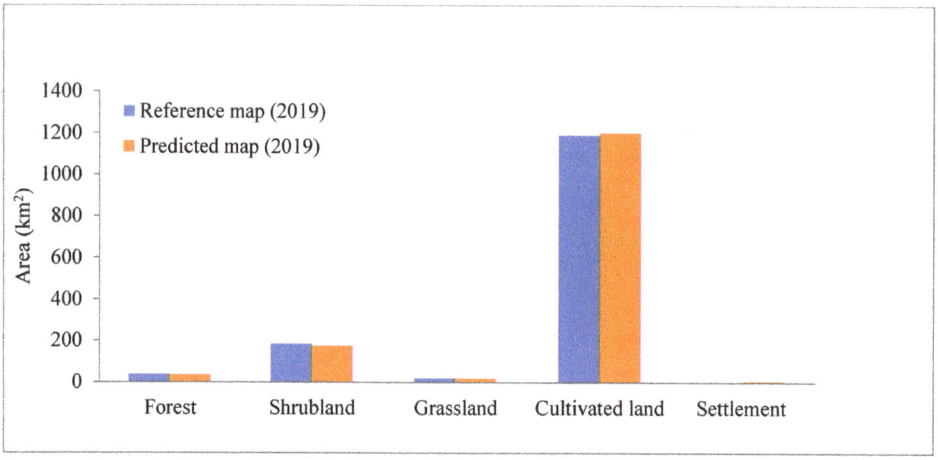

Figure 16. Comparison of the reference (baseline) and predicted areas of the LULC classes in the Gumara watershed in 2019.

3.9. LULC Prediction

The future LULC conditions of the Gumara watershed were predicted considering the historical change trends under two distinct scenarios. Figure 17 and Table 9 present the future LULC dynamics of the Gumara watershed in 2035 and 2065 under the BAU and GOV scenarios simulated based on the CA–Markov model. Under the BAU scenario, consistent increases in cultivated land and settlement will occur in the watershed relative to those in the reference year (2019). Based on these results, the cultivated land area, which accounted for 1187.48 km^2 (83.08%) in 2019, will increase to 1240.61 km^2 (86.77%) and 1272.62 km^2 (89.01%) in 2035 and 2065, respectively (Table 9). In these years, cultivated land is expected to expand in the lower part adjacent to Lake Tana and in the upper part of the watershed near Debre Tabor town due to the suitability of the slope of the land surface for crop production (Figure 14a,b). In particular, fragmented grassland covers found in the lower part of the watershed adjacent to Lake Tana (wetland part) are highly likely to be converted to cultivated land because this area is known for its potential for rice production. Similarly, under the BAU scenario, the settlement area will increase from 2.63 km^2 (0.18%) in 2019 to 8.06 km^2 (0.56%) and 11.90 km^2 (0.83%) in 2035 and 2065, respectively (Table 9). As presented in Figure 17a,b, significant expansion of urban settlements in and around Debre Tabor town was predicted in the future years 2035 and 2065. Forests, shrublands, and cultivated land found near towns were the main contributors to the expansion of settlements. In contrast to the expansion of cultivated land and settlements, consistent decreases in forest, shrubland, and grassland areas will occur in the Gumara watershed relative to those in the reference year (2019). Based on the results, the BAU scenario predicts forests to be 20.53 km^2 (1.44%) and 15.50 km^2 (1.08%), shrublands to 149.5 km^2 (10.46%) and 119.70 (8.37%), and grasslands to be 11.12 km^2 (0.78%) and 10.10 km^2 (0.71%) in 2035 and 2065, respectively. Under this scenario, the forest and shrubland cover found near Debre Tabor town was predicted to be converted to settlement and cultivated land; only a small percentage of the forests were reserved at high elevations, while the shrubland was reserved at mid-elevation areas (Figure 17a,b).

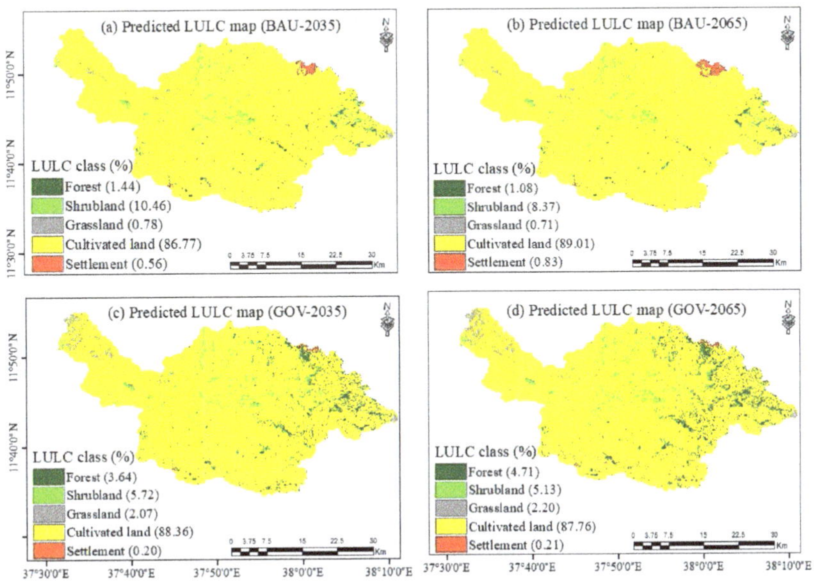

Figure 17. Predicted LULC maps of the Gumara watershed: (**a**) for 2035 and (**b**) for 2065 under the BAU scenario; (**c**) for 2035 and (**d**) for 2065 under the GOV scenario.

Table 9. Predicted area and percentage of LULC classes in the Gumara watershed in 2019 and in 2035 and 2065 under the BAU and GOV scenarios.

LULC Class	Reference (2019)		BAU (2035)		BAU (2065)		GOV (2035)		GOV (2065)	
	Area (km^2)	%	Area (km^2)	%	Area (km^2)	%	Area (km^2)	%	Area (km^2)	%
Forest	37.01	2.59	20.53	1.44	15.50	1.08	52.05	3.64	67.30	4.71
Shrubland	182.81	12.79	149.50	10.46	119.70	8.37	81.74	5.72	73.28	5.13
Grassland	19.47	1.36	11.12	0.78	10.10	0.71	29.65	2.07	31.44	2.20
Cultivated land	1184.48	83.08	1236.61	86.77	1268.62	89.01	1259.44	88.36	1250.86	87.76
Settlement	2.63	0.18	8.06	0.56	11.90	0.83	2.93	0.20	2.96	0.21

Unlike in the BAU scenario, forest and grassland are expected to increase at the expense of a decrease in cultivated land under the GOV scenario (Table 9). In this scenario, forest cover is expected to reach 52.05 km^2 (3.64%) and 67.30 km^2 (4.71%), and grassland cover is expected to reach 29.65 km^2 (2.07%) and 31.44 km^2 (2.2%) in 2035 and 2065, respectively. Under this scenario, shrubland will cover a smaller percentage (5.72% in 2035 and 5.12% in 2065) of the watershed due to the conversion of shrubland to forest through the implementation of afforestation programs. Under the GOV scenario, cultivated land is expected to decline slightly from 1263.44 km^2 (88.36%) in 2035 to 1254.86 km^2 (87.76%)

in 2065. This scenario also promotes controlled or limited settlement expansion to reduce competition for various natural and economic resources.

3.10. Change Detection (2019–2065)

Table 10a,b shows the gains and losses of each LULC class in the Gumara watershed for the future periods under the two different scenarios. The results revealed contrasting changes under the different scenarios. For instance, under the BAU scenario, for the period from 2019 to 2065, cultivated land increased by 60.45 km^2, while forest and grassland decreased by 9.86 and 8.29 km^2, respectively (Table 10a). In contrast, for the same period, under the GOV scenario, forest and grassland increased by 42.02 and 13.01 km^2, respectively, while cultivated land decreased by 26.30 km^2 (Table 10b). Figure 18a,b also show the net change values for the future periods for different LULC classes under different scenarios. Based on the results, positive net change values of cultivated land and settlement and negative net change values of other classes were found under the BAU scenario (Figure 18a). For instance, net changes of 43.44, 8.00, and 3.85 km^2 in cultivated land and 5.14, 3.85, and 8.88 km^2 in settlement were found for the three periods. In contrast, under the GOV scenario, positive net change values for forestland and grassland and negative net change values for the other classes were found in the three respective periods (Figure 18b). For instance, the largest positive net change in forest area was detected in the period from 2019–2065, with a value of 42.01 km^2, while a negative net change in forest area (loss) was detected for shrubland, with a value of −28.58 km^2. In general, in the BAU scenario, larger values of losses than gains (negative net change) for forest, shrubland and grassland and larger values of gains than losses (positive net changes) for cultivated land and settlement were found for the future periods (Table 10a, Figure 18a). Under the GOV scenario, larger values of gains than losses (positive net changes) for forest, shrubland and grassland and larger values of losses than gains (negative net change) for cultivated land were found for the future periods (Table 10b, Figure 18b).

Table 10. (**a**) Gains and losses of LULC classes between 2019 and 2065 under the BAU scenario. (**b**) Gains and losses of LULC classes between 2019 and 2065 under the GOV scenario.

(a)

LULC	2019–2035		2035–2065		2019–2065	
	Gain	Loss	Gain	Loss	Gain	Loss
Forest	0.00	6.72	0.00	3.14	0.00	9.86
Shrubland	3.43	37.87	2.57	10.29	2.57	44.73
Grassland	0.00	7.29	0.00	1.00	0.00	8.29
Cultivated land	48.59	5.15	11.86	3.86	60.45	9.00
Settlement	5.15	0.02	3.86	0.05	9.00	0.02

(b)

LULC	2019–2035		2035–2065		2019–2065	
	Gain	Loss	Gain	Loss	Gain	Loss
Forest	26.73	0.00	15.29	0.00	42.02	0.00
Shrubland	0.01	20.15	0.05	8.43	0.09	28.58
Grassland	11.15	0.02	1.86	0.03	13.01	0.00
Cultivated land	0.00	17.72	0.00	8.58	0.00	26.30
Settlement	0.00	0.00	0.00	0.00	0.01	0.00

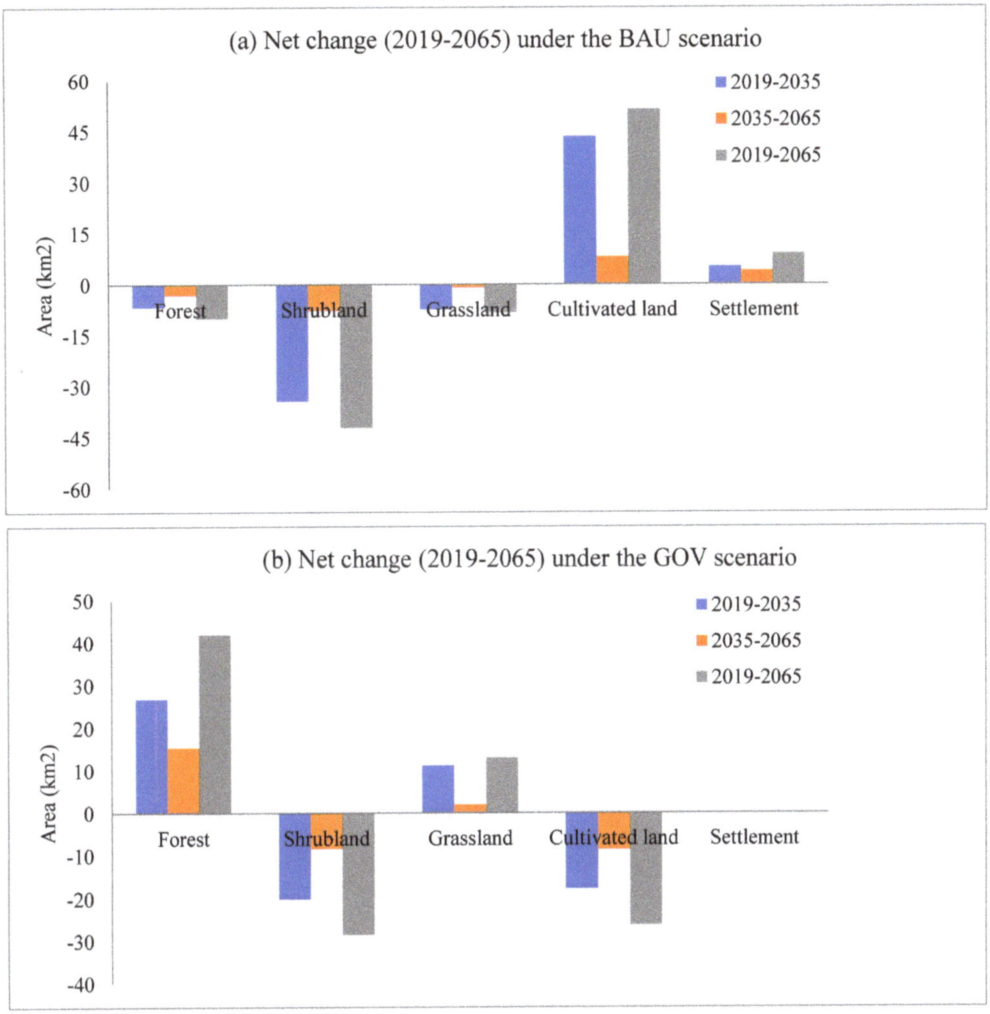

Figure 18. Net changes (gain-losses): (**a**) net change (2019–2065) under the BAU scenario and (**b**) net change (2019–2065) under the GOV scenario.

4. Discussion

4.1. LULC Classification and Change Detection

In this study, LULC classification was performed using the RF algorithm available in the GEE environment, and a total of five classes were identified, including forest, shrubland, grassland, cultivated land, and settlement. Based on the generated LULC maps of the four years, cultivated land covers the largest area percentage, while settlement covers the smallest. According to the classification accuracy assessment, the overall accuracies were 94.39%, 94.84%, 93.13%, and 91.13% for the years 1985, 2000, 2010, and 2019, respectively. In terms of the kappa coefficients, 0.92, 0.94, 0.90, and 0.88 were found for the four corresponding years. According to the criteria provided in Congalton [75], all of the historical LULC maps showed perfect agreement, with kappa coefficient values greater than 0.81. The results of the classification accuracy assessment found in this study showed slight improvement relative to previous studies conducted in the Gumara watershed. For example, Chakilu and Moges [32] reported an overall accuracy of 88.33% and a kappa coefficient of 0.73 for 2013

LULC map classified using the ML classifier. These improvements in classification accuracy for this study may be due to the use of vegetation indices and topographic factors as additional variables in refining the classification. In this regard, consistently with this study, a number of studies have shown the importance of vegetation indices and topographic factors in improving LULC classification accuracy [55,76,77].

In the study watershed, significant expansion of cultivated land and settlement and declines in forest, shrubland, and grassland were observed in the historical years (1985–2019). For this, deforestation for housing and charcoal consumption, overgrazing, extensive rice cultivation and other natural and human-induced factors contributed to the decline of forests, grasslands, and shrublands [31,43,78]. Similarly to this study, studies conducted in the Lake Tana subbasin, where the Gumara watershed is located, have indicated the expansion of cultivated land in recent decades [36,49]. Other local studies conducted in the Didessa [35], Andassa [79], Koga [80], and Finchaa [34] watersheds have reported significant expansion of cultivated land and settlement areas in recent decades, which is consistent with the findings of this study.

4.2. Impacts of LULC Change on Socioeconomic and Environmental Conditions

In this study, LULC prediction for the years 2035 and 2065 was performed under the BAU and GOV scenarios. Under the BAU scenario, an increase in cultivated land and settlements is highly expected. More specifically, under this scenario, the expansion of cultivated land will occur in areas with gentle slopes suitable for agricultural crop production. Consistently with these findings, similar scenario-based studies have reported a significant increase in cultivated land and settlements under the BAU in future years [20,24,81]. This significant LULC change has a number of implications for the socioeconomic and environmental conditions of the study watershed. From a socioeconomic perspective, agricultural cultivation practiced by local farmers plays an important role in ensuring food security for the local community. Local farmers can also benefit from selling vegetables, fruits, and cereal crops (for example, rice, wheat, barley, and millet) to the surrounding urban community. However, intensive agricultural cultivation has had a significant negative impact on reducing the quality of land resources and future productivity. It can also lead to excessive erosion of topsoil (fertile soil), which is crucial for agricultural production. Rapid population growth in the future could create competition for land resources and agricultural inputs (for example, fertilizers, insecticides, pesticides, improved seeds, etc.) among households. In the area, the significant conversion of grassland to cultivated land is also currently affecting indigenous cattle production [43,82]. This problem is most common in rural areas adjacent to the shore of Lake Tana [43]. Other similar studies also indicated the negative consequences of intensive cultivation in reducing the economic value of livestock production [83,84]. Furthermore, land degradation and loss of forest have negative impacts on livelihoods and poverty levels, and can lead to migration and displacements.

From an environmental perspective, if intensive agricultural cultivation continues in the future under the BAU scenario, the Gumara watershed may face serious environmental issues, such as soil erosion and degradation, deforestation, flooding, landslides, and contamination of surface and groundwater resources due to intensive use of fertilizer for cultivation, loss of biodiversity, and depletion of natural resources [32,36,41]. Intensive agricultural cultivation and settlement expansions also have adverse effects on the ecosystem service values and hydrology of the studied watershed. As the study watershed is the main flow contributor to Lake Tana, sediments and contaminated surface runoff from cultivated lands pollute the lake water. Recent studies have indicated that the deterioration of the water quality of Lake Tana is the result of soil erosion and contaminant flows from the contributing watersheds [50,85].

In contrast, under the GOV scenario, the expansion of cultivated land settlement areas will remain steady, and forests and other vegetation cover are expected to improve in the future. Compared to the BAU scenario, the GOV scenario is environmentally sustainable and economically profitable [24]. Environmentally, the GOV scenario can be considered sustainable, as it can reduce the adverse impact of climate change on a local or regional scale by increasing carbon sequestration and preserving biodiversity [46]. Economically, it can provide revenue for the local community from timber and nontimber products, firewood, spices, ecotourism, and medicinal plants [44]. In general, forests play a significant role in the livelihood of rural communities and can be considered a pathway for green economic growth [46].

As indicated in several studies [20,81], LULC dynamics are complex, and future LULC predictions are prone to uncertainty due to the uncertainty of drivers in the future. In this study, topographic, distance-to-socioeconomic development, and evidence of likelihood of change (disturbance) driver variables were considered for future LULC prediction. However, driver variables related to population density were not considered due to a lack of accurate census data. Therefore, further research can be conducted by considering these drivers when the government releases recent and reliable census data.

4.3. *Relevance of Scenario-Based LULC Change Detection and Prediction to Policy and Practice*

The scenario-based LULC change detection predictions covered in this study can provide significant input for land use policy and practice. The results of change detection under the BAU and GOV scenarios can help land use planners, policymakers, and watershed managers develop sustainable land use management plans for promoting sustainable development goals. According to the results of this study, an expansion in cultivated land and a reduction in forest and grassland over the past 34 years were observed. This could create competition for crop cultivation and livestock grazing among households. To handle such competition for land resources, introducing a more integrated sustainable land resource management policy that accounts for the socioeconomic values of crop and livestock production is recommended [84,86]. In this regard, the government and other stakeholders should work to combat excessive land resource utilization and rapid population growth to ensure the food security of local farmers [87]. Stakeholders should also work cooperatively to address issues related to land tenure security, land use rights, and land certification. Furthermore, to reduce the adverse effects of land use change on the environment, stakeholders should encourage communities to participate in the GLI program. It is also recommended that various land management practices, such as agroforestry, polyculture, crop rotation, and smart agriculture, be implemented to increase the economic productivity and environmental sustainability of the GLI program.

5. Conclusions

Long-term LULC change detection and prediction studies are crucial for providing significant information on the current and future hydrological conditions of watersheds and large river basins. For this reason, this study focused on change detection and prediction of LULC in the Gumara watershed using the cloud-based platform Google Earth Engine (GEE) and the geospatial prediction model cellular automata–Markov (CA–Markov). The first part of the study focused on LULC classification and change detection in historical periods. For LULC classification, Landsat-5/OLI and 8/TM images from 1985, 2000, 2010, and 2019 were used as inputs for the RF classification algorithm built into the GEE environment. To improve the classification accuracy and distinguishability of complex classes, other variables, such as vegetation indices (NDVI and SAVI) and topographic factors (elevation and slope), were used as inputs for the RF classifier. Using this cloud-based image classification approach, five major LULC classes, namely, forest, shrubland, grassland, cultivated land, and settlements, were identified for the study watershed. The results of the change analysis for the four historical periods (1985–2000, 200–2010, 2010–2019, and 1985–2019) revealed significant LULC changes in the study watershed. Over the past 34 years (1985–2019), expansions of cultivated land and settlements and decreases in forest, shrubland, and grassland were detected due to increasing anthropogenic disturbances. From the analysis, a significant increase in cultivated land from 58.60 to 83.08% and a decrease in forest from 5.22 to 2.59% were detected between 1985 and 2019.

In the second part of this study, future LULC maps of the Gumara watershed were generated for the years 2035 and 2065 using the CA–Markov model available in the land change modeler (LCM) module of IDRISI Selva 17.0. To predict future LULC maps, earlier and later images, the Markovian transition probability matrix (TPM), and transition suitability maps (TSMs) were used as inputs for the CA–Markov model. To generate the TSMs, a multilayer perceptron (MLP) artificial neural network (ANN) algorithm was employed for this study. Two LULC scenarios, namely, the business-as-usual (BAU) and governance (GOV), were considered. For the first scenario (BAU), the most dominant LULC TSMs were generated based on the current trend of LULC change using six LULC driver variables, such as elevation, slope, distance from streams, distance from roads, distance from towns, and evidence likelihood. Under this scenario, significant expansions of cultivated land and settlements were predicted for the years 2035 and 2065 due to human-induced anthropogenic disturbances. Therefore, cultivated land and settlements are expected to increase from 83.08 to 89.01% and 0.18 to 0.21%, respectively, between 2019 and 2065.

On the other hand, the GOV scenario considers the Green Legacy Initiative program, which is currently under implementation. For this scenario, topographic factors (slope and elevation) were considered to generate TSMs. Under this scenario, forest and grassland areas are expected to increase from 2.59% in 2019 to 4.71% in 2065 due to the implementation of afforestation and reforestation programs and controlled anthropogenic disturbances. Overall, the scenario-based findings of this study can provide information for government policymakers, land use planners, and watershed managers to developing sustainable land management policies.

In this study, LULC classification was performed using spectral bands, vegetation indices, and topographic factors as inputs for the RF algorithm. In this regard, further research can be performed by considering texture information following recent studies [10,88]. In this study, future LULC prediction was performed considering topographic and proximity-related factors. Hence, future predictions of LULC maps can also be further researched by considering population density data when the government releases accurate and updated census data. Moreover, further studies could be performed by studying the impacts of LULC on the hydrology and ecosystem service values of the studied watershed.

Supplementary Materials: The following supporting information can be downloaded at: https://www.mdpi.com/article/10.3390/land13030396/s1, S1: Total number of training and validation points; S2: Confusion matrix; S3: Transition potential maps (2010–2019).

Author Contributions: H.B.: designed the methodology, scripted the Google Earth Engine code, analyzed the data, interpreted the results, and wrote the manuscript. A.M.M. and G.T. conceived and supervised the study and edited and wrote the manuscript. All authors have read and agreed to the published version of the manuscript.

Funding: This research work was financially supported by the Africa Center of Excellence for Water Management (ACEWM), Addis Ababa University, Ethiopia (Project code: 57940-ET).

Data Availability Statement: The original contributions presented in the study are included in the article/Supplementary Material. The data and Google Earth Engine scripts used for this research are available from the corresponding author upon request (email: haile.belay@aau.edu.et).

Acknowledgments: The authors would like to acknowledge the developers of Google Earth Engine for their free cloud-based platform. The authors would also like to acknowledge the U.S. Geological Survey for providing the Landsat images.

Conflicts of Interest: The authors of this manuscript declare no conflicts of interest.

References

1. Navin, M.S.; Agilandeeswari, L. Comprehensive review on land use/land cover change classification in remote sensing. *J. Spectr. Imaging* **2020**, *9*, a8. [CrossRef]
2. Tadese, S.; Soromessa, T.; Bekele, T. Analysis of the Current and Future Prediction of Land Use/Land Cover Change Using Remote Sensing and the CA-Markov Model in Majang Forest Biosphere Reserves of Gambella, Southwestern Ethiopia. *Sci. World J.* **2021**, *2021*, 6685045. [CrossRef] [PubMed]
3. Dibaba, W.T.; Demissie, T.A.; Miegel, K. Watershed hydrological response to combined land use/land cover and climate change in highland ethiopia: Finchaa catchment. *Water* **2020**, *12*, 1801. [CrossRef]
4. Yang, L.; Feng, Q.; Yin, Z.; Deo, R.C.; Wen, X.; Si, J.; Li, C. Separation of the Climatic and Land Cover Impacts on the Flow Regime Changes in Two Watersheds of Northeastern Tibetan Plateau. *Adv. Meteorol.* **2017**, *2017*, 6310401. [CrossRef]
5. Otukei, J.R.; Blaschke, T. Land cover change assessment using decision trees, support vector machines and maximum likelihood classification algorithms. *Int. J. Appl. Earth Obs. Geoinf.* **2010**, *12*, 27–31. [CrossRef]
6. Thanh Noi, P.; Kappas, M. Comparison of Random Forest, k-Nearest Neighbor, and Support Vector Machine Classifiers for Land Cover Classification Using Sentinel-2 Imagery. *Sensors* **2017**, *18*, 18. [CrossRef] [PubMed]
7. Abijith, D.; Saravanan, S. Assessment of Land Use and Land Cover Change Detection And Prediction Using Remote Sensing And CA Markov in the Northern Coastal Districts of Tamil Nadu, India Devanantham abijith National Institute of Technology Tiruchirappalli. *Res. Sq.* **2021**, *29*, 86055–86067.
8. Arfasa, G.F.; Owusu-Sekyere, E.; Doke, D.A. Predictions of land use/land cover change, drivers, and their implications on water availability for irrigation in the Vea catchment, Ghana. *Geocarto Int.* **2023**, *38*, 2243093. [CrossRef]
9. Chang, X.; Zhang, F.; Cong, K.; Liu, X. Scenario simulation of land use and land cover change in mining area. *Sci. Rep.* **2021**, *11*, 12910. [CrossRef]
10. Shafizadeh-Moghadam, H.; Khazaei, M.; Alavipanah, S.K.; Weng, Q. Google Earth Engine for large-scale land use and land cover mapping: An object-based classification approach using spectral, textural and topographical factors. *GIScience Remote Sens.* **2021**, *58*, 914–928. [CrossRef]
11. Noi Phan, T.; Kuch, V.; Lehnert, L.W. Land cover classification using google earth engine and random forest classifier-the role of image composition. *Remote Sens.* **2020**, *12*, 2411. [CrossRef]
12. Xiong, J.; Thenkabail, P.S.; Tilton, J.C.; Gumma, M.K.; Teluguntla, P.; Oliphant, A.; Congalton, R.G.; Yadav, K.; Gorelick, N. Nominal 30-m cropland extent map of continental Africa by integrating pixel-based and object-based algorithms using Sentinel-2 and Landsat-8 data on google earth engine. *Remote Sens.* **2017**, *9*, 1065. [CrossRef]
13. Stromann, O.; Nascetti, A.; Yousif, O.; Ban, Y. Dimensionality Reduction and Feature Selection for Object-Based Land Cover Classification based on Sentinel-1 and Sentinel-2 Time Series Using Google Earth Engine. *Remote Sens.* **2020**, *12*, 76. [CrossRef]
14. Camargo, F.F.; Sano, E.E.; Almeida, C.M.; Mura, J.C.; Almeida, T. A comparative assessment of machine-learning techniques for land use and land cover classification of the Brazilian tropical savanna using ALOS-2/PALSAR-2 polarimetric images. *Remote Sens.* **2019**, *11*, 1600. [CrossRef]
15. Neetu; Ray, S.S. Exploring machine learning classification algorithms for crop classification using sentinel 2 data. *Int. Arch. Photogramm. Remote Sens. Spat. Inf. Sci. ISPRS Arch.* **2019**, *42*, 573–578. [CrossRef]
16. Adam, E.; Mutanga, O.; Odindi, J.; Abdel-Rahman, E.M. Land-use/cover classification in a heterogeneous coastal landscape using RapidEye imagery: Evaluating the performance of random forest and support vector machines classifiers. *Int. J. Remote Sens.* **2014**, *35*, 3440–3458. [CrossRef]
17. Breiman, L. Random forests. *Mach. Learn.* **2001**, *45*, 5–32. [CrossRef]
18. Tassi, A.; Vizzari, M. Object-oriented lulc classification in google earth engine combining snic, glcm, and machine learning algorithms. *Remote Sens.* **2020**, *12*, 3766. [CrossRef]
19. Kadri, N.; Jebari, S.; Augusseau, X.; Mahdhi, N.; Lestrelin, G.; Berndtsson, R. Analysis of Four Decades of Land Use and Land Cover Change in Semiarid Tunisia Using Google Earth Engine. *Remote Sens.* **2023**, *15*, 3257. [CrossRef]
20. Hamad, R.; Balzter, H.; Kolo, K. Predicting land use/land cover changes using a CA-Markov model under two different scenarios. *Sustainability* **2018**, *10*, 3421. [CrossRef]
21. Leta, M.K.; Demissie, T.A.; Tränckner, J. Modeling and prediction of land use land cover change dynamics based on land change modeler (Lcm) in nashe watershed, upper blue nile basin, Ethiopia. *Sustainability* **2021**, *13*, 3740. [CrossRef]
22. Sisay, G.; Gesesse, B.; Fürst, C.; Kassie, M.; Kebede, B. Modeling of land use/land cover dynamics using artificial neural network and cellular automata Markov chain algorithms in Goang watershed, Ethiopia. *Heliyon* **2023**, *9*, e20088. [CrossRef] [PubMed]
23. Halmy, M.W.A.; Gessler, P.E.; Hicke, J.A.; Salem, B.B. Land use/land cover change detection and prediction in the north-western coastal desert of Egypt using Markov-CA. *Appl. Geogr.* **2015**, *63*, 101–112. [CrossRef]
24. Ruben, G.B.; Zhang, K.; Dong, Z.; Xia, J. Analysis and projection of land-use/land-cover dynamics through scenario-based simulations using the CA-Markov model: A case study in guanting reservoir basin, China. *Sustainability* **2020**, *12*, 3747. [CrossRef]
25. Shoyama, K. Assessment of land-use scenarios at a national scale using intensity analysis and figure of merit components. *Land* **2021**, *10*, 379. [CrossRef]
26. Van Vuuren, D.P.; Kok, M.T.J.; Girod, B.; Lucas, P.L.; de Vries, B. Scenarios in global environmental assessments: Key characteristics and lessons for future use. *Glob. Environ. Change* **2012**, *22*, 884–895. [CrossRef]

27. O'Neill, B.C.; Kriegler, E.; Ebi, K.L.; Kemp-Benedict, E.; Riahi, K.; Rothman, D.S.; Van Ruijven, B.J.; Van Vuuren, D.P.; Birkmann, J.; Kok, K. The roads ahead: Narratives for shared socioeconomic pathways describing world futures in the 21st century. *Glob. Environ. Change* **2017**, *42*, 169–180. [CrossRef]
28. Popp, A.; Calvin, K.; Fujimori, S.; Havlik, P.; Humpenöder, F.; Stehfest, E.; Bodirsky, B.L.; Dietrich, J.P.; Doelmann, J.C.; Gusti, M.; et al. Land-use futures in the shared socio-economic pathways. *Glob. Environ. Change* **2017**, *42*, 331–345. [CrossRef]
29. Wang, R.; Hou, H.; Murayama, Y. Scenario-based simulation of Tianjin city using a cellular automata-Markov model. *Sustainability* **2018**, *10*, 2633. [CrossRef]
30. Gebresellase, S.H.; Wu, Z.; Xu, H.; Muhammad, W.I. Scenario-Based LULC Dynamics Projection Using the CA–Markov Model on Upper Awash Basin (UAB), Ethiopia. *Sustainability* **2023**, *15*, 1683. [CrossRef]
31. Berihun, M.L.; Tsunekawa, A.; Haregeweyn, N.; Meshesha, D.T.; Adgo, E.; Tsubo, M.; Masunaga, T.; Fenta, A.A.; Sultan, D.; Yibeltal, M. Exploring land use/land cover changes, drivers and their implications in contrasting agro-ecological environments of Ethiopia. *Land Use Policy* **2019**, *87*, 104052. [CrossRef]
32. Chakilu, G.G.; Moges, M.A. Assessing the Land Use/Cover Dynamics and its Impact on the Low Flow of Gumara Watershed, Upper Blue Nile Basin, Ethiopia. *Hydrol. Curr. Res.* **2017**, *8*, 268. [CrossRef]
33. Yesuph, A.Y.; Dagnew, A.B. Land use/cover spatiotemporal dynamics, driving forces and implications at the Beshillo catchment of the Blue Nile Basin, North Eastern Highlands of Ethiopia. *Environ. Syst. Res.* **2019**, *8*, 21. [CrossRef]
34. Dibaba, W.T.; Demissie, T.A.; Miegel, K. Drivers and implications of land use/land cover dynamics in Finchaa catchment, northwestern Ethiopia. *Land* **2020**, *9*, 113. [CrossRef]
35. Tolessa, T.; Dechassa, C.; Simane, B.; Alamerew, B.; Kidane, M. Land use/land cover dynamics in response to various driving forces in Didessa sub-basin, Ethiopia. *GeoJournal* **2020**, *85*, 747–760. [CrossRef]
36. Bogale, A. Review, impact of land use/cover change on soil erosion in the Lake Tana Basin, Upper Blue Nile, Ethiopia. *Appl. Water Sci.* **2020**, *10*, 235. [CrossRef]
37. Faichia, C.; Tong, Z.; Zhang, J.; Liu, X.; Kazuva, E.; Ullah, K.; Al-Shaibah, B. Using rs data-based ca–markov model for dynamic simulation of historical and future lucc in Vientiane, Laos. *Sustainability* **2020**, *12*, 8410. [CrossRef]
38. Lukas, P.; Melesse, A.M.; Kenea, T.T. Prediction of Future Land Use/Land Cover Changes Using a Coupled CA-ANN Model in the Upper Omo–Gibe River Basin, Ethiopia. *Remote Sens.* **2023**, *15*, 1148. [CrossRef]
39. Mohamed, A.; Worku, H. Simulating urban land use and cover dynamics using cellular automata and Markov chain approach in Addis Ababa and the surrounding. *Urban Clim.* **2020**, *31*, 100545. [CrossRef]
40. Anteneh, M.; Mohammed, W. Effects of land cover changes and slope gradient on soil quality in the Gumara watershed, Lake Tana basin of North—West Ethiopia. *Model. Earth Syst. Environ.* **2020**, *6*, 85–97. [CrossRef]
41. Wubie, M.A.; Assen, M.; Nicolau, M.D. Patterns, causes and consequences of land use/cover dynamics in the Gumara watershed of lake Tana basin, Northwestern Ethiopia. *Environ. Syst. Res.* **2016**, *5*, 8. [CrossRef]
42. Szuster, B.W.; Chen, Q.; Borger, M. A comparison of classification techniques to support land cover and land use analysis in tropical coastal zones. *Appl. Geogr.* **2011**, *31*, 525–532. [CrossRef]
43. Desta, M.A.; Zeleke, G.; Payne, W.A.; Shenkoru, T.; Dile, Y. The impacts of rice cultivation on an indigenous Fogera cattle population at the eastern shore of Lake Tana, Ethiopia. *Ecol. Process.* **2019**, *8*, 19. [CrossRef]
44. Beyene, A.; Shumetie, A. *Green Legacy Initiative for Sustainable Economic Development in Ethiopia*; Ethiopian Economic Association (EEA): Addis Ababa, Ethiopia, 2023.
45. Fikreyesus, D.; Gizaw, S.; Mayers, J.; Barrett, S. Mass Tree Planting: Prospects for a Green Legacy in Ethiopia. 2022. Available online: https://opendocs.ids.ac.uk/opendocs/handle/20.500.12413/17524 (accessed on 1 September 2023).
46. Kassa, H.; Abiyu, A.; Hagazi, N.; Mokria, M.; Kassawmar, T.; Gitz, V. Forest landscape restoration in Ethiopia: Progress and challenges. *Front. For. Glob. Change* **2022**, *5*, 796106. [CrossRef]
47. Tegegne, G.; Melesse, A.M.; Asfaw, D.H.; Worqlul, A.W. Flood frequency analyses over different basin scales in the Blue Nile River Basin, Ethiopia. *Hydrology* **2020**, *7*, 44. [CrossRef]
48. Tefera, B.; Kassa, H. Trends and driving forces of Eucalyptus plantation by smallholders in the Lake Tana Watershed of Ethiopia. In *Social and Ecological System Dynamics: Characteristics, Trends, and Integration in the Lake Tana Basin, Ethiopia*; Springer: Cham, Switzerland, 2017; pp. 563–580.
49. Getachew, B.; Manjunatha, B.R.; Bhat, H.G. Modeling projected impacts of climate and land use/land cover changes on hydrological responses in the Lake Tana Basin, upper Blue Nile River Basin, Ethiopia. *J. Hydrol.* **2021**, *595*, 125974. [CrossRef]
50. Tikuye, B.G.; Gill, L.; Rusnak, M.; Manjunatha, B.R. Modelling the impacts of changing land use and climate on sediment and nutrient retention in Lake Tana Basin, Upper Blue Nile River Basin, Ethiopia. *Ecol. Model.* **2023**, *482*, 110383. [CrossRef]
51. Saah, D.; Johnson, G.; Ashmall, B.; Tondapu, G.; Tenneson, K.; Patterson, M.; Poortinga, A.; Markert, K.; Quyen, N.H.; San Aung, K.; et al. Collect Earth: An online tool for systematic reference data collection in land cover and use applications. *Environ. Model. Softw.* **2019**, *118*, 166–171. [CrossRef]
52. Farr, T.G.; Rosen, P.A.; Caro, E.; Crippen, R.; Duren, R.; Hensley, S.; Kobrick, M.; Paller, M.; Rodriguez, E.; Roth, L. The shuttle radar topography mission. *Rev. Geophys.* **2007**, *45*. [CrossRef]
53. Reid, R.S.; Kruska, R.L.; Muthui, N.; Taye, A.; Wotton, S.; Wilson, C.J.; Mulatu, W. Land-use and land-cover dynamics in response to changes in climatic, biological and socio-political forces: The case of southwestern Ethiopia. *Landsc. Ecol.* **2000**, *15*, 339–355. [CrossRef]

54. Abdi, A.M. Land cover and land use classification performance of machine learning algorithms in a boreal landscape using Sentinel-2 data. *GIScience Remote Sens.* **2020**, *57*, 1–20. [CrossRef]
55. Mananze, S.; Pôças, I.; Cunha, M. Mapping and assessing the dynamics of shifting agricultural landscapes using google earth engine cloud computing, a case study in Mozambique. *Remote Sens.* **2020**, *12*, 1279. [CrossRef]
56. Talukdar, S.; Singha, P.; Mahato, S.; Shahfahad; Pal, S.; Liou, Y.A.; Rahman, A. Land-use land-cover classification by machine learning classifiers for satellite observations-A review. *Remote Sens.* **2020**, *12*, 1135. [CrossRef]
57. Rouse, J.W.; Haas, R.H.; Schell, J.A.; Deering, D.W. Monitoring vegetation systems in the Great Plains with ERTS. *NASA Spec. Publ.* **1974**, *351*, 309.
58. Huete, A.R. A soil-adjusted vegetation index (SAVI). *Remote Sens. Environ.* **1988**, *25*, 295–309. [CrossRef]
59. Nasiri, V.; Deljouei, A.; Moradi, F.; Sadeghi, S.M.M.; Borz, S.A. Land Use and Land Cover Mapping Using Sentinel-2, Landsat-8 Satellite Images, and Google Earth Engine: A Comparison of Two Composition Methods. *Remote Sens.* **2022**, *14*, 1977. [CrossRef]
60. Congalton, R.G.; Green, K. *Assessing the Accuracy of Remotely Sensed Data: Principles and Practices*; CRC Press: Boca Raton, FL, USA, 2019.
61. Foody, G.M. Status of land cover classification accuracy assessment. *Remote Sens. Environ.* **2002**, *80*, 185–201. [CrossRef]
62. Braimoh, A.K. Random and systematic land-cover transitions in northern Ghana. *Agric. Ecosyst. Environ.* **2006**, *113*, 254–263. [CrossRef]
63. Yuan, D. Survey of multispectral methods for land-cover change analysis. In *Remote Sensing Change Detection: Environmental Monitoring Methods and Application*; Taylor and Francis Ltd.: Abingdon, UK, 1999.
64. Pontius, R.G.; Shusas, E.; McEachern, M. Detecting important categorical land changes while accounting for persistence. *Agric. Ecosyst. Environ.* **2004**, *101*, 251–268. [CrossRef]
65. Asif, M.; Kazmi, J.H.; Tariq, A.; Zhao, N.; Guluzade, R.; Soufan, W.; Almutairi, K.F.; Sabagh, A.E.; Aslam, M. Modelling of land use and land cover changes and prediction using CA-Markov and Random Forest. *Geocarto Int.* **2023**, *38*, 2210532. [CrossRef]
66. Beroho, M.; Briak, H.; Cherif, E.K.; Boulahfa, I.; Ouallali, A.; Mrabet, R.; Kebede, F.; Bernardino, A.; Aboumaria, K. Future Scenarios of Land Use/Land Cover (LULC) Based on a CA-Markov Simulation Model: Case of a Mediterranean Watershed in Morocco. *Remote Sens.* **2023**, *15*, 1162. [CrossRef]
67. Kafy, A.A.; Naim, M.N.H.; Subramanyam, G.; Faisal, A.A.; Ahmed, N.U.; Rakib, A.A.; Kona, M.A.; Sattar, G.S. Cellular Automata approach in dynamic modelling of land cover changes using RapidEye images in Dhaka, Bangladesh. *Environ. Chall.* **2021**, *4*, 100084. [CrossRef]
68. Liping, C.; Yujun, S.; Saeed, S. Monitoring and predicting land use and land cover changes using remote sensing and GIS techniques—A case study of a hilly area, Jiangle, China. *PLoS ONE* **2018**, *13*, e0200493. [CrossRef] [PubMed]
69. Vu, T.T.; Shen, Y. Land-use and land-cover changes in dong trieu district, vietnam, during past two decades and their driving forces. *Land* **2021**, *10*, 798. [CrossRef]
70. Eastman, J.R. *IDRISI Taiga: Guide to GIS and Image Processing Volume—Manual Version 16.02*; Clark Labs Clark University: Worcester, MA, USA, 2009; p. 325.
71. Cramér, H. *Mathematical Methods of Statistics*; Princeton University Press: Princeton, NJ, USA, 1999; Volume 43.
72. Cohen, J. *Statistical Power Analysis for the Behavioral Sciences*; Academic Press: Cambridge, MA, USA, 2013.
73. Abbas, S.; Yaseen, M.; Latif, Y.; Waseem, M.; Muhammad, S.; Leta, M.K.; Sher, S.; Imran, M.A.; Adnan, M.; Khan, T.H. Spatiotemporal Analysis of Climatic Extremes over the Upper Indus Basin, Pakistan. *Water* **2022**, *14*, 1718. [CrossRef]
74. Kamaraj, M.; Rangarajan, S. Predicting the future land use and land cover changes for Bhavani basin, Tamil Nadu, India, using QGIS MOLUSCE plugin. *Environ. Sci. Pollut. Res.* **2022**, *29*, 86337–86348. [CrossRef] [PubMed]
75. Congalton, R.G. Accuracy assessment and validation of remotely sensed and other spatial information. *Int. J. Wildland Fire* **2001**, *10*, 321–328. [CrossRef]
76. Abida, K.; Barbouchi, M.; Boudabbous, K.; Toukabri, W.; Saad, K.; Bousnina, H.; Sahli Chahed, T. Sentinel-2 Data for Land Use Mapping: Comparing Different Supervised Classifications in Semi-Arid Areas. *Agriculture* **2022**, *12*, 1429. [CrossRef]
77. Amini, S.; Saber, M.; Rabiei-Dastjerdi, H.; Homayouni, S. Urban Land Use and Land Cover Change Analysis Using Random Forest Classification of Landsat Time Series. *Remote Sens.* **2022**, *14*, 2654. [CrossRef]
78. Regasa, M.S.; Nones, M.; Adeba, D. A review on land use and land cover change in Ethiopian basins. *Land* **2021**, *10*, 585. [CrossRef]
79. Gashaw, T.; Tulu, T.; Argaw, M.; Worqlul, A.W.; Tolessa, T.; Kindu, M. Estimating the impacts of land use/land cover changes on Ecosystem Service Values: The case of the Andassa watershed in the Upper Blue Nile basin of Ethiopia. *Ecosyst. Serv.* **2018**, *31*, 219–228. [CrossRef]
80. Sewnet, A.; Abebe, G. Land use and land cover change and implication to watershed degradation by using GIS and remote sensing in the Koga watershed, North Western Ethiopia. *Earth Sci. Inform.* **2018**, *11*, 99–108. [CrossRef]
81. Samie, A.; Deng, X.; Jia, S.; Chen, D. Scenario-based simulation on dynamics of land-use-land-cover change in Punjab province, Pakistan. *Sustainability* **2017**, *9*, 1285. [CrossRef]
82. Amsalu, T.; Addisu, S. Assessment of Grazing Land and Livestock Feed Balance in Gummara-Rib Watershed, Ethiopia. *Curr. Agric. Res. J.* **2014**, *2*, 114–122. [CrossRef]
83. Aklile, Y.; Beyene, F. Examining drivers of land use change among pastoralists in Eastern Ethiopia. *J. Land Use Sci.* **2014**, *9*, 402–413. [CrossRef]

84. Mekuria, W.; Mekonnen, K.; Thorne, P.; Bezabih, M.; Tamene, L.; Abera, W. Competition for land resources: Driving forces and consequences in crop-livestock production systems of the Ethiopian highlands. *Ecol. Process.* **2018**, *7*, 30. [CrossRef]
85. Goshu, G.; Koelmans, A.A.; de Klein, J.J.M. Water quality of Lake Tana basin, Upper Blue Nile, Ethiopia. A review of available data. In *Social and Ecological System Dynamics: Characteristics, Trends, and Integration in the Lake Tana Basin, Ethiopia*; Springer: Cham, Switzerland, 2017; pp. 127–141.
86. Cochrane, L.; Hadis, S. Functionality of the land certification program in Ethiopia: Exploratory evaluation of the processes of updating certificates. *Land* **2019**, *8*, 149. [CrossRef]
87. Agidew, A.m.A.; Singh, K.N. The implications of land use and land cover changes for rural household food insecurity in the Northeastern highlands of Ethiopia: The case of the Teleyayen sub-watershed. *Agric. Food Secur.* **2017**, *6*, 56. [CrossRef]
88. Cuypers, S.; Nascetti, A.; Vergauwen, M. Land Use and Land Cover Mapping with VHR and Multi-Temporal Sentinel-2 Imagery. *Remote Sens.* **2023**, *15*, 2501. [CrossRef]

land

MDPI

Article

'Kesho' Scenario Development for Supporting Water-Energy Food Security under Future Conditions in Zanzibar

Rebecca J. S. Newman [1,*], Charis Enns [2], Claudia Capitani [1], Jessica P. R. Thorn [3], Colin J. Courtney-Mustaphi [4], Sam J. Buckton [1], Eugyen Suzanne Om [1], Ioan Fazey [1], Tahir A. Haji [5], Aziza Y. Nchimbi [5], Rebecca W. Kariuki [6] and Robert A. Marchant [1]

1 Department of Environment and Geography, University of York, York YO10 5DD, UK; claudia.capitani@gmail.com (C.C.); sam.buckton@york.ac.uk (S.J.B.); eso513@york.ac.uk (E.S.O.); ioan.fazey@york.ac.uk (I.F.); robert.marchant@york.ac.uk (R.A.M.)
2 Global Development Institute, University of Manchester, Manchester M13 9PL, UK; charis.enns@manchester.ac.uk
3 School of Geography and Sustainable Development, University of St Andrews, St Andrews KY16 9AJ, UK; jessica.thorn@st-andrews.ac.uk
4 Department of Environment Science, University of Basel, 4001 Basel, Switzerland; colin.courtney-mustaphi@unibas.ch
5 Department of Forestry and Renewable and Non-Renewable Resources, Mjini Magharibi Region, Zanzibar V646+9P9, Tanzania; twahir.haji@yahoo.com (T.A.H.); azizanchimbi@gmail.com (A.Y.N.)
6 School of Sustainability, Arizona State University, Tempe, AZ 85281, USA; rebecca.kariuki@asu.edu
* Correspondence: rebecca.newman@york.ac.uk

Abstract: Social-ecological interactions mediate water–energy–food security in small developing islands, but community-scale insights are underrepresented in nexus research. These interactions are dynamic in their response to environmental and anthropogenic pressures and need to be understood to inform sustainable land use planning into the future. This study centered on bringing together diverse stakeholders to explore water–energy–food futures using the "*Kesho*" (meaning "tomorrow" in Kiswahili) scenario tool for two of the largest islands that comprise the Zanzibar Archipelago. The methodology comprised four core stages: (1) exploration of how past drivers of change impacted water–energy–food security; (2) modeling of a *Business as Usual Scenario* for land cover change; (3) narrative development to describe alternative futures for 2030 based on themes developed at the community scale; and (4) predictions about how narratives would shape land cover and its implications for the nexus. These results were used to model alternate land cover scenarios in TerrSet IDRISI (v. 18.31) and produce visual representations of expected change. Findings demonstrated that deforestation, saltwater incursion, and a reduction in permanent waterbodies were projected by 2030 in a *Business as Usual Scenario*. Three alternative scenario narratives were developed, these included *Adaptation, Ecosystem Management*, and *Settlement Planning*. The results demonstrate that the effectiveness of actions under the scenario options differ between the islands, indicating the importance of understanding the suitability of national policies across considered scales. Synergies across the alternative scenario narratives also emerged, including integrated approaches for managing environmental change, community participation in decision making, effective protection of forests, cultural sensitivity to settlement planning, and poverty alleviation. These synergies could be used to plan strategic action towards effectively strengthening water–energy–food security in Zanzibar.

Keywords: livelihoods; nexus; climate change; development; ecosystems; western Indian ocean

1. Introduction

The interactions amongst land use, climate, and socio-economic changes impact the resilience and sustainability of social-ecological systems [1]. In small islands within the Western Indian Ocean (WIO), shifts in these dynamics have significant impacts on livelihoods and resource security—particularly as they are concentrated over smaller spatial extents [2]. Small islands in this region are undergoing rapid changes related to transitions to industrial fishing, exploitation of gas and oil, expansion of tourism, and urbanization—all of which degrade ecosystems and water resources that support local livelihoods [3].

Alongside social changes, climate change poses significant threats to local economies, many of which are informal. Communities on small islands are vulnerable to impacts such as rising sea levels, coastal erosion, drought, high wind speeds, erratic rainfall, and changes in sea temperature and acidity. Climate change also poses a threat to the unique biodiversity of the WIO, including its coral reefs, seagrasses, and mangroves, which are vital for coastal protection and supporting marine life [4].

As well as facing more incremental changes associated with climate change, communities living on small islands are exposed to natural hazards that include heavy rainfall, droughts, extremely high temperatures, storm surges, cyclones, and tsunamis [5]. All these pressures impact people and the way they interact with their environment to meet their needs [6].

There is a growing awareness that in these contexts' knowledge exchange and integration of local experience into decision making is key to building adaptive capacity [2]. Participatory scenarios can be used to consider how to manage the causes and consequences of complex social-ecological challenges, as they offer an opportunity to explore multiple and interacting pressures alongside effects on livelihoods [1,7,8]. By simulating different future conditions, scenarios can inform strategies, investments, and plans that are robust under a range of potential futures [1]. They can also help to pre-empt conflicting agendas and prepare for external influences [9–11]. This could support building the adaptive capacity of communities on small islands in response to social-ecological changes.

In a small island context, a water–energy–food nexus approach is critical for evaluating synergies and trade-offs of different land management decisions, given that they operate at tight spatial scales and effects can be rapid [12]. The water–energy–food nexus recognizes that systems are interconnected and interdependent. For example, in coastal island communities, water is essential for hydropower dams and irrigating agricultural areas, while energy is needed to pump, desalinate, treat, and distribute water as well as produce, process, preserve, and transport food [6,12]. Another example is forests and their role in protecting coastlines from inundation, which results in the salinization of agricultural soils [13] and available groundwater resources [14].

Analysis of the resource system allows the identification of effective policies for improving adaptive capacity [15,16]. For instance, previous research in Zanzibar identified solar as an important energy transition to disentangle from expensive dependencies for electricity from mainland Tanzania and ensure consistent pumping of water [17]. An integrated approach to exploring resource security across sectors could, therefore, potentially better inform where priorities are set and shifted.

Despite this need, there are a limited number of emergent studies exploring future options to support a sustainable water–energy–food nexus in small islands [12,18,19]. Even then, the studies tended to focus on broad system levels and did not capture more in-depth local interactions. This is an important gap to consider, given that social-ecological changes have multiple impacts across the water–energy–food nexus.

Considering that several layers of understanding are needed to fully unveil opportunities for addressing sustainability challenges, multiple perspectives from diverse stakeholders across scales and levels are needed. These include power hierarchies, access to knowledge and capacity, and can be nested across scales [20,21]. This would improve not only the feasibility and validity but also the uptake and concreteness of scenarios [22,23]. By positioning future scenarios around the water–energy–food security nexus, this study attempts to ensure that discussions respond appropriately to local needs.

The purpose of this research was to create scenario alternatives to address actual and emerging challenges for water–energy–food security experienced by local communities in a small island context. There were four core objectives: (1) to explore how key drivers of change operate spatially and temporally; (2) to develop a model for land cover change in a *Business as Usual* scenario for 2030 based on community-based insights; (3) to develop coherent and tangible scenario narratives using alternative themes generated by communities, *Ecosystem Management*, *Adaptation*, and *Settlement Planning*; and (4) to model land cover implications of alternative scenarios for water–energy–food security towards 2030.

Findings show that drivers of change might change the land cover and its potential impact on the water–energy–food nexus under alternative scenario trajectories. They also unveil alignments in strategic areas of focus for responding to and preparing for change across scenario narratives. Consequently, results could be used to help inform planned action towards local, regional, and national poverty alleviation and sustainability agendas.

2. Methods

2.1. Study Area

Zanzibar, a semi-autonomous territory that forms a political union with mainland Tanzania, is an archipelago in the Indian Ocean (Figure 1). Zanzibar's population stands at 1,889,773 [24]. Most of this population growth has been in urban areas, with growing disparities in welfare between the islands of Pemba and Unguja. Poverty levels based on household consumption stand at 25.7% [25].

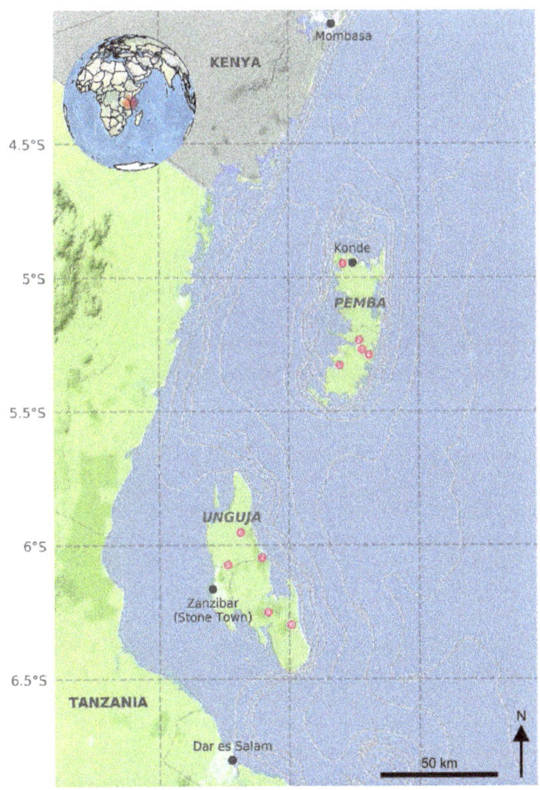

Figure 1. Location of Unguja and Pemba on the East African coastline, original map provided by [26] and adapted by [13] to include community-based workshop sites in Unguja (**left**) and Pemba (**right**).

Members of each of these communities were included as stakeholders in this study. Coordinates provided by the Department of Renewable and Non-Renewable Resources, Zanzibar. Pemba sites include villages: (1) Mji Mpya, (2) Macho Mane, (3) Mfikiwa, (4) Pujini, and (5) Chumbageni. Unguja sites include villages: (6) Kinyasini, (7) Pongwe, (8) Kizimbani, (9) Pete, and (10) Jambiani Kikadini.

The islands have a humid tropical monsoon climate with 1600–1900 mm annual rainfall and an average annual temperature of 27.5 °C [27]. There are four main seasons: "kaskazi" (hot season) between December and February; "masika" (long rainy season) between March and May; "kipupwe" (cold season with high winds) between June and September; and "vuli" (short rains) between October and December. However, climate variability is changing these more predictable patterns [28].

Key livelihoods include farming, fishing, and seaweed aquaculture [29]. The islands are experiencing fast rates of change, stimulated by population growth, migration, urbanization, tourism, and increased land demand for agriculture and forestry products [30]. Tourism contributed 29.2% of the 2022 GDP and resulted in investment in key services. For some villages, high levels of tourism have resulted in the loss of access to beaches and the sea, which has resulted in a loss of livelihoods [31].

Freshwater lenses above saltwater are the main freshwater sources in Zanzibar. Only 24% of rainwater is captured into groundwater, as approximately 44% evaporates and 32% is surface runoff [32]. Only 1.3% is captured for domestic and irrigation purposes through rainwater harvesting and remains underutilized on the island [33]. Recent research has indicated that groundwater abstraction in Unguja Island is higher than the rate of recharge. This means that groundwater is overexploited and exposed to pollution and saltwater intrusion [34]. Seasonal imbalances occur during the dry season when recharge is at its lowest, but abstraction rates are at their highest due to demands from tourism [34,35]. Saltwater intrusion is already widespread due to the pumping of groundwater from coastal aquifers [36]. Water is abstracted from springs and caves and supplied to communities through pipelines and public taps [35].

Zanzibar's electric grid is reliant on hydroelectricity generated on mainland Tanzania, which is carried through two underwater cables, one with a capacity of 100 megawatts to Unguja and another 25-megawatt cable to Pemba island [12]. Each island has substations to redistribute this electricity with significant wattage loss [17]. Demand for electricity is set to exceed the cable capacity in the next few years, with development outpacing capacity [17]. Because Zanzibar's electricity is generated in Tanzania, ZECO, the Zanzibari electrical distributor, pays the Tanzania Electric Supply Company (TANESCO, Tanzania's electrical provider) each month for the megawattage received [17]. To lessen the reliance on Tanzania for energy needs, Zanzibar has set a target of using 30% local renewables by 2030 [17].

Food security is a major issue in Zanzibar, with approximately 65% of households experiencing poor food consumption and 32% being severely food insecure [37]. Many households in Zanzibar depend upon subsistence activities, such as fishing and farming, to meet their food needs and are highly exposed to food insecurity because of stressors such as climate change alongside increased food demands from tourism [38].

2.2. Research Approach

The approach for research was based on principles outlined by a scenario analysis tool called "*Kesho*" (meaning "tomorrow" in Kiswahili) [39]. *Kesho* provides a structured framework allowing the participation of diverse stakeholders across scales to connect their insights directly into land cover modeling. The method has been applied in the Southern Agricultural Growth Corridor Kilombero, southwest Tanzania [23], in the Serengeti Landscape on pastoral transitions [1], to assess natural capital in Tanzania [40] and evaluate mountain social-ecological system transitions in Kenya and Ethiopia [41].

In the case of this research, the objectives of working towards water–energy–food security were developed through ethnographic research and confirmed after focus groups with village leaders and elders (number of participants (n) = 40). Transcripts and field notes of these discussions were inductively coded, and emergent themes were used to guide the rest of the research process [6]. Boundary conditions were spatially set to island level in Unguja and Pemba and temporally to 2030 to coincide with the 17 Sustainable Development Goals and the first-time horizon for Zanzibar's Development Vision for 2050 [42].

In the second stage, which took place from April to August 2019, scenario themes were co-developed by communities in community-based workshops in ten villages across the two islands (n = 142) (see Table 1). Communities went through a process of identifying key drivers of change and the impacts of land cover change on water, energy, and food. They made predictions for how drivers might contribute to challenges for water–energy–food security in the future. They then suggested several strategies that would be used to mediate these issues. The clustering and prioritization of these strategies lead to three core themes to frame future scenarios: *Ecosystem Management*, *Adaptation*, and *Settlement Planning* [13].

Table 1. Community workshop locations for the pre multi-stakeholder workshop preparation [13]. Locations represent diverse land cover types across both islands. The first five locations are on Pemba Island, and the last five locations are on Unguja Island.

Island	Workshop Group Location	Main Land Cover Type Represented
Pemba	Macho Mane	Peri-urban
	Mfikiwa	Commercial farming
	Pujini	Commercial farming and mangrove cover
	Chumbageni	Coastal area with mangrove cover and some tourism
	Mji Mpya	Protected forest
Unguja	Jambiani Kikadini	Coastal with high levels of tourism
	Pongwe	Coastal with medium levels of tourism and some mangrove cover
	Kinyasini	Peri-urban and commercial farming
	Kizimbani	Commercial farming (spice farming)
	Pete	Protected forest reserve and mangrove cover

The third phase, the focus of this article, involved scenario modeling, starting with land cover modeling of a *Business as Usual* scenario based on insights from the community-based workshops to inform variable section for causes of land cover change [13]. This was followed by two multistakeholder scenario workshops, which took place in October 2019 and involved both expert stakeholders and community representatives (n = 67). Stakeholders created narratives for alternative sustainable scenarios andpredicted how each of the alternative scenarios would alter land cover and the impact this might have on water–energy–food security for 2030. Land cover predictions made by participants were modeled to create spatially explicit visualizations of the scenarios centered on stakeholder perspectives.

2.3. Participant Selection

Two full-day scenario workshops were undertaken with local stakeholder participants, three Kiswahili-speaking facilitators, and the author, one in Macho Mane Pemba (n = 44) and one in Stonetown Unguja (n = 23). Stakeholders involved community representatives from all ten previous sites (one man, one woman, and the "*Sheha*" (village leader)). These stakeholders had been originally recruited by each village leader across the village sample sites to represent a diversity of ages, genders, and occupations. In each community workshop, participants selected two representatives to advocate for their views in the multi-stakeholder scenarios.

Members from Zanzibar's agriculture, forestry, environment, water, energy, and tourism departments attended alongside two non-governmental organizations, Milele Foundation (a sustainable livelihood organization) and Wildlife Conservation Society. These stakeholders were recruited through the support of fieldwork facilitators based at the Department of Forestry and Renewable and Non-Renewable Resources in Zanzibar.

At the Pemba workshop, 37% of the participants were women, and 63% were men. In the Unguja workshop, 48% of the participants were women, and 52% were men. This was achieved by specifying to community leaders and ministries that we were aiming for equal gender representation. At the village level, three out of ten village leaders were women, suggesting a move to more equal participation of women in leadership roles.

The participants were split into three working groups in each scenario workshop. These were organized to mix both community representatives and representatives from institutions together. Three Kiswahili-speaking facilitators supported the groups throughout the day, assisting those who were unable to read or write and explaining activity instructions.

All facilitators had undergone a one-day training with the lead author during the week preceding the workshops to become familiar with the workshop activities. During this training, the need to identify participant needs and encourage participation across age, gender, and background was discussed. Facilitators were asked to keep a record of any imbalances in contributions that they became aware of during the workshops.

2.4. Water–Energy–Food Context

Timelines were produced in which drivers of land use and land cover change influencing water–energy–food were identified from 40 years ago to the present day. Participants were asked to consider environmental, social, environmental, technical, and political drivers of change to avoid sectoral biases. The timeline was used on a horizontal axis, and on the vertical axis, trends in water–energy–food security were drawn. Each group discussed how the events on the timeline contributed to these trends and made notes on points made.

2.5. Spatial Mapping of Drivers

Groups reflected on the scale at which different drivers of change might be felt across the island (Unguja or Pemba) by 2030. Each group was given a blank map which just detailed "*Shehia*" (ward) boundaries. Group one colored and labeled areas in the island (Pemba or Unguja) where population growth would be highest and explained why. Group two did the same for climate change, and group three for socio-economic development.

2.6. Scenario Narrative Development

The third activity focused on the development of three alternate pathway trajectories based on themes that emerged during community-based scenario planning [13]. Group one explored improving the adaptive capacity of communities, group two focused on the protection and regeneration of ecosystems (including places of cultural and spiritual importance), and group three discussed sustainable settlement planning (summarized as *Adaption*, *Ecosystem Management* and *Settlement Planning*).

Participants were asked to describe what the islands would look like, how different elements interact, and what the day-to-day implications might be. These scenarios considered drivers that were both impactful and uncertain. Narratives were plausible and internally consistent, even when exploring extreme or unlikely futures. Each group had several guiding questions to prompt discussion (see Table S1).

2.7. Land Cover Change Predictions

Once pathway narratives were created, participants were asked to reflect on how pathways might influence land use and land cover by 2030. These included predictions of land cover conversions, the possible percentage change, and the likelihood of this occurring on a scale of 0 (not possible) to 4 (very likely), alongside explanations of where such change might occur. Participants were given a range of photographs of land cover types taken from the relevant island and land cover maps for Zanzibar from 2019 as guides.

Next, participants were then asked to reflect on how much land use or land cover change might impact livelihood aspects, including water, energy, food, health, and shelter (health and shelter added by participant request) on a scale from extremely positive to extremely negative for the year 2030.

3. Land Cover Modelling

3.1. *Business as Usual Scenario*

To model land use and land cover changes in a *Business as Usual* scenario, baseline temporal scope was defined for 2015 and 2017, and a future prediction for the year 2030. Land cover predictions for the 2030 baseline scenarios were produced using the Land Cover Modeler in the TerrSet Package of IDRISI (v. 18.31). Towo land cover maps for the years 2015 and 2017 were used as a baseline reference for change. Land cover maps obtained from *Copernicus* Global Land Service were used at a raster resolution of 100 m [43]. Land cover classes were modified in accordance with the local community and expert knowledge of the area. Change analysis was then applied to calculate transitions between land cover types (see Table S2 and Figure S1).

Transitions were grouped into four sub-models for Unguja and three sub-models in Pemba based on the learned understanding of drivers of land cover change developed through community-based workshops [13]. Within the transition sub-model structure, eleven independent spatial variables were included to potentially explain spatial changes in land cover; and included elevation, slope, distance from nearest protected areas, distance from the sea, distance from nearest roads, distance from settlements, distance from closed forest, distance from wetlands, distance from built-up areas, and soil composition (see Table S3). A Boolean layer was used to calculate the relative frequency of pixels that have undergone change, i.e., the evidence likelihood transformation [44]. Transitions of less than 250 pixels were excluded from the model to improve model accuracy (see Tables S4 and S5).

Multi-layer perceptron (MLP) analysis was then performed to calibrate each sub-model using dynamic learning rates. MLP analysis predicts the potential of a pixel to transition based on the explanatory power of the selected independent spatial variables. MLP then develops a model based on samples of pixels that went through the different transitions shown in the sub-model alongside samples that were eligible for transition but did not change [45,46]. Finally, Markov chain analysis was used to analyze the 2015 and 2017 land cover images and produce a transition probability matrix, transition areas matrix, and conditional probability images for the specified future date of 2030 [47].

Because the transitions to the built-up land class were few between 2015 and 2017, it was impossible to model this adequately using the MLP analysis. As settlement expansion and increased tourism infrastructure came up as major themes for land use and land cover change in earlier scoping research, it was decided that this needed to be sufficiently reflected in the baseline maps to provide a benchmark of potential change for the year 2030 [6]. Therefore, the transition to built-up land cover was projected separately and added to the BAU land cover maps (see Table S6).

Other *Kesho* applications explored the *Business as Usual* scenario in tangent to sustainable alternatives during the workshop process [1,23,40,41]. In this case, the *Business as Usual* model was created using insights from community-based insights [13] and then used as a reference point for predicting land cover change in other scenarios (i.e., "30% less built up transitions than a BAU scenario"). This was decided so that collaboration and co-creation amongst diverse stakeholders could focus on solutions and how to work towards positive outcomes.

3.2. *Alternative Scenario Modelling*

The next step involved spatially allocating land cover changes for each scenario alternative based on the perceptions, likelihood, and location of change identified during workshops. Boolean constraint maps were produced using the Reclass function in the TerrSet Package of IDRISI (v. 18.31) using perceptions of predicted land cover transitions. Spatial decision variable maps were created in ArcMap to include specific areas or conditions where change might occur. Additional variables that could affect the location of transitions were derived from both previous research employing scoping focus groups and community-based workshops [6,13] as well as wider literature [17,28,29,48].

A fuzzy module was used to convert all the decision variables into factor weights from 0 to 1. Within the fuzzy module, the parameters of each variable (i.e., monotonically increasing or decreasing) were specified. Using the Multi Criteria Evaluation module, the weight of influence of each decision variable was calculated based on the authors' understanding of the area.

The Multi-Objective Land Allocation (MOLA) function was then used to select the most spatially viable parcels of land for predicted transitions based on both the Boolean maps and likelihood indicators specified by communities. Within the MOLA function, the likelihood of change is area demand predictions specified by stakeholders were applied when calculating the objective weight of each land cover transition. The resultant spatial allocation maps were brought into ArcMap to model land cover for 2030 under each scenario using the conditional function in the spatial analyst toolbox (see Tables S7–S12). Land classes were simplified with the guidance of the Department of Forestry and Renewable and Non-Renewable resources to reflect local understandings of land cover types (see Tables S13–S15).

4. Results

4.1. Past Influences on Water–Energy–Food

In both islands there were several interrelated factors that caused land use and land cover changes, and ultimately affected water–energy–food security in the last 40 years. Participants detailed events that mediated the wider context in which the nexus operated; they also made specific links between their perceptions of water–energy–food and the events on their timelines, which are summarised below (Figure 2a,b).

In accompanying timeline notes, Pemba participants wrote that from the year 2000:

> *"Water has raised due to strong electric power and the strengthening of the water supply projects"* (Pemba, October 2019).

They also stated that food security was variable due to drought, floods, and inflation of imports seen from 1970 to the 1990s. Moreover, fuelwood supplies became increasingly degraded due to poor planning for road and building developments, excavation for water and electricity infrastructure, quarries for aggregates to make blocks for new buildings, and the creation of salt pools between 1990 and the 2000s.

In parallel to this, from the 2000s onwards, in Pemba, agriculture, trade, tourism, and urbanization were perceived to increase income and provide alternative routes for attaining water–energy–food security. Formal institutions and legal instruments were also introduced to inform land planning, which was thought to improve the sustainability of land use. However, from 2005 onwards, climate change, decreased soil fertility, and degradation of the natural environment hindered further improvements in nexus outcomes.

In Unguja, since the 2000s, participants commented that energy needs changed with the increase in trade and manufacturing, which saw greater electricity provision. Also, from the 2000s onwards, water supplies were thought to have become *"more modern"* as people transitioned from well water to piped water through the addition of new water infrastructure.

It was further mentioned that there was a lack of knowledge about water sanitation in the past and that greater awareness has helped to promote improvements. It was recognized that land shortages put pressure on farmers to increase yields, which led to a focus from the government to introduce inputs such as fertilizers and pesticides. Such land shortages also led to land conflicts and deforestation. Environmental degradation from 2005 was thought to have greatly reduced fuelwood supply (Figure 3a,b).

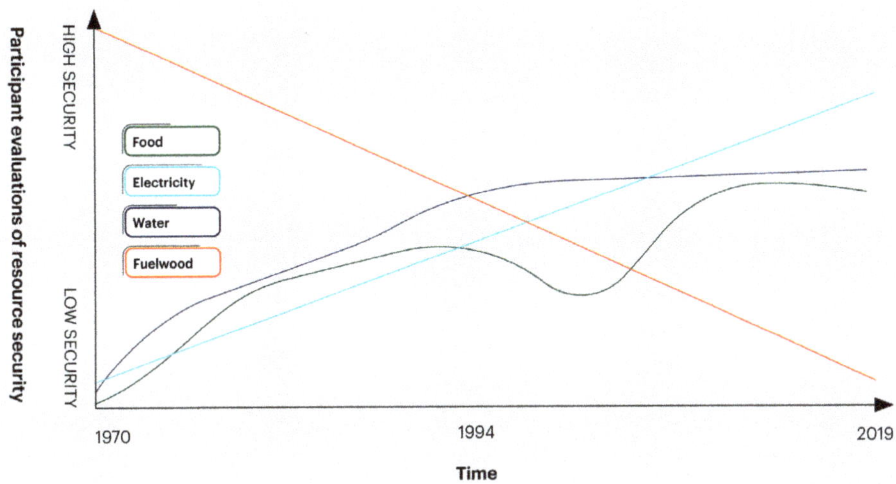

(a)

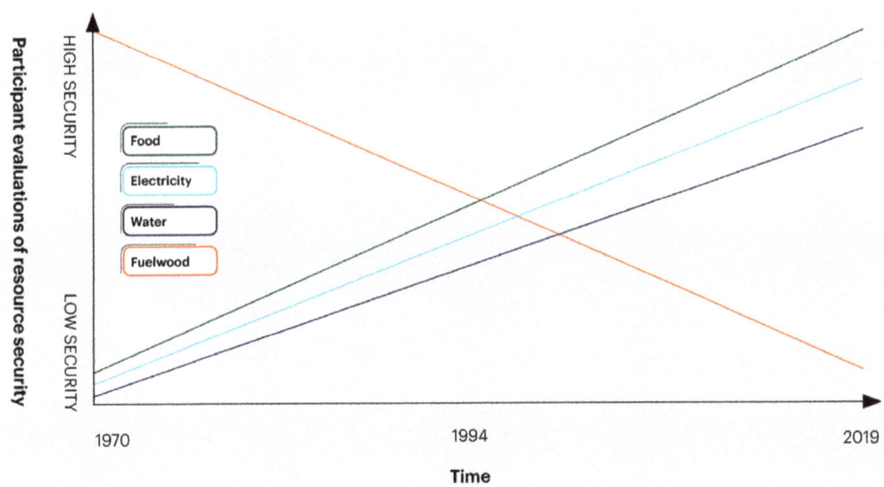

(b)

Figure 2. (**a**) Pemba water–energy–food security perceptions from participants; (**b**) Unguja water–energy–food perceptions from participants. In the workshops, participants drew timelines from 1970 onwards and added events to those timelines that impacted water–energy–food security. Above those timelines they drew line graphs depicting trends for water–energy–food security over that time. They then wrote notes to explain any connections they made between events on the timeline and their interpretation of water–energy–food security (see Figures S2 and S3 for original drawings).

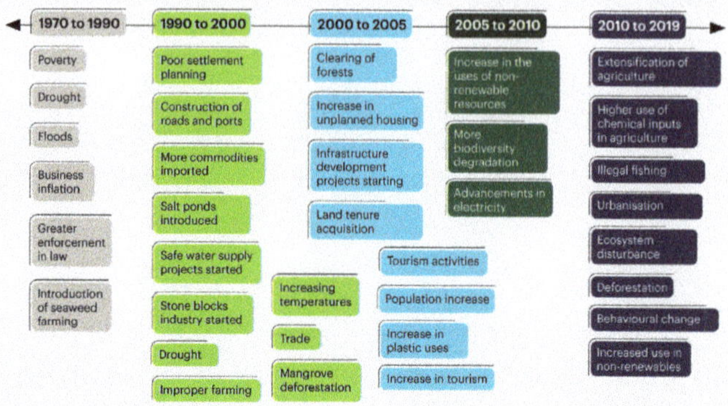

(**a**)

(**b**)

Figure 3. (**a**) Pemba timeline of events identified by participants which shaped water–energy–food security; (**b**) Unguja timeline of events which shaped water–energy–food security identified by participants.

4.2. Spatial Distribution of Drivers of Change

Three key drivers of change were identified in community-based workshops held before the multi-stakeholder workshops: population increase through both migration and internal migration, socio-economic development, and climate change [13]. In the community-based workshops, participants focused on their locality, so to understand these at the island scale, participants were asked to code areas of impact from low to high.

In Pemba, population increase from internal movement and economic migration focused on urban centers, areas with natural capital for tourism, adequate land, productive fishing, or access to the port. Socio-economic development was expected in areas adjacent to existing urban centers, around the fishing port, and in specific sites of greater natural capital that could attract tourism. Both were thought to contribute to deforestation:

> *"Increase in population leads to scarcity of land for settlement and forest area decreases. Road constructions leads to disappearance of forest and people's settlement"* (Unguja, October 2019).

Climate change impacts were predicted in specific coastal areas where flooding events have been experienced previously or where there were greater levels of deforestation. Interestingly, stakeholders graded climate change impacts as medium intensity rather than high due to ongoing mitigation in place, including afforestation of mangroves, construction of seawalls, establishment of community forests, increased awareness, and enforcement of forest regulations (Figure 4a).

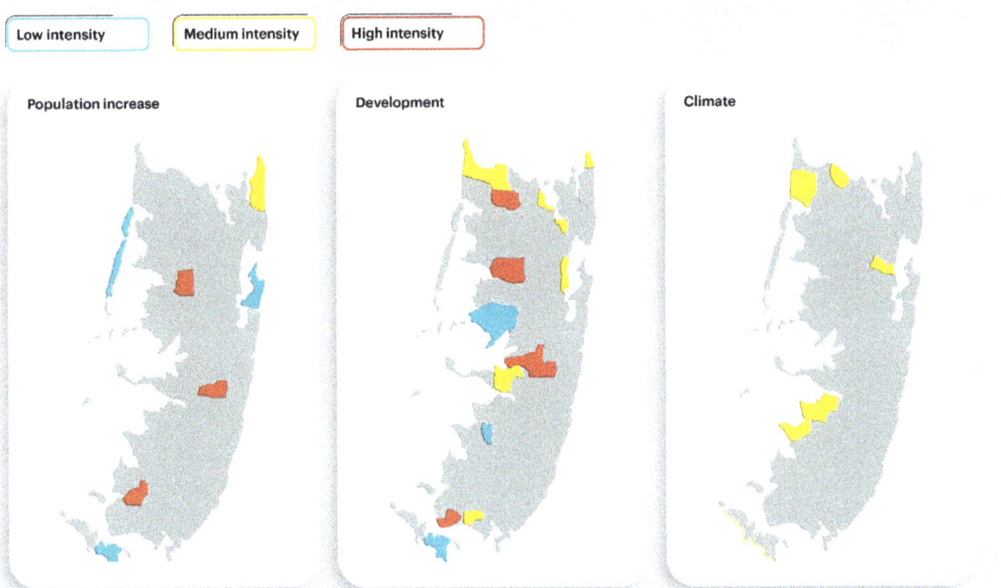

(**a**)

Figure 4. *Cont.*

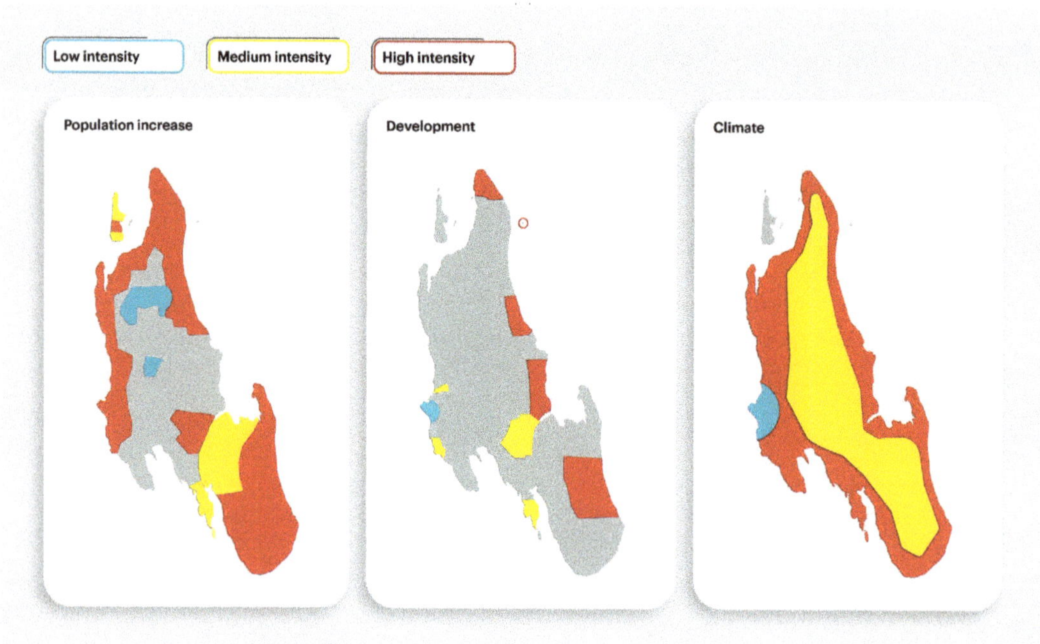

(**b**)

Figure 4. (**a**) Pemba drivers; (**b**) Unguja drivers. Population increase refers to internal migration, development is associated with socio-economic factors, and climate encompasses sea-level rise, severe weather events, and climate variability. Participants colored maps with their predictions and made notes to explain any specific connections. For development in Unguja, participants also predicted socio-economic development on one of the islets, circled in red.

Conversely, in Unguja, population increase from economic migration was expected to be more intense in the north and south coastal zones, areas that already experience high levels of tourism. This was thought to extend across the north-west coast, as coastal zones have been steadily developed for tourism infrastructure alongside other development projects. Development, on the other hand, was predicted to increase on the east coast in areas with current low levels of tourism. This was because of the establishment of development projects initiated as part of the Zanzibar Investment Promotion Authority (ZIPA) zones.

Unlike Pemba, where climate change impacts were predicted to be quite concentrated in certain places, participants in Unguja expressed concerns about the high intensity of pressures associated with wave overwash and salination of water wells across the entire coastal zone, due to deforestation on the coastline and frequent flooding (Figure 4b).

4.3. *Alternative Scenario Narratives*

Scenario narratives were summarised and written in Kiswahili by participants and then later translated into English (Table 2). There are some interesting commonalities between the three alternate scenarios. For instance, all scenarios recognize that sustainability requires an integrated response, social change, and poverty reduction. As such, water–energy–food impacts were often indirect and related to economic factors. One group explained that:

Table 2. Pathway narratives for 2030 scenarios based on themes generated from community-based workshops. Narratives are based on the translation of written narratives produced by subgroups in the multi-stakeholder workshops.

Scenario	Narrative Unguja	Narrative Pemba
Adaption	Adaptive capacity is enhanced through education and training across three key areas: entrepreneurship, cooperative group establishment, and community development projects. Such training recognizes how needs change over temporal scales and explores resource scarcity across spatial scales. This requires a coordinated response from a wide range of stakeholders across government sectors, development organizations, and private companies. To support this scenario, government bodies would review policies, laws, and guidance.	This scenario builds on existing innovative efforts to provide new ways to obtain income and encourage entrepreneurship and community environmental education. Existing barriers to adaptation are addressed through acknowledgment of the impact of poverty on readiness. Communities and individuals are now empowered to make changes to livelihoods to improve their outcomes. Education for communities is practical and involves multiple stakeholders across government sectors, NGOs, and community-based organizations. Policy between the health, education, and agriculture sectors is joined up to support adaptive capacity.
Ecosystem management	Natural forests are protected in small islands, wetlands, and areas supporting water reserves via gazettement involving communities from the beginning. To support this process, policies that connect land, forests, and fisheries are put in place, combined with continual awareness campaigns. Due to poverty and the current dependence on natural forests, efforts are made to ensure access to alternative resources and livelihoods.	This scenario emphasizes a multi-stakeholder response to strengthen laws around conservation, environmental education, and suitable land use planning. To conserve natural forests and water bodies, governing bodies have a broad overview of driving factors, such as development activities, population increase, lack of education, and poverty. Initiatives are underway to address how entrenched beliefs and traditional practices influence communities' motivation to change.
Settlement planning	This scenario focuses on youth employment, industry, and local investments. A national land use plan is developed, which considers settlement planning guidance and how economic growth relating to tourism coincides with settlements. This should be developed and supported by a coordinated team across sectors relating to land and settlements. Sanitation is continually improved through better access to cleaning facilities (i.e., waste management, sewage systems, and recycling facilities).	Settlements experience substantive changes associated with the growth of industries of agriculture, tourism, and extraction of gas and oil. This results in increased employment, business opportunities, community service, and infrastructure improvements. Spatially, land planning considers these impacts by enforcing the recommendations of environmental impact assessments and encouraging multi-story homes to make efficient use of the space available. Concurrently, this scenario protects customs, traditions, and norms for people in Zanzibar -to avoid disintegration, security challenges, and lifestyles. Cooperation between stakeholders, such as government, institutions, and civil society, is important to navigate the transitions appropriately.

"Land is the resource which contributes much in improving people's income through economic activities like agriculture, industries as well as investment in tourism" (Unguja, October 2019).

Education through continual engagement of communities was seen as important for protecting forests and improving adaptive capacity. Settlement planning scenarios emphasized the need to plan for how economic transitions relating to tourism might socially impact communities.

But there were also calls for political attention to be given to certain actions; in Unguja, participants stated that:

"We beg the government for industries and investments in the country. We beg the government to improve cleaning facilities in our villages. The government should ensure employment for the youth" (Unguja, October 2019).

Whereas in Pemba, participants called upon the government to

"Follow up the implementation of the policies and rules on urban planning" (Pemba, October 2019).

This indicates that thought needs to go not only into policy creation according to needs but also into how that policy can be acted upon in a meaningful way.

4.4. Business as Usual and Alternative Scenario Modelling

Land cover modeling drew out the strong linkages between forests across the water–energy–food nexus. In Pemba, land cover transitions to forested wetlands occurred in the coastal fringes. In Unguja, transitions to wetland forest occurred on slow-draining soils further inland. This indicates that transitions to forested wetlands are not likely to be associated with reforestation but with flooding existing forest spaces through coastal flooding and heavy rainfall.

Deforestation was expected across all scenarios. For Pemba, the highest occurrence of deforestation (across natural forests) occurred in the *Business as Usual* scenario, followed by *Settlement Planning*, *Adaption*, and *Ecosystem Management*. Unsurprisingly, the *Ecosystem Management* scenario, which had the lowest expected rates of deforestation, was the only scenario that did not reduce permanent waterbody cover.

For Unguja, the highest rates for deforestation were seen in the *Business as Usual* scenario, but this was followed by *Ecosystem Management*, *Settlement Planning*, and then *Adaption*. The *Adaptation* scenario also saw the lowest amount of losses for permanent waterbodies.

In Pemba, the *Adaptation* and *Ecosystem Management* scenarios included the expansion of cropland, whereas *Settlement Planning* involved more plantation forests. In Unguja, cropland expansion was not demonstrated in the alternative scenarios, however there was a focus on plantation forest expansion in both the *Ecosystem Management* and *Settlement Planning* scenarios (see Figures 5a,b and 6a,b and Table 3).

Table 3. Predicted land cover change for the *Business as Usual Scenario* and the three alternative scenarios (S1—*Adaption;* S2—*Ecosystem management;* S3—*Settlement planning*). The *Business as Usual* scenario in Pemba predicted sea level rise and encroachment onto land. The *Ecosystem Management* scenario in Unguja predicted a slight reduction in seaward extent due to mangrove reforestation.

	Pemba					Unguja				
Land classes	**2019 (ha)**	**BAU (ha)**	**S1 (ha)**	**S2 (ha)**	**S3 (ha)**	**2019 (ha)**	**BAU (ha)**	**S1 (ha)**	**S2 (ha)**	**S3 (ha)**
Shrubland	2.2	1.1	1.6	1.1	1.6	2.5	1.8	2.4	1.9	2.0
Cropland	1.4	1.7	1.5	1.8	1.4	1.1	1.1	1.0	1.0	1.0
Built up	0.8	0.9	0.9	0.8	0.8	2.3	2.4	2.4	2.4	2.4
Bare sparse vegetation	0.0	0.0	0.0	0.0	0.1	0.0	0.0	0.0	0.0	0.0
Permanent water bodies	0.4	0.0	0.2	0.2	0.4	0.4	0.1	0.3	0.2	0.2
Forested wetland	4.8	8.9	5.4	5.4	5.4	3.5	8.4	4.2	5.0	5.0
Natural forest	21.5	17.8	20.9	21.2	20.4	36.3	32.3	35.8	35.4	35.5
Plantation forest	1.0	1.0	1.0	1.0	1.3	3.2	3.2	3.2	3.2	3.1
Sea (gains and losses)	-	0.5	-	-	-	-	-	-	-0.2	-

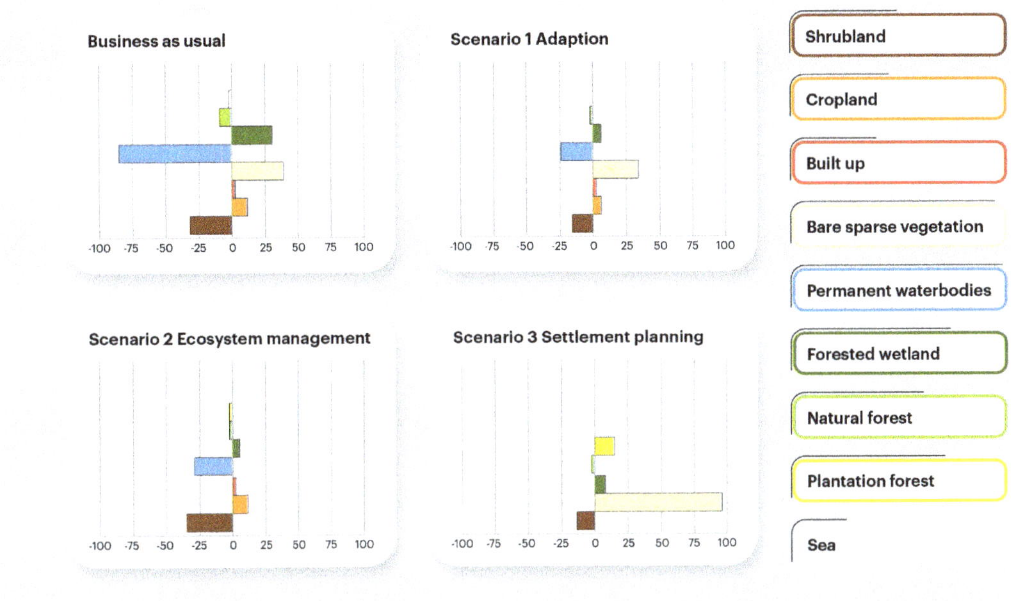

(**a**)

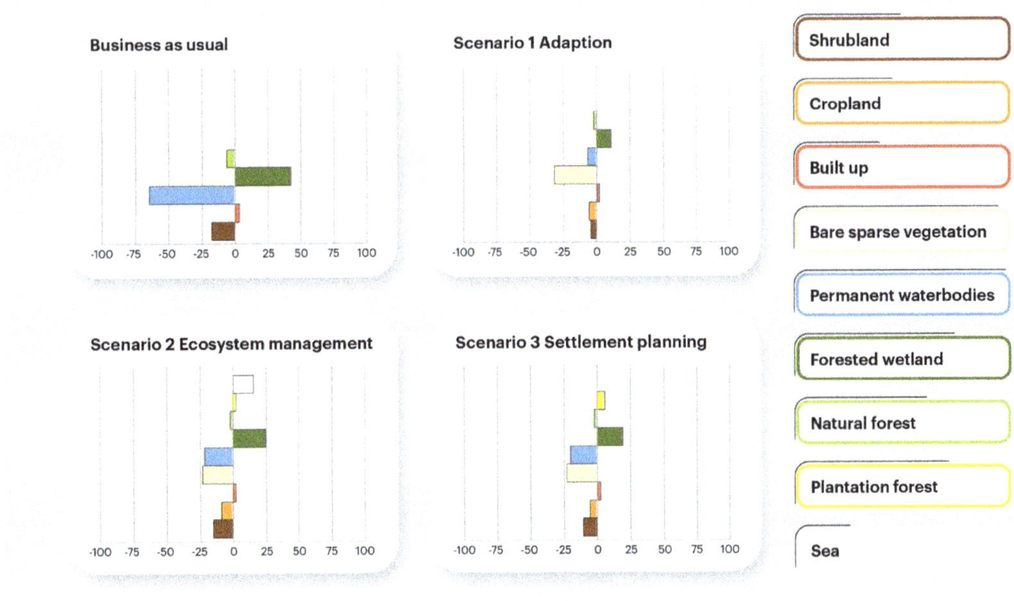

(**b**)

Figure 5. (**a**) Pemba land change; (**b**) Unguja land change. Percentage of land cover change for each land class between the present day and 2030 for the *Business as Usual* scenario, plus the three alternative scenarios, *Adaption*, *Ecosystem Management*, and *Settlement Planning*, developed in the workshops according to the modeling outcomes.

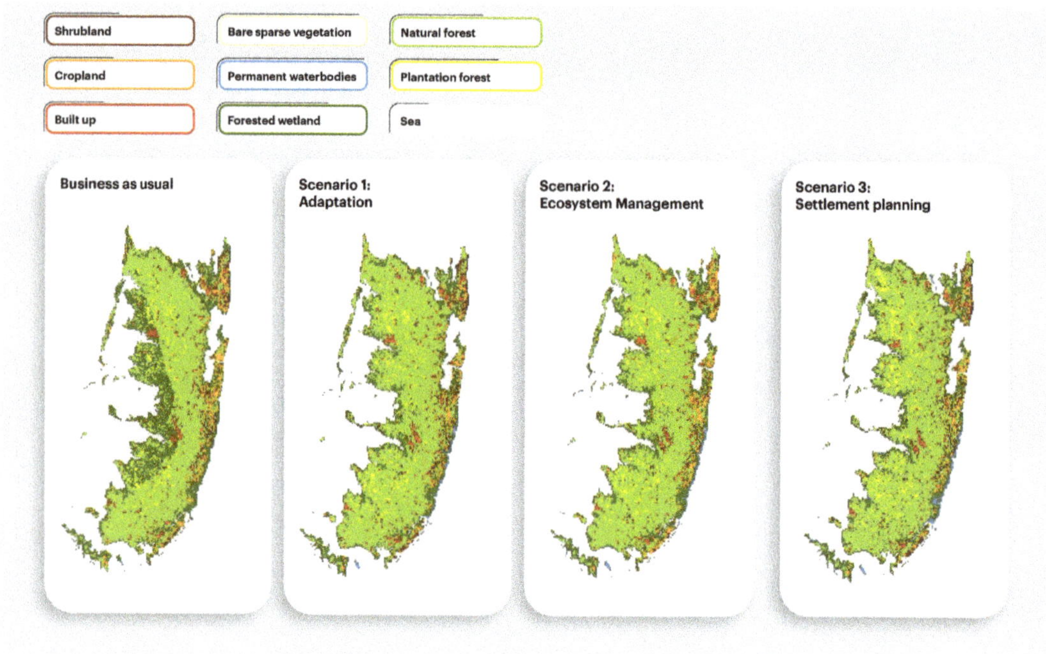

(**a**)

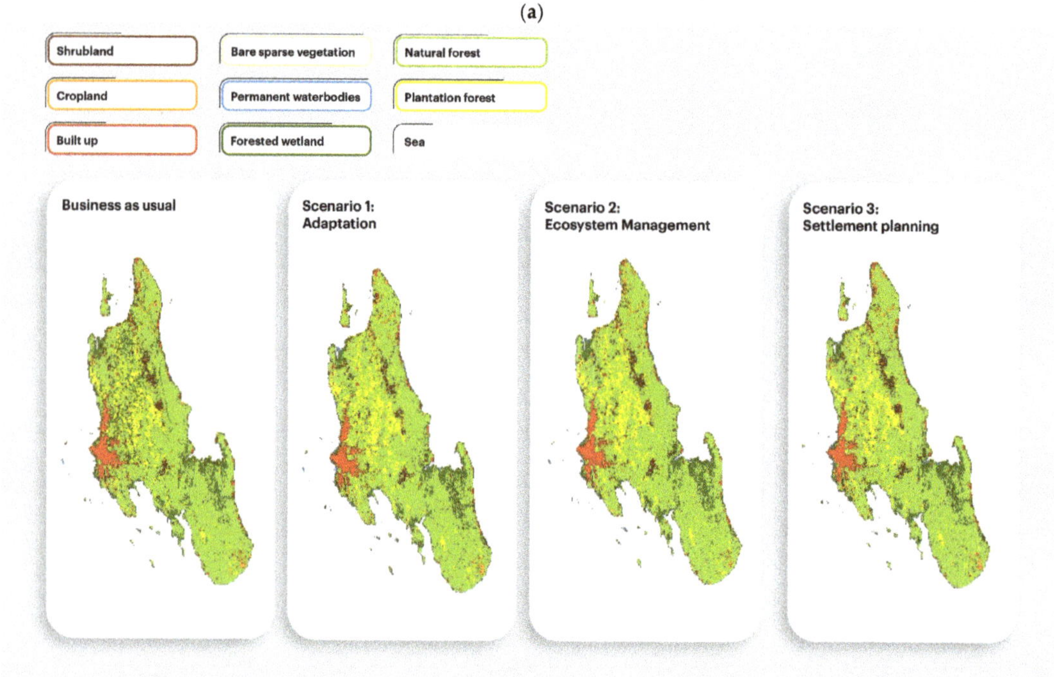

(**b**)

Figure 6. (**a**) Pemba land change map; (**b**) Unguja land change map. Maps depict land cover associated with the *Business as Usual* scenario and the three alternative scenarios, *Adaption*, *Ecosystem Management*, and *Settlement Planning*, according to modeling outcomes.

4.5. Overall Impacts of Scenarios on Water–Energy–Food

Considering the overall water–energy–food evaluation for each scenario, the *Settlement Planning* scenario for Pemba came out as the most effective at supporting resource security. In Unguja, the overall mean for water–energy–food evaluation was similar, but the *Ecosystem Management* scenario was predicted to support water–energy–food security most effectively (Table 4).

Table 4. Evaluation of how each scenario will impact livelihood aspects, starting with water, energy and food, and additional concerns, health, and shelter, by request of the participants on a scale of −4, very bad, to 4, very good.

		Water	Electricity	Gas	Firewood	Food	Health	Shelter
Pemba	Adaption	1	2	1	−1	1	1	1
	Ecosystem Management	0	0	1	−1	0	1	1
	Settlement Planning	2	3	1	−1	0	1	1
Unguja	Adaption	2	3	3	1	3	3	2
	Ecosystem Management	4	3	2	2	3	3	2
	Settlement Planning	4	3	1	−1	3	3	1

Across nearly all the scenarios, except for the *Adaption* and *Ecosystem Management* scenario for Unguja, the security of fuelwood supplies was thought to likely reduce. Participants in Pemba explained that:

> *"Clean and affordable energy seems to be weak due to large number of people using firewood which is its affordable and accessible"* (Pemba, October 2019).

Participants predicted that as fuelwood supplies become more unsustainable, other forms of energy use (predominantly gas) would increase. At the same time, barriers to its accessibility were mentioned:

> *"Gas energy is not strong due to lack of awareness among the community as well as the high price"* (Pemba, October 2019).

Participants conveyed a wish to protect forests for ecosystem benefits, including coastal protection, air quality, and soil structure. But they also recognized the challenge this posed for meeting their needs for energy, construction, and extraction of other non-wood resources, such as medicinal plants.

In Unguja, the *Ecosystem Management* and *Settlement Planning* scenarios both predicted high levels of water security despite the loss of permanent waterbodies shown in the land cover predictions. This potentially arises from an assumption that positive improvements would continue from witnessing the previous introduction of water infrastructure. There could also be a disconnect in understanding the extreme pressures regarding over-extraction and lack of foresight about climate change impacts.

When evaluating the overall effectiveness of the alternative scenarios to meet resource needs, participants felt it was important to consider health and shelter due to the implicit effects resource changes have on these two factors.

5. Discussion

The *Business as Usual* scenario gives a strong indication that saltwater incursion is likely to increase, as well as flooding of inland spaces where soil composition and land use do not promote adequate infiltration. Alongside this, the spatial analysis shows deforestation and a likely reduction in permanent water bodies. This aligns with previous research, indicating that small WIO islands face serious freshwater shortages resulting from environmental change and deforestation [49]. These risks are then often compounded by social factors such as inadequate funding for adequate water infrastructure, inequality of access, and rapid changes to watersheds and groundwater [48]. Although all alternative scenarios envisaged some continued deforestation of natural forests, some were more effective than others in mediating this. As well as pointing out areas of concern, scenarios offer an insight into where cropland expansion or plantation forest could take place, which could strengthen water–energy–food security.

By bringing together diverse stakeholders, scenarios provide more holistic evaluation strategies to secure water–energy–food into the future [1]. While no one scenario offered an ideal set of solutions for addressing environmental challenges and meeting water–energy–food security, several cross-cutting themes for future planning that support resource security became apparent. These included integrated approaches for managing environmental change, community participation in decision-making, effective protection of forests, cultural sensitivity to settlement planning and development, and poverty alleviation. Because of these emergent alignments, this discussion will focus on examining how scenarios could support strategic action toward water–energy–food security, and, more broadly, the 17 Sustainable Development Goals. Although the scenarios did not explore beyond the 2030 time horizon, they offer insights that can contribute to the achievement of Zanzibar's Development Vision for 2050 towards their 2030 targets [42] and wider agendas for sustainable land management, such as the African Agenda for 2063 [50].

5.1. Integrated Planning across Sectors

Within workshops, participants emphasized the value of the cross-cutting nexus government approach, which explicitly facilitates cross-linking between water–energy–food issues. This aligns with recommendations from the most recent SDG progress report, which state that "*epistemic communities need to reflect the diversity of society, and their interactions will need to be far more multi-directional and multi-disciplinary, so they can effectively address complex and interlinked challenges and goals*" (p. 91) [51]. Zanzibar's 2050 Development Vision also sets out to develop linkages across all sectors [42]. More generally, wider research states that an integrated approach for water–energy–food is seen as essential for future security [52]. Efforts should, therefore, pay attention to how meaningful collaboration happens.

Supporting collaborative practices across sectors might require the designation of specific roles to identify overlapping areas of interest where different groups might be able to contribute more integratively [53]. Resourcing aimed at cross-sector working groups might also be needed to stimulate more coherent action. Further to this, multi- and trans-disciplinary action requires critical reflections on power distribution and accountabilities [54]. Consequently, reflexivity is needed to surface participants' positionality (including their values and beliefs) and challenge underlying assumptions and power relations [55,56]. This can provide a more accurate and nuanced understanding of the impacts of actions undertaken and avoid a culture of silence where fundamental issues remain unresolved [57].

5.2. Community Participation in Decision Making

Participants advocated for participation in decision making. Integrating bottom-up experiences of all segments of populations and subnational identities into policy decision making has been shown to result in tangible synergistic solutions that are actionable [58] and key for the delivery of the SDGs [51]. While horizontal linkages for cohered responses to change can be enabled through common peer-to-peer understanding, vertical dialogues connecting experience to policy are often limited [59]. Moreover, while community-level insights are often acknowledged as important to developing a holistic understanding of complex issues, they are not drawn upon directly but through advocates such as NGOs, as in the case of the development of Zanzibar's 2050 vision [42]. This said, there is a movement towards people-focused processes for development; for example, the Africa 2063 vision outlines key stakeholder groups for consultation, including women and youth [50].

Specific to this research focus, there is an increasing awareness that social-ecological understandings for water–energy–food systems are missing in nexus analysis, which risks inadequate management [60,61]. But for effective elicitation, attention needs to be given to what prevents effective participation [20,21,62,63]. For instance, communities may lack trust in authorities due to past experiences of exploitation or broken promises. This mistrust can lead to reluctance to participate in new initiatives. Self-esteem, capacity, gender dynamics, and differences in education can also impact the dynamics and communication in participatory methods [21].

In this study, preparatory work was carried out at the community level to inform the framing of the multi-stakeholder workshops to try to enhance agency. Nonetheless, there were still barriers to participation in the multi-stakeholder workshops, which could have influenced the results. These were predominantly around differences in literacy and education, which affected the confidence of individuals to participate. Facilitators were aware of and mediated these challenges through supportive discussion and confidence building around the value of insights from community perspectives. But, this observation serves as a reminder that there can still be an imbalance of community voice representation even when represented in multi-stakeholder workshops.

5.3. Protection of Remaining Forests

Scenario narratives advocated for more regulation and protection of forests, especially those known to play a critical role in protecting coastlines from rising sea levels and the increased frequency of severe storms [64]. Key elements in this regard include restoring natural barriers like mangroves, which can help protect against erosion and flooding, as well as creating buffer zones and regulating coastal development and construction away from high water marks [65,66]. Awareness campaigns and education around certain themes, such as deforestation, were highlighted as central to aligning communities to land use policies. For instance, previous research in Zanzibar found that education about the ecological links between mangrove ecosystems and resources such as fish was key to obtaining management support [67].

Concurrently, it was recognized that alternative energy provision interventions are needed so that people are not adversely affected by restricted access to forest resources, which they depend on for energy needs [68]. As part of the Zanzibar Vision for 2050, there is a focus on extracting offshore oil and gas, which has tensions with international goals for reducing the use of fossil fuels [69]. Transitioning to gas usage for cooking also ultimately depends on people's capacity to afford this as a fuel source.

There are broader links to energy that need to be considered within energy security transitions. Zanzibar's strong dependence on mainland Tanzania remains a constraint to development as demand is expected to exceed cable capacity within the next few years due to the introduction of new hotels [17]. Lack of energy results in frequent power cuts and water extraction and supply challenges. There is also a constant tension between Zanzibar and TANESCO (Tanzania's electrical provider) because of excessive unpaid electricity bills, resulting in threats to cut off power supply [17]. This would have extremely negative implications for the provision of water, which depends upon electricity for pumping.

Renewable energy has some capacity to reshape energy relationships and enhance resilience. For instance, ZAWA (Zanzibar's water authority) has begun installing solar water pumps [17]. Solar is viewed as a more feasible alternative for households who cannot afford to connect their homes to the national grid. When this research was conducted, connection costs were variable, and the onus for expanding electric infrastructure was placed on the user. Since then, ZENCO (Zanzibar's electricity provider) has announced a flat rate of 200,000 TSH (around 85 USD). While this is a fairer approach, it is still out of reach to many citizens. As a result, solar remains a more viable option, especially for those without land ownership, as it is an investment that could be relocated.

5.4. Centralising Community Needs in Planning

Findings revealed a need for more consideration of diverse needs to design the most effective use of the space. Whilst there is an acknowledgment of unplanned settlements and challenges regarding poor sanitation and energy infrastructure, Zanzibar's 2050 Vision plans do not yet explicitly address this, especially in rural areas. At present, there is a limited understanding of how the built environment impacts the water–energy–food nexus [70]. Houses in rural settlements are mostly built without formal planning and are socially constructed in relation to how people organize their lives [71]. More interrogation is needed to explore how the current nature of settlements aligns with opportunities for strengthening water–energy–food security, but also how it might constrain it.

What also came through within the research process was the concern that participants had about development trajectories in terms of their suitability with social values and livelihoods. There was a strong advocation for settlements to have a degree of separation from tourism centers. Given that tourism is a vital part of Zanzibar's economy, it is essential to implement sustainable tourism practices that do not adversely affect the vital ecosystems or cultural heritage that local communities value [72,73]. At present, the benefits of tourism are not distributed amongst the local communities. Conversely, they have seen a reduction in quality and availability, increases in the price of locally caught fish, and conflicts with coastal zone use [31,38,72].

Looking forward, there are also several considerations that need to be made to improve community resilience to outlined drivers. Improved drainage systems, erosion control measures, and construction of infrastructure that is resilient to climate change impacts will be key in adapting to and mitigating the impacts of environmental changes [8]. Building guidelines should be revised to encourage the construction of energy-efficient buildings, incorporating renewable energy sources such as solar and wind power, thereby reducing dependency on non-renewable energy and forest products. Green infrastructure spaces and rain gardens could also aid in water management, improve water and air quality, provide local space for food production, and mitigate urban heat island effects [8].

5.5. Poverty Alleviation

Communities further sought to harness opportunities to strengthen their livelihood outcomes and, in turn, water–energy–food security. There was a call for more support for entrepreneurship, which could also see Zanzibari's connection to tourism-related opportunities. There was also a demand for adopting more innovation, for instance, in the farming sector. Community development, together with education, was also seen as important for stimulating skills and broadening opportunities from urban to rural communities. Further to this, there were ideas around how communities could organize themselves to develop co-operatives, which might stimulate funding streams for development activities. Zanzibar's Development Plan for 2050 has a strong focus on diverse income opportunities and seeks to stimulate opportunities across agriculture, finance, trade, tourism, blue economy, creative and digital, and oil and gas sectors [42]. It also seeks to create better links between tourism and local produce as well as support training to increase people's capabilities to enter the workforce [42]. More in-depth planning is needed to ensure people can harness potential opportunities through subsidized training programs and advisory groups to connect local people to employment opportunities.

5.6. Limitations

Whilst findings contribute to understanding how land cover change could impact water-energy-food security in Zanzibar, there are limitations to the research. Firtsly, the *Business as Usual* scenario modeling used land cover maps over a close time series to project land cover changes for the future. It would be more suitable to analyze changes over a longer time interval, but this was not available at the time of analysis. The *Business as Usual* map, alongside the alternative scenario maps, should not be interpreted as accurate representations of the future but as comparative visuals of alternative approaches for managing drivers of change into the future to guide decision making.

Moreover, though there was a concerted effort to bring together diverse stakeholders, it is worth reflecting on who was not involved. For instance, migratory groups from the Masai Mara periodically travel to and from Zanzibar for work during the high tourism season. They interact with the landscape to meet their needs and, therefore, have an influence on the nexus—but were not included in visioning for the future. Similarly, international migrants who have settled in Zanzibar were not involved but often have an impact on land cover through building or tourism-related business.

Moreover, while efforts were made to reduce bias for the sampling of participants by setting out specific guidance for the recruitment, there is some likely bias evident in the results in using village leaders as gatekeepers. For instance, a likely over-representation of the current political party.

Finally, ideally, all stakeholders should have had the opportunity to give feedback on the results and further modify the land cover maps. In this case, it was not possible due to the global pandemic in 2020.

5.7. Future Considerations

Though the research process was able to draw out planning approaches that could support water–energy–food security through the mediation of drivers of change, there was no one set of solutions that completely addressed resource challenges, and deforestation was still expected to occur to some extent across all the scenarios modeled.

Whilst scenario methodologies are useful for working with the future and challenging assumptions behind societal patterns, they can be seen as limited as they tend to focus on plausible futures. This can mean that they don't always sufficiently challenge systematic failures that might require more transformative change [74]. This study focused on 2030, but having a longer time horizon could help participants explore more ambitious interventions, but at the same time, it means working with greater uncertainty. Future research could valuably employ more "pathways"-type future approaches such as Three Horizons, working with local participants to collectively envision their desired futures and creatively identify necessary actions for supporting transformation [74].

Another consideration here is the lack of consideration towards global influences when exploring water–energy–food security to local land use. Communities in small islands are net importers of food and, therefore, are highly exposed to food price spikes [75]. While subsistence farming is an important aspect of meeting food needs, climate change, land use competition, and poor soil fertility contribute to reduced yield [76]. Though results are useful for building a picture of place-based water–energy–food security understandings, there is a need to capture global influences and their impacts across these more local scales. This might be achieved by involving trade, geopolitical, and global conservation experts in the workshop dialogues.

Further work is also needed to explore the wider implications of informal migrant settlements and the social-ecological interactions of migrations within the water–energy–food nexus. Unplanned urban sprawl from both local populations and migrant settlements can lead to expansions into natural environments or agricultural land, as in the case of Chuini Zanzibar [77]. Social-ecological relationships underpinning water–energy–food security might be especially precarious for migrant populations because of land tenure insecurity. For instance, those renting property or land can be more reluctant to invest in infrastructure such as electricity because they could be forced to move [17]. Therefore, what might be happening is a divergent set of social-ecological relationships for migrant populations compared with Zanzibari communities. These need to be understood so that they can be captured in future scenarios.

6. Conclusions

As communities in small islands within the Western Indian Ocean have strong social-ecological relationships within the water–energy–food nexus, they are well-positioned to explain the implications of different types of pressures and how this mediates these relationships. This study provides an example of how community insights can be integrated with active decision makers to evaluate the impact of alternative interventions more effectively. Findings reveal important understandings with regard to where drivers of change might most impact water–energy–food security and offer socially acceptable interventions to mediate such.

Results could inform near-term land use planning that can enhance climate-resilient rural livelihoods, better conservation outcomes, and sustainable tourism development. Specific policy recommendations arising from this work include: (1) focusing conservation and forest regeneration efforts around areas predicting a loss in permanent water bodies; (2) working with communities and hoteliers to reforest coastlines to provide protective buffers against storm surges at a landscape scale; (3) invest in solar infrastructure for water pumps to limit disruptions in supply caused power outages; (4) actively support the establishment of woodlots for producing fuelwood at scale, while also investing in long term sustainable energy transitions; (5) a focus on local enterprise and the development of skills to harness emergent opportunities; and (6) enhance adaptive capacity of communities through training in innovative agricultural methods.

As well as more specific policy outcomes, the results from the study also advise on processes for how policy decisions should arise, with community involvement and integration between sectors being key. Enhanced participation from communities at the onset of landscape planning was thought to provide a better opportunity for protecting community values and culture. Productive discourse and co-development of strategic action involving relevant organizations were thought to be important for increasing Zanzibar's capacity to support better water–energy–food outcomes.

Understandings from communities and experts about social-ecological interactions within the water–energy–food nexus also indicate that scenario alternatives would operate differently for the different islands, demonstrating the importance of understanding the social, environmental, and economic contexts and how they might connect to national policies.

Supplementary Materials: The following supporting information can be downloaded at https://www.mdpi.com/article/10.3390/land13020195/s1, Figure S1. Map of Unguja and Pemba islands and land use and land cover for the year 2019 (when fieldwork was conducted); Figure S2. Pemba; Figure S3. Unguja; Table S1. Guiding questions for scenario narrative formation in Pemba and Unguja multistakeholder workshops; Table S2. Land cover class descriptions used in initial modelling process (note these were later simplified further); Table S3. The data sources for variable land cover modelling; Table S4. Sub model details for transition potential maps explaining land cover change between 2015 and 2017 in Unguja; Table S5. Sub model details for transition potential maps explaining land cover change between 2015 and 2017 in Pemba; Table S6. Area of built up land cover class; Table S7. Modelling conditions for the adaptive capacity scenario in Pemba. Likelihood of land transition ranges from 1 (least likely) to 4 (most likely). Relationship of the influencing variables to the pattern of transition described as monotonically increasing (the further away from the variable the more likely the transition), monotonically decreasing (the further away the variable is the less likely the transition); Table S8. Modelling conditions for the adaptive capacity scenario in Unguja. Likelihood of land transition ranges from 1 (least likely) to 4 (most likely). Relationship of the influencing variables to the pattern of transition described as monotonically increasing (the further away from the variable the more likely the transition), monotonically decreasing (the further away the variable is the less likely the transition); Table S9. Modelling conditions for the ecosystem management scenario for Pemba. Likelihood of land transition ranges from 1 (least likely) to 4 (most likely). Relationship of the influencing variables to the pattern of transition described as monotonically increasing (the further away from the variable the more likely the transition), monotonically decreasing (the further away the variable is the less likely the transition); Table S10.

Modelling conditions for the ecosystem management scenario in Unguja. Likelihood of land transition ranges from 1 (least likely) to 4 (most likely). Relationship of the influencing variables to the pattern of transition described as monotonically increasing (the further away from the variable the more likely the transition), monotonically decreasing (the further away the variable is the less likely the transition); Table S11. Modelling conditions for the settlement planning scenario in Pemba. Likelihood of land transition ranges from 1 (least likely) to 4 (most likely). Relationship of the influencing variables to the pattern of transition described as monotonically increasing (the further away from the variable the more likely the transition), monotonically decreasing (the further away the variable is the less likely the transition); Table S12. Modelling conditions for the settlement planning scenario in Unguja. Likelihood of land transition ranges from 1 (least likely) to 4 (most likely). Relationship of the influencing variables to the pattern of transition described as monotonically increasing (the further away from the variable the more likely the transition), monotonically decreasing (the further away the variable is the less likely the transition); Table S13. Percentage differences in land cover area between the present day actual land cover and the predicted values for the BAU scenario for 2030 in Pemba; Table S14. Percentage differences in land cover area between the present day actual land cover and the predicted values for the BAU scenario for 2030 in Unguja; Table S15. Details of land cover consolidation for final map images and calculations.

Author Contributions: Conceptualisation by R.J.S.N., R.A.M., C.E., A.Y.N., T.A.H., C.C., J.P.R.T., C.J.C.-M., R.W.K., S.J.B., I.F. and E.S.O. Data collection R.J.S.N., A.Y.N. and T.A.H. and analysis by R.J.S.N. Writing by R.J.S.N. Editing R.J.S.N., J.P.R.T., C.E., C.J.C.-M. and R.A.M. Supervision by C.E and R.A.M. All authors have read and agreed to the published version of the manuscript.

Funding: This document has been produced with the financial assistance of the Economic and social research council (Grant no. Economic and Social Research Council: ES/J500215/1) as well as the European Union (Grant no. DCI-PANAF/2020/420-028), through the African Research Initiative for Scientific Excellence (ARISE), pilot programme. ARISE is implemented by the African Academy of Sciences with support from the European Commission and the African Union Commission.

Data Availability Statement: Land cover maps are available by email request to Rebecca J. S. Newman at rebecca.newman@york.ac.uk.

Acknowledgments: Special thanks to the Department of Forestry and Renewable and Non-Renewable Resources in Zanzibar, which supported this project. Thanks to all those stakeholders who contributed, from communities, government sectors, and non-governmental organizations in Zanzibar. Thanks also to the reviewers whose thoughtful feedback helped to improve the contribution of this paper.

Conflicts of Interest: Authors can confirm there are no conflicts of interest with regards to this publication.

References

1. Kariuki, R.W.; Capitani, C.; Munishi, L.K.; Shoemaker, A.; Courtney, C.J.; William, N.; Lane, P.J.; Marchant, R. Serengeti's futures: Exploring land use and land cover change scenarios to craft pathways for meeting conservation and development goals. *Front. Conserv. Sci.* **2022**, *3*, 920143. [CrossRef]
2. Poti, M.; Hugé, J.; Shanker, K.; Koedam, N.; Dahdouh-Guebas, F. Learning from small islands in the Western Indian Ocean (WIO): A systematic review of responses to environmental change. *Ocean Coast. Manag.* **2022**, *227*, 106268. [CrossRef]
3. Laffoley, D.; Baxter, J.M.; Hassoun, A.E.R.; Spalding, M.; Osborn, D.; Oliver, J.; Andrew, T.G. Towards a western Indian Ocean regional ocean acidification action plan. In *Workshop Report*; IUCN: Gland, Switzerland, 2020; 29p.
4. IPCC. *Climate Change 2022: Impacts, Adaption, and Vulnerability. Contribution of Working Group II to the Sixth Assessment Report of the Intergovernmental Panel on Climate Change*; Pörtner, H.-O., Roberts, D.C., Tignor, M., Poloczanska, E.S., Mintenbeck, K., Alegría, A., Craig, M., Langsdorf, S., Löschke, S., Möller, V., et al., Eds.; Cambridge University Press: Cambridge, UK; New York, NY, USA, 2022; 3056p. [CrossRef]
5. Mycoo, M.M.; Wairiu, D.; Campbell, V.; Duvat, Y.; Golbuu, S.; Maharaj, J.; Nalau, P.; Nunn, J.; Pinnegar, J.; Warrick, O.; et al. Small Islands. In *Climate Change 2022: Impacts, Adaption, and Vulnerability. Contribution of Working Group II to the Sixth Assessment Report of the Intergovernmental Panel on Climate Change*; Pörtner, H.-O., Roberts, D.C., Tignor, M., Poloczanska, E.S., Mintenbeck, K., Alegría, A., Craig, M., Langsdorf, S., Löschke, S., Möller, V., et al., Eds.; Cambridge University Press: Cambridge, UK, 2022.
6. Newman, R.J.S.; Thorn, J.P.R.; Haji, T.A.; Nchimbi, A.Y.; Musa, I.; Enns, C.; Marchant, R.A. A people-centred framework for exploring water-energy-food security in a small developing island. *Popul. Environ.* **2023**, *45*, 14. [CrossRef]

7. Marchant, R. Using the Past to Chart Future Pathways? In *East Africa's Human Environment Interactions*; Palgrave Macmillan: Cham, Switzerland, 2021. [CrossRef]
8. Thorn, J.P.R.; Klein, J.A.; Steger, C.E.; Hopping, K.A.; Capitani, C.; Tucker, C.M.; Reid, R.S.; Marchant, R. Scenario archetypes reveal risks and opportunities for mountain futures. *Glob. Environ. Chang.* **2021**, *69*, 102291. [CrossRef]
9. Haasnoot, M.; Kwakkel, J.H.; Walker, W.E.; ter Maat, J. Dynamic adaptive policy pathways: A method for crafting robust decisions for a deeply uncertain world. *Glob. Environ. Chang.* **2013**, *23*, 485–498. [CrossRef]
10. Hermans, L.M.; Haasnoot, M.; ter Maat, J.; Kwakkel, J.H. Designing monitoring arrangements for collaborative learning about adaption pathways. *Environ. Sci. Policy* **2017**, *69*, 29–38. [CrossRef]
11. Lavorel, S.; Colloff, M.J.; Locatelli, B.; Gorddard, R.; Prober, S.M.; Gabillet, M.; Devaux, C.; Laforgue, D.; Peyrache-Gadeau, V. Mustering the power of ecosystems for adaption to climate change. *Environ. Sci. Policy* **2019**, *92*, 87–97. [CrossRef]
12. Winters, Z.S.; Crisman, T.L.; Dumke, D.T. Sustainability of the Water-Energy-Food Nexus in Caribbean Small Island Developing States. *Water* **2022**, *14*, 322. [CrossRef]
13. Newman, R.J.S.; Capitani, C.; Haji, T.A.; Nchimbi, A.Y.; Enna, C.; Marchant, R. Community-based scenario's framework for supporting water, energy, and food security in Zanzibar. *Sustain. Sci.* **2024**. [CrossRef]
14. Barbier, E.B. Valuing the storm protection service of estuarine and coastal ecosystems. *Ecosyst. Serv.* **2015**, *11*, 32–38. [CrossRef]
15. Kurian, M. The water-energy-food nexus. *Environ. Sci. Policy* **2017**, *68*, 97–106. [CrossRef]
16. Mpandeli, S.; Naidoo, D.; Mabhaudhi, T.; Nhemachena, C.; Nhamo, L.; Liphadzi, S.; Hlahla, S.; Modi, A.T. Climate Change Adaption through the Water-Energy-Food Nexus in Southern Africa. *Int. J. Environ. Res. Public Health* **2018**, *15*, 2306. [CrossRef] [PubMed]
17. Dean, E. The "Department of Human Needs": Renewable energy and the water–energy–land nexus in Zanzibar. *Econ. Anthropol.* **2023**, *10*, 246–255. [CrossRef]
18. Martin del Campo, F.; Singh, S.; Mijts, E. The resource (in)sufficiency of the Caribbean: Analysing socio-metabolic risks (SMR) of water, energy, and food. *Front. Clim.* **2023**, *5*, 1085740. [CrossRef]
19. Crisman, T.L.; Winters, Z.S. Caribbean small island developing states must incorporate water quality and quantity in adaptive management of the water-energy-food nexus. *Front. Environ. Sci.* **2023**, *11*, 1212552. [CrossRef]
20. Cleaver, F. The inequality of social capital and the reproduction of chronic poverty. *World Dev.* **2005**, *33*, 893–906. [CrossRef]
21. Gustavsson, M.; Lindström, L.; Jiddawi, N.S.; de la Torre-Castro, M. Procedural and distributive justice in a community-based managed Marine Protected Area in Zanzibar, Tanzania. *Mar. Policy* **2014**, *46*, 91–100. [CrossRef]
22. Saito, O.; Kamiyama, C.; Hashimoto, S.; Matsui, T.; Shoyama, K.; Kabaya, K.; Uetake, T.; Taki, H.; Ishikawa, Y.; Matsushita, K.; et al. Co-design of national-scale future scenarios in Japan to predict and assess natural capital and ecosystem services. *Sustain. Sci.* **2018**, *14*, 5–21. [CrossRef]
23. Thorn, J.P.R.; Wijesinghe, A.; Tam, C.; Nakei, J.B.; Magesa, R.; Mbwambo, L.; Mtega, G.; Marchant, R. *Kesho Mpya (New Tomorrow): Envisioning a Sustainable Future for the Southern Agricultural Growth Corridor of Tanzania (SAGCOT)*; UNEP-WCMC, SAGCOT: Cambridge, UK; Dar es Salaam, Tanzania, 2022. Available online: https://developmentcorridors.org/wp-content/uploads/2021/12/SAGCOT-Brochure.pdf (accessed on 15 December 2023).
24. Office of the Chief Government Statistician Zanzibar (OCGS). *Zanzibar Socio Economic Survey 2022*; Preliminary Statistical Report; OCGS: Zanzibar, Tanzania, 2023.
25. World Bank. *Towards a More Inclusive Zanzibar Economy: Zanzibar Poverty Assessment 2022*; World Bank: Washington, DC, USA, 2022. Available online: https://documents1.worldbank.org/curated/en/099755011022231706/pdf/P1753400df8a050240810307f8599187b46.pdf (accessed on 1 December 2023).
26. Leclair, M. Aquaculture and Marine Conservation: Case Study of Zanzibar. 2020. Available online: https://www.iucn.org/sites/dev/files/content/documents/zanzibar_case_study_2020.pdf (accessed on 2 July 2022).
27. Department of Environment (DoE). *The Status of Zanzibar Coastal Resources—Towards the Development of Integrated Coastal Management Strategies and Action Plan*; The Zanzibar Revolutionary Government: Zanzibar, Tanzania, 2009.
28. Makame, M.O.; Shackleton, S. Perceptions of climate variability and change in relation to observed data among two east coast communities in Zanzibar, East Africa. *Clim. Dev.* **2019**, *12*, 801–813. [CrossRef]
29. Suckall, N.; Tompkins, E.; Stringer, L. Identifying trade-offs between adaptation, mitigation and development in community responses to climate and socio-economic stresses: Evidence from Zanzibar, Tanzania. *Appl. Geogr.* **2014**, *46*, 111–121. [CrossRef]
30. Kukkonen, M.; Käyhkö, N. Spatio-temporal analysis of forest changes in contrasting land use regimes of Zanzibar, Tanzania. *Appl. Geogr.* **2014**, *55*, 193–202. [CrossRef]
31. Lange, G.M. Tourism in Zanzibar: Incentives for sustainable management of the coastal environment. *Ecosyst. Serv.* **2015**, *11*, 5–11. [CrossRef]
32. Haji, H.S. *Water Balance Assessment in Unguja Island, Zanzibar, Tanzania*; Faculty of Geo-Information and Earth Observation (ITC), University of Twente: Enschede, The Netherlands, 2010.
33. The Revolutionary Government of Zanzibar. An Assessment of Rainwater harvesting Potential in Zanzibar. 2007. Available online: https://apps.worldagroforestry.org/downloads/O1E_Assessment_of_Rainwater_harvesting_Potential_in_Zanzibar.pdf (accessed on 3 September 2020).

34. Pembe-Ali, Z.; Mwamila, T.B.; Lufingo, M.; Gwenzi, W.; Marwa, J.; Rwiza, M.J.; Lugodisha, I.; Qi, Q.; Noubactep, C. Application of the Kilimanjaro Concept in Reversing Seawater Intrusion and Securing Water Supply in Zanzibar, Tanzania. *Water* **2021**, *13*, 2085. [CrossRef]
35. Gössling, S. The consequences of tourism for sustainable water use on a tropical island: Zanzibar, Tanzania. *J. Environ. Manag.* **2001**, *61*, 179–191. [CrossRef] [PubMed]
36. Mato, R.A. Groundwater quality degradation due to saltwater intrusion in Zanzibar Municipality. *Afr. J. Environ. Sci. Technol.* **2015**, *9*, 734–740. [CrossRef]
37. Nyangasa, M.A.; Buck, C.; Kelm, S.; Sheikh, M.; Hebestreit, A. Exploring Food Access and Sociodemographic Correlates of Food Consumption and Food Insecurity in Zanzibari Households. *Int. J. Environ. Res. Public Health* **2019**, *16*, 1557. [CrossRef] [PubMed]
38. Makame, O.M.; Richard, Y.M.K.; Layla, A.S. Climate change and household food insecurity among fishing communities in the eastern coast of Zanzibar. *J. Dev. Agric. Econ.* **2015**, *7*, 131–142. [CrossRef]
39. Capitani, C.; Marchant, R. KESHO: A Participatory Scenario Modelling Tool. 2019. Available online: http://www.real-project.eu/wp-content/uploads/2019/05/KESHO_English-1.pdf (accessed on 20 January 2024).
40. Capitani, C.; Garedew, W.; Mitiku, A.; Berecha, G.; Hailu, B.T.; Heiskanen, J.; Hurskainen, P.; Platts, P.J.; Siljander, M.; Pinard, F.; et al. Views from two mountains: Exploring climate change impacts on traditional farming communities of Eastern Africa highlands through participatory scenarios. *Sustain. Sci.* **2019**, *14*, 191–203. [CrossRef]
41. Capitani, C.; Van Soesbergen, A.; Mukama, K.; Malugu, I.; Mbilinyi, B.; Chamuya, N.; Kempen, B.; Malimbwi, R.; Mant, R.; Munishi, P.; et al. Scenarios of land use and land cover change and their multiple impacts on natural capital in Tanzania. *Environ. Conserv.* **2019**, *46*, 17–24. [CrossRef]
42. Zanzibar Planning Commission. Zanzibar Development Vision 2050. 2020, pp. 1–86. Available online: https://zarip.planningznz.go.tz/handle/123456789/18 (accessed on 4 November 2023).
43. Buchhorn, M.; Smets, B.; Bertels, L.; De Roo, B.; Lesiv, M.; Tsendbazar, N.E.; Linlin, L.; Tarko, A. *Copernicus Global Land Service: Land Cover 100 m: Version 3 Globe 2015–2019: Product User Manual*; Zenodo: Geneve, Switzerland, September 2020. [CrossRef]
44. Clark Labs. *TerrSet Help System*; Clark University: Worcester, MA, USA, 2017.
45. Civco, D.L. Artificial neural networks for land cover classification and mapping. *Int. J. Geogr. Inf. Syst.* **1993**, *7*, 173–186. [CrossRef]
46. Sangermano, F.; Eastman, J.R.; Zhu, H. Similarity weighted instance based learning for the generation of transition potentials in land change modeling. *Trans. GIS* **2010**, *14*, 569–580. [CrossRef]
47. Takada, T.; Miyamoto, A.; Hasegawa, S.F. Derivation of a yearly transition probability matrix for land-use dynamics and its applications, *Landsc. Ecol.* **2010**, *25*, 561–572.
48. Makame, M.O.; Kangalawe, R.Y.M. Water Security and Local People Sensitivity to Climate Variability and Change among Coastal Communities in Zanzibar. *J. Sustain. Dev.* **2018**, *11*, 23. [CrossRef]
49. Comte, J.-C.; Cassidy, R.; Obando, J.; Robins, N.; Ibrahim, K.; Melchioly, S.; Mjemah, I.; Shauri, H.; Bourhane, A.; Mohamed, I.; et al. Challenges in groundwater resource management in coastal aquifers of East Africa: Investigations and lessons learnt in the Comoros Islands, Kenya, and Tanzania. *J. Hydrol. Reg. Stud.* **2016**, *5*, 179–199. [CrossRef]
50. DeGhetto, K.; Gray, J.R.; Kiggundu, M.N. The African Union's Agenda 2063: Aspirations, challenges, and opportunities for management research. *Afr. J. Manag.* **2016**, *2*, 93–116. [CrossRef]
51. United Nations. Global Sustainable Development Report. 2023. Available online: https://sdgs.un.org/sites/default/files/2023-09/FINAL%20GSDR%202023-Digital%20-110923_1.pdf (accessed on 15 December 2023).
52. Daher, B.T.; Mohtar, R.H. Water–energy–food (WEF) Nexus Tool 2.0: Guiding integrative resource planning and decision-making. In *Sustainability in the Water Energy Food Nexus*; Routledge: Abingdon, UK, 2017.
53. Oborn, E.; Dawson, S. Knowledge and practice in multidisciplinary teams: Struggle, accommodation, and privilege. *Hum. Relat.* **2010**, *63*, 1835–1857. [CrossRef]
54. Comeau-Vallée, M.; Langley, A. The Interplay of Inter- and Intraprofessional Boundary Work in Multidisciplinary Teams. *Organ. Stud.* **2019**, *41*, 017084061984802. [CrossRef]
55. Bilella, P.R.; Mulder, S.S.; Zaveri, S. To Be or Not to Be an Evaluator for Transformational Change: Perspectives from the Global South. In *Transformational Evaluation for the Global Crises of Our Times*; Van den Berg, R.D., Magro, C., Adrien, M.-H., Eds.; IDEAS: Bloomington, MN, USA, 2021; pp. 157–175.
56. Gates, E.F.; Page, G.; Crespo, J.M.; Oporto, M.N.; Bohórquez, J. Ethics of evaluation for socio-ecological transformation: Case-based critical systems analysis of motivation, power, expertise, and legitimacy. *Evaluation* **2023**, *29*, 23–49. [CrossRef]
57. Fetterman, D. Transformative empowerment evaluation and freirean pedagogy: Alignment with an emancipatory tradition. *New Dir. Eval.* **2017**, *2017*, 111–126. [CrossRef]
58. Pereira, L.; Asrar, G.R.; Bhargava, R.; Fisher, L.H.; Hsu, A.; Jabbour, J.; Nel, J.; Selomane, O.; Sitas, N.; Trisos, C.; et al. Grounding global environmental assessments through bottom-up futures based on local practices and perspectives. *Sustain. Sci.* **2021**, *16*, 1907–1922. [CrossRef]
59. Pittore, K.; Debons, P. Operationalizing Food System Governance: The Case of Fort Portal Food Change Lab. *Sustainability* **2023**, *15*, 3527. [CrossRef]
60. Yung, L.; Louder, E.; Gallagher, L.A.; Jones, K.; Wyborn, C. How Methods for Navigating Uncertainty Connect Science and Policy at the Water-Energy-Food Nexus. *Front. Environ. Sci.* **2019**, *7*, 37. [CrossRef]

61. Hibbett, E.; Rushforth, R.R.; Roberts, E.; Ryan, S.M.; Pfeiffer, K.; Bloom, N.E.; Ruddell, B.L. Citizen-Led Community Innovation for Food Energy Water Nexus Resilience. *Front. Environ. Sci.* **2020**, *8*, 571614. [CrossRef]
62. Cleaver, F. Paradoxes of participation: Questioning participatory approaches to development. *J. Int. Dev.* **1999**, *11*, 597–612. [CrossRef]
63. Brown, G.; Kyttä, M. Key issues and priorities in participatory mapping: Toward integration or increased specialization? *Appl. Geogr.* **2018**, *95*, 1–8. [CrossRef]
64. Chunga, D.D. Including Stakeholder's Perspectives on Mangrove Ecosystems Degradation and Restoration to Support Blue Carbon in Jozani-Chwaka Bay National Park, Zanzibar. Master's Thesis, World Maritime University, Malmö, Sweden, 2023; pp. 1–17. Available online: https://storage.googleapis.com/dissertations/2023/OSGM_2023_CHUNGA.pdf (accessed on 7 November 2023).
65. Monga, E.; Mangora, M.M.; Trettin, C.C. Impact of mangrove planting on forest biomass carbon and other structural attributes in the Rufiji Delta, Tanzania. *Glob. Ecol. Conserv.* **2022**, *35*, e02100. [CrossRef]
66. Nyangoko, B.P.; Berg, H.; Mangora, M.M.; Shalli, M.S.; Gullström, M. Community perceptions of climate change and ecosystem-based adaptation in the mangrove ecosystem of the Rufiji Delta, Tanzania. *Clim. Dev.* **2022**, *14*, 896–908. [CrossRef]
67. Shunula, J. Public awareness, key to mangrove management and conservation: The case of Zanzibar. *Trees* **2001**, *16*, 209–212. [CrossRef]
68. Okello, J.A.; Alati, V.M.; Kodikara, S.; Kairo, J.; Dahdouh-Guebas, F.; Koedam, N. The status of Mtwapa Creek mangroves as perceived by the local communities. *West. Indian Ocean J. Mar. Sci.* **2019**, *18*, 67. [CrossRef]
69. UNFCCC. Conference of the Parties Serving as the Meeting of the Parties to the Paris Agreement, Fifth Session: First Global Stocktake. *United Nations Framework Convention on Climate Change, Dubai.* 2023. Available online: https://unfccc.int/sites/default/files/resource/cma2023_L17_adv.pdf (accessed on 14 November 2023).
70. Heard, B.R.; Miller, S.A.; Liang, S.; Xu, M. Emerging challenges and opportunities for the food–energy–water nexus in urban systems. *Curr. Opin. Chem. Eng.* **2017**, *17*, 48–53. [CrossRef]
71. Myers, G.A. Social construction of peri-urban places and alternative planning in Zanzibar. *Afr. Aff.* **2010**, *109*, 575–595. [CrossRef]
72. De Jong Cleyndert, G.; Newman, R.J.S.; Cuni-Sanchez, A.; Marchant, R. Adaption of Seaweed Farmers in Zanzibar to the Impacts of Climate Change. In *African Handbook of Climate Change Adaptation*; Springer: Cham, Switzerland, 2021; pp. 3–28. [CrossRef]
73. Baloch, Q.B.; Shah, S.N.; Iqbal, N.; Sheeraz, M.; Asadullah, M.; Mahar, S.; Khan, A.U. Impact of tourism development upon environmental sustainability: A suggested framework for sustainable ecotourism. *Environ. Sci. Pollut. Res.* **2023**, *30*, 5917–5930. [CrossRef] [PubMed]
74. Sharpe, B.; Hodgson, A.; Leicester, G.; Lyon, A.; Fazey, I. Three horizons: A pathways practice for transformation. *Ecol. Soc.* **2016**, *21*, 47. [CrossRef]
75. Connell, J.; Lowitt, K.; Saint Ville, A.; Hickey, G.M. Food Security and Sovereignty in Small Island Developing States: Contemporary Crises and Challenges. In *Food Security in Small Island States*; Connell, J., Lowitt, K., Eds.; Springer: Singapore, 2020. [CrossRef]
76. Paddock, J.; Smith, A. What role for trade in food security? Insights from a small island archipelago. *J. Peasant Stud.* **2018**, *45*, 368–388. [CrossRef]
77. Suomela, M. Developing an Informal Settlement into a Resilient Urban Centre in Zanzibar, Tanzania. Master's Thesis, Aalto University, Espoo, Finland, 2019; pp. 1–184. Available online: https://aaltodoc.aalto.fi/server/api/core/bitstreams/2a71b57f-0fbf-459f-85e8-c59bb9b01460/content (accessed on 8 November 2023).

land

MDPI

Article

Key Drivers of Land Use Changes in the Rural Area of Gargano (South Italy) and Their Implications for the Local Sustainable Development

Nazgul Esengulova [1], Pasquale Balena [2], Caterina De Lucia [3], Antonio Lopolito [3,*] and Pasquale Pazienza [3]

[1] Department of Humanities, Letters, Cultural Heritage, Education Sciences, University of Foggia, 71121 Foggia, Italy; nazgul.esengulova@unifg.it
[2] Department of Civil, Environmental, Land, Building Engineering, and Chemistry (DICATECh), Polytechnic University of Bari, 70125 Bari, Italy; pasquale.balena@poliba.it
[3] Department of Economics, Management and Territory, University of Foggia, 71121 Foggia, Italy; caterina.delucia@unifg.it (C.D.L.); pasquale.pazienza@unifg.it (P.P.)
* Correspondence: antonio.lopolito@unifg.it

Abstract: This study examines the dynamics of land use and land cover change (LULCC) in the Gargano area (Southern Italy) to reveal crucial insights into the socio-economic and environmental impacts on its unique natural and cultural resources. This analysis was conducted using a mixed approach of GIS data and expert interviews to investigate significant changes in the Gargano area, from 2000 to 2018, and their drivers. Artificial surfaces gained 22% of their original surfaces, while heterogeneous areas and pastures lost 25% and 78%, respectively. Urbanization and deforestation emerged as major concerns, reflecting heightened sensitivity to these transformative processes. Agricultural intensification and support policies were perceived as potential pressure sources on specific natural components. Conversely, these drivers counteracted land abandonment. Drivers such as education level and agricultural extensification were seen as levers for a more desirable land cover dynamic. Identified actions include providing targeted support for agriculture within environmental constraints, addressing land ownership fragmentation, supporting agricultural extensification, and promoting environmental awareness.

Keywords: LULCC dynamics; Gargano; land use drivers; geospatial analysis; expert interviews

1. Introduction

Land use/land cover change (LULCC) has become a great concern in recent years worldwide, due to its adverse impacts on the environment and the consequent effect on economies and social dynamics. LULCC inhibits sustainable development [1]. An analysis of global LULCC drivers for the period of 1982–2016 carried out by Song et al. [2] states that approximately 60% of global land changes are attributed to direct human activity, with the remainder associated with indirect factors such as climate change. Changes in land use and land cover are the result of intricate interactions between people and the environment [3,4]. Hence, a complex and holistic approach incorporating both human and environmental dimensions is required for a comprehensive analysis. The identification of the key drivers of LULC changes is challenging due to the strong interdependencies and interconnectedness in socio-ecological systems (SESs) [5–7]. SESs are dynamic, and their dynamism implies that we can only identify the main drivers for a brief period of time with a need for systematic monitoring and updates. Remote sensing and a geographic information system analysis play an important role in collecting a sufficient amount of data on LULC changes [8,9].

Spatio-temporal analysis of LULCC along with on-site observations, statistics, and quantitative and qualitative data from local people and experts can provide for comprehensive research and a full picture of the ongoing and possible future scenarios, which are essential for effective decision- and policy-making targeted at sustainability in present and future times. An analysis of LULCC drivers is important for designing sustainable development strategies [10].

The scientific literature states demographic, socio-economic, institutional, technological, and biophysical factors as the main factors shaping land use and land cover change in different countries across the globe, with some of these factors prevailing more or less in certain regions depending on the specifics of the study area and the local context [11,12].

Many of these studies analyze developing countries with high rates of urbanization. Calzada et al. [13] examine land use changes in two contrasting tropical dry regions of Mexico, identifying a swift decline in natural areas due to urban expansion. A future land use simulator predicts reduced agricultural and pasture lands and increased human settlement areas by 2038. Other similar studies have analyzed LULCC in countries such as Ethiopia [14], Tanzania [15], and China [16]. These studies have identified factors such as intensive agriculture, socio-economic developments, population growth, and the quantity of exported goods as determinants of declining agricultural yield, biodiversity loss, increased aridity, land degradation, and water resource decline, exacerbating the gap in land and water supply–demand. Demographic and socio-economic factors emerged as crucial drivers in changes leading to a rise in impervious surfaces.

As argued by these studies, various approaches are employed in the analysis of the driving forces behind LULCC and landscape dynamics. Commonly, GIS and remote sensing techniques are utilized for scrutinizing LULCC changes, offering valuable insights into the spatial distribution and temporal evolution of land use patterns. An assessment of LULCC frequently involves the use of satellite imagery, such as Landsat images captured at different time points [8,14]. Landsat satellites, equipped with multispectral sensors, provide a comprehensive view of the Earth's surface, enabling the monitoring of changes in land use and land cover over specific periods. Moreover, to analyze spatiotemporal changes, some studies adopt integrated approaches like the cellular automata–artificial neural network (CAANN) methodology [17], which combines cellular automata and artificial neural networks to model and simulate complex land use dynamics, offering a comprehensive understanding of landscape transformations over time. Additionally, researchers often integrate future land use simulation models into their analyses to predict potential scenarios based on current trends and identified drivers, adopting such techniques as Dinamica [18], Markov-FLUS [19], SLEUTH cellular automata [20], artificial neural network–Markov chains [21], and CLUE-S [7]. These models serve as valuable tools for understanding and predicting the dynamics of land use across different spatial scales.

On a global basis, technological innovation, demographic change, globalization, urbanization, and climate change are the primary determinants. These principal forces influence the demand and supply of land assets and engage in complex interactions [22,23]. The other important driving factor of land use change according to Lambin is also the history of land use [24,25], which shapes land practices. Regionally, contextual factors such as cultural preferences, biophysical characteristics, institutional arrangements, prevailing socio-economic conditions, and policy initiatives may mitigate the impacts of global drivers in rural areas and specifically in marginal territories [23]. For instance, Mack et al. [26], in their recent research, found that indirect drivers such as migration and migrant money flow of remittances have a strong impact on land systems and consequently on land change. Another study of local drivers carried out by Tahmasebi [27] suggests that, in rural areas, other often neglected factors such as psychological drivers can also have a strong impact on how land users make their decisions about land practices. Land practices and their management further create various LULCC patterns that can either be harmful or beneficial to other ecosystems [28]. This is critical if we consider areas with fragile ecosystems, which are less resistant in the face of external influence and have less ability to quickly recover from any negative disturbances and, if pushed forward beyond the threshold, would not sustain any damage. Semi-arid and arid areas are considered fragile systems (they have inherent properties of fragility). However, under increased or unpredictable disturbances caused by natural calamities or anthropogenic impact, they may demonstrate a high level of species alteration in number and composition [29–31]. Thus, designing complex, holistic land use strategies and plans is of special significance when it comes to sustainable development, in particular for inner or marginal areas where different land use types often overlap.

This task has become a major challenge, especially in the case of European countries where there exists a multilayer policy-making process along with a multi-faceted wide rural territory. To ease the designing process of tailored local development policies, a 'place-based vision' approach has been developed in the EU rural policy practice, as a method to favor sustainable regional development [32], in specific territorial conditions such as remote and/or marginalized regions [33]. To develop a strategic place-based vision, it is important to map the current situation in the territory, and often, this starts with mapping the natural resources at hand, the main land use transformations, and their socio-economic determinants. The Mediterranean area has seen several transformations in recent decades, particularly driven by demographic changes leading to considerable variations in LULCC in terms of urban sprawl, abandoned rural areas, and the decrease in agricultural land [34,35]. Also, many pastoral regimes have been abandoned, thus causing an increase in natural vegetation of unfertile and unproductive land or changes in wildfire regimes [36,37]. The above changes have contributed to worsening global warming and drying climate conditions, particularly in Italy, Spain, Greece and Portugal [38].

In the context of the fragile and marginalized risky rural area of the Mediterranean basin, our analysis is focused on the micro-transformations that occurred in the area of Gargano, a mountain area in the south of Italy with semi-arid conditions and a predominantly agricultural economy [39], in the period 2000–2018.

The Gargano area (approximately 170 thousand inhabitants) relies on the key sectors of tourism, agriculture, and industry. Tourism is a major contributor to the local income, with some problems of sustainability linked to seasonality. Despite challenges like abandonment in some areas, agriculture remains crucial, with a multitude of small-scale agricultural activities taking place, particularly in olive tree and vegetable farming. However, the region faces the issue of younger generations migrating in search of better job opportunities elsewhere. The industrial sector experiences limited development, with most companies being smaller than the provincial average.

The purpose of this paper is to assess LULCC in Gargano (South Italy) and provide adequate policy recommendations based on the above transformations. The main research aims that the present study intends to pursue are (1) to identify the major LULCC transformations and (2) to investigate key drivers of these transformations.

The first question is addressed by means of a geographical information system (GIS) analysis, while the related drivers are qualitatively assessed based on the experts' views. This approach is illustrated in Section 2, which includes a focus on the case study area before introducing the GIS analysis and the method to assess the influence of drivers. In Section 3, the results of the analysis are presented. Section 4 returns to the research question and provides some policy implications and some concluding remarks.

2. Materials and Methods

The information employed in this research was collected from diverse sources: the gray literature for the socio-economic, demographic, climatic, and physiographic data of the case study area, GIS data to observe LULCC, and the scientific literature and expert interviews to identify and assess the major drivers of land use change. To ease the reading, this information has been organized into three subsections, related respectively to the case study, the LULCC analysis, and the assessment of the drivers.

2.1. Case Study Area

Gargano is a historical and geographical sub-region in Foggia, Apulia, southeast Italy, consisting of a wide isolated mountain massif and the Gargano National Park. The region is rich in culture, art, and spirituality, and 35% of all Italian botanical species can be found in the park. The climate is semi-arid, with average temperatures of around 15 °C, with summers between 25 °C and 30 °C, and peaks over 40 °C on the hottest days. Winters are relatively temperate, with temperatures not dropping below 0 °C.

The Gargano region is characterized by sedimentary rocks, particularly limestone and dolomites, created during the Cretaceous and Jurassic periods. The biological landscape is distinguished by its calcareous composition, karst characteristics, caves, sinkholes, and a substantial abundance of woods. The Gargano Promontory constitutes an island separated from the rest of the peninsula by the Tavoliere delle Puglie.

The study area streams are radial and hierarchical, originating from torrential sources and flowing near coastlines. Some flow into Lake Lesina and others towards Lake Varano. Inclines and rocky substrates in river valleys cause frequent flood events and long-term low water levels, transporting solid materials.

This research focuses on 13 municipalities representing more than 80% of the Gargano extension, including Cagnano Varano, Carpino, Ischitella, Manfredonia, Mattinata, Monte Sant'Angelo, Peschici, Rignano Garganico, San Giovanni Rotondo, San Marco in Lamis, Sannicandro Garganico, Vico Del Gargano, and Vieste.

The total area of the study area is 150,857 hectares. There are two commercial ports, in Manfredonia and Rodi Garganico, and the closest airports are located in Foggia (104 km) and Bari (143 km). The rail network there is the least developed infrastructure network in the area, with a couple of regional operators providing short-length connections. There is a fairly good road network connecting small towns and cities, but bus transportation does not fully satisfy transportation needs due to its low frequency (Report on Mediterranean Adventure Tourism, 2020).

The Gargano area, with a population of 172,207 people as of 2019, is characterized by three key sectors: tourism, industrial activities, and agriculture. Tourism accounts for 11% of the national GDP, which is worth 900 million euros [40], while the agricultural sector plays a significant role in shaping land use patterns. However, some agricultural areas have experienced abandonment due to water scarcity, inadequate infrastructure, and insufficient services.

Most of the local population is employed in small-scale agricultural activities (7367 people), particularly related to olive tree and vegetable farming, and especially in the tourist industry (10,305 people). The area is heavily affected by the migration of the youngest generations to northern Italy and other countries, motivated by the search for better job opportunities [41].

The industry still has poor development, with low industrialization rates and most companies having a size less than the provincial average. The manufacturing industry contributes 7.5% to the provincial total, with only rare exceptions. The construction industry occupies a small percentage of the provincial GDP (4.40%). Specialized production, such as electrical machinery, electronic and optical equipment, motors, generators, and transformers, enjoys a certain vitality, with workers making up about 91% of the province's total for this category [41].

The Rodi Garganico Citrus Oasis is a natural area in the northern part of Gargano, known for its centuries-old production of citrus fruits, such as "Duretta del Gargano" and "Bionda del Gargano" oranges, as well as "Femminello" lemon, which received IGP acknowledgment [42].

2.2. Geospatial Analysis

Data on land cover, land use and transitions between different categories are some of the most frequently requested pieces of information for the formulation of sustainable land management and planning strategies, needed to provide information to support decision-making processes at community, national and local levels and to verify the effectiveness of environmental policies [43,44].

The ever-increasing availability of high-resolution geographic data provided by land government agencies, research infrastructures and environmental agencies is contributing significantly to the ability to study recent soil transformations at unprecedented scales of detail [45].

As a result, technical–scientific and dissemination initiatives are flourishing, including the periodic reports on land consumption produced by the Higher Institute for Environmental Protection and Research (ISPRA) and the Research Centre on Soil Consumption (CRCS), which provide a very detailed statistical picture of the transformations underway in Italy without, however, going into the merits of the territorial dynamics that triggered, favored, or counteracted them. In this study, therefore, we aim to contextualize the observations on changes in land use and cover in the study area, investigating the determining factors and discussing their possible implications.

The CORINE (Coordination of Information on the Environment) program is mainly based on the remote sensing technique, using classification techniques that depend on the automatic and semi-automatic interpretation of images.

In this context, for the analysis of transformations, of particular importance is the CORINE Land Cover (CLC) initiative, created at the European level specifically for the detection and monitoring of the characteristics of land cover and use, with particular attention to protection needs. The main aim of the initiative is to dynamically verify the state of the environment in the community area, to provide support for the development of common policies, monitor their effects, and propose possible corrective measures. Between 1985 and 1990, the European Commission promoted and financed the CORINE program and set up an information system on the state of the environment in Europe. Nomenclature systems and working methodologies for the creation of the CORINE Land Cover (CLC) database have also been developed and approved at the European level. It was initially implemented in 1990, while subsequent updates refer to the years 2000, 2006, 2012, and 2018.

CLC products are based on the photointerpretation of satellite images following the standard methodology and nomenclature with the following characteristics:

- A total of 44 land cover classes at the third hierarchical level of the CORINE nomenclature;
- Minimum mapping unit (MMU) for the coverage of 25 hectares;
- Minimum amplitude of linear elements of 100 m;
- MMU for change (LCC) of 5 hectares.

For the analysis of the dynamics of agroforestry land uses, we adapted the methodology proposed in the Territorial Coordination Plan of the Province of Foggia [40]. Through the application of a land cover change classification matrix that simplifies the interpretation of possible changes between any two classes to a limited number of processes, we obtained the following:

- *Persistences*:
  - PU: artificial surfaces;
  - RA: arable soil;
  - PC: permanent crop;
  - PP: pastures;
  - PH: heterogeneous area;
  - PN: natural areas.
- *Agricultural Conversions*:
  - HE: arable to heterogeneous areas;
  - ER: permanent crop to arable areas;
  - AP: pasture to other agricultural use;
  - TC: other agricultural use to permanent crops;
  - OC: heterogeneous areas to permanent crops;
  - CP: other agricultural use to pastures.
- *Net Transformation*:
  - UR: urbanization of natural or agricultural areas;
  - DC: deforestation for cropping (e.g., from woods and shrubs to meadows and pastures);

- DP: deforestation for pastures (e.g., from woods and shrublands to complex particulate cropping systems);
- DU: de-urbanization and de-impermeabilization of soils;
- NA: recolonization by natural areas (processes of secondary recolonization by spontaneous vegetation, e.g., from arable land in non-irrigated areas to woods and shrubs).

Despite the consolidated validity of the CORINE system, a limitation of this methodology lies in the fact that the classification adopted by the CORINE program is unable to capture detailed ecosystem variations, especially in natural areas [46]. It is also important to consider that the CORINE Land Cover (CLC) dataset presents an MMU of 25 hectares for areal phenomena [47], thus preventing the detection of small territorial fragmentations with dimensions smaller than 25 hectares and with different classifications compared to surrounding surfaces that exceed this threshold. This problem takes on particular relevance in the territorial characteristics of the study area, mainly involving the lands classified as "2.2.1 Vineyards" and "2.2.2 Plantations of fruit trees and berries". Their extension, especially in coastal areas, often exceeds 25 hectares, highlighting the need for ancillary data for a more accurate estimate [48].

2.3. Assessment of Land Use Drivers

For the identification of the potential drivers of the observed LULCC in the area studied, we performed a literature analysis based on Science Direct and Scopus. We considered very recent papers (2017–2023) and selected 31 drivers which apply to the study area among those analyzed in these papers [9,49–52]. Table 1 reports the selected drivers grouped as (i) physical and anthropogenic or (ii) socio-economic factors.

Table 1. Drivers of land use change drivers.

ID	Physical and Anthropogenic	ID	Socio-Economic
D1	Population growth rural	D17	Migration
D2	Population growth urban	D18	Labor shortage
D3	Agricultural intensification: modern	D19	Cultural Change in religious patterns
D4	Agricultural intensification: traditional	D20	Level of education
D5	Agricultural extensification	D21	Customary land tenure system
D6	Irrigation (dams, rivers)	D22	Economic rising living standard
D7	Improved crop varieties	D23	Financial capital of rural farmers (poverty)
D8	Mining	D24	Foreign agricultural medium-scale investments
D9	Bushfire	D25	International funding/development aid
D10	Livestock	D26	Credits by family, bank, government or NGO
D11	Road network	D27	Science and research
D12	Soil type and fertility	D28	Service offered by extension officers
D13	Topography	D29	Governmental laws
D14	Rainfall variability	D30	National agricultural programs
D15	Temperature variability	D31	Fertilizer subsidies
D16	Wind intensity		

Demographic aspects and agricultural practices are the main anthropogenic factors, while soil type, topography and atmospheric phenomena are among the physical ones. The social factors include migration, unemployment, and level of education. The availability of financial resources is mentioned as economic factors.

To assess their relevance as LULCC drivers and their impact for the sustainable development of the rural Gargano, we individually interviewed 11 experts by means of an ad hoc questionnaire. Table 2 reports the affiliation, scientific background, and research or activity field of the experts interviewed.

Table 2. Affiliation, scientific background, and research or activity field of the experts for interviews.

ID	Affiliation	Background	Field of Research or Activity
E1	Polytechnic University of Bari, Italy	Territorial, Urban and Environmental Planning	Territorial, Urban and Environmental Planning
E2	University of Foggia, Italy	Social and Political Philosophy	Cultural Heritage
E3	University of Ferrara, Italy	Geological Sciences	Stratigraphic Geology and Sedimentology
E4	Ministry of Education, Italy	Agricultural Science	Rural Development
E5	Gargano National Park, Italy	Communication Studies	Protected Areas Management
E6	Gargano National Park, Italy	Economics	Ecotourism, Destination Management
E7	University of Foggia, Italy	Economics	Water Management, Regional and Local Land Use planning
E8	University of Foggia, Italy	Agricultural Science	Remote Sensing, Land Use Change and Natural Resources Management
E9	University of Foggia, Italy	Economics	Bioeconomy and Renewable Energy
E10	University of Foggia, Italy	Economics	Rural Development
E11	University of Foggia, Italy	Natural Science	Land Planning

The experts were chosen through a process involving the review of published papers, recommendations from fellow scientists, and consultations with local institutions, including the University of Foggia and the National Park of Gargano. An essential criterion was their affiliation with scientific institutions in Italy and their thorough understanding of the causal relationships between driving forces and LULCC in the region.

The questionnaire was based on the information obtained from the geospatial analysis and the literature review. It was formed of two core sections, reserved, respectively, in the assessment and by means of Likert scales, of (i) the land use changes and (ii) their drivers. The LULCCs were assessed with respect to their potential effects on the sustainable development of the case study area. To this end, the experts were asked to consider the LULCC that occurred in the period analyzed (2000–2018), using a geographical representation and a table reporting their extension in hectares and % in terms of the surface in the year 2000. Subsequently, the respondents were provided with a list of the LULCC, divided into three different categories according to the classification described in Section 2.2 (i.e., persistences, agricultural conversions and net transformations), and asked the following question:

Q1. What potential effects on the sustainable development of the Gargano region can be anticipated as a result of the following LULCC which occurred between 2000 and 2018? (Possible answers: very negative; negative; slightly negative; none; slightly positive; positive; very positive).

It should be borne in mind that Q1 is not a simple question as it requires the respondents to perform an assessment of each single LULCC listed.

Subsequently, the respondents were asked to assess the LULCC drivers with respect to their ability to generate some land use changes in abstract (Q2) and their capability to produce specific LULCC (Q3):

Q2. To what extent have these factors contributed to causing one or more of the described 'changes' (persistences, agricultural conversions, or net transformations)? (Possible answers: not at all, they were a collateral cause of marginal importance, they were a relevant contributing factor, they were the main cause, they were almost the sole cause).

Q3. If the following factors have, in some way, contributed to causing the changes indicated in the columns, check the boxes in the corresponding intersections.

As with Q1, Q2 and Q3 are complex questions requiring a reiterated task to the respondent. Q2 has 31 responses (the number of the drivers identified), and Q3 has even a greater number as it requires the comparison of each driver with each LULLC reported from the analysis.

3. Results

3.1. Geospatial Analysis

Table 3 summarizes the geospatial analysis whose results have been aggregated in three land cover macro categories, namely, artificial surfaces, agricultural areas, and natural areas, with a breakdown of agricultural areas into three main sub-categories. The information reported relates to the extension of the categories' surface in the year 2000 (column a), the loss and the gain of the surface (respectively, columns b and c) in the period 2000–2018, the balance of that period (column d) and the new extension in 2018 (column e).

Table 3. Land use change in the period 2000–2018.

	2000 Extension	Loss (%)	Gain (%)	Balance (%)	2018 Extension
	a	*b*	*c*	$d = b - c$	$e = a + d$
Artificial Surfaces	4183.20	1100.71 (0.26)	2025.56 (0.48)	924.84 (0.22)	5108.04
Agricultural Areas	71,107.90	15,317.59 (0.21)	11,678.15 (0.16)	−3639.43 (−0.05)	67,468.46
Arable Soil	*31,625.69*	*2734.72 (0.08)*	*3175.48 (0.10)*	*440.76 (0.01)*	*32,066.45*
Permanent Crops	*22,064.49*	*1716.62 (0.07)*	*3307.86 (0.14)*	*1591.23 (0.07)*	*23,655.73*
Pastures	*2308.52*	*2198.82 (0.95)*	*385.94 (0.16)*	*−1812.87 (−0.78)*	*495.64*
Heterogeneous Areas	*15,109.20*	*8667.42 (0.57)*	*4808.85 (0.31)*	*−3858.56 (−0.25)*	*11,250.64*
Natural Areas	111,153.99	41,757.74 (0.37)	44,472.34 (0.4)	2714.59 (0.02)	113,868.59

The first result relates to the remarkable increase in the artificial surface, which, in 2018, was 22% more extended than in 2000, with a net gain of almost one thousand ha. This is especially due to the expansion of urban areas, industrial facilities, port construction, extraction activity and the building of sport and leisure facilities. The loss of 26% of the surface is largely an internal conversion especially related to the transition from different kinds of urban areas and the transformation of some discontinuous urban areas in industrial sites or leisure facilities.

On the contrary, the agricultural areas lost about 3.6 thousand hectares passing from 71 to 67 and reducing by about 5%. This reduction is mainly due to a sharp decline in pastureland, which lost more than 78% of its original surface, and due to the downgrading of heterogenous areas. These two categories declined by more than 5.6 thousand hectares overall, especially in favor of natural areas. At the same time, there has been a significant increase in the areas with more intensive agricultural activity, such as that pertaining to arable soil and permanent crops, gaining more than 1.5 thousand ha.

Another relevant result concerns the extension of the natural areas with a net gain of 2%, corresponding to 2.700 hectares, which have been recolonized by spontaneous vegetation. A large part of the loss and gain for this macro category occurred between internal subcategories and can therefore be conceived as internal conversions. Most of the transformations that affected natural areas in the period 2000–2018 involved surfaces used for agricultural purposes (4.8 thousand ha). Of these, almost half, about 2.4 thousand ha, underwent a sort of hybridization, transitioning from natural areas to heterogeneous surfaces with the following CORINE Land Cover label: "Land principally occupied by agriculture, with significant areas of natural vegetation". Another thousand hectares were converted into arable land. On the other hand, during the same period, natural areas withdrew 7.6 thousand hectares from agricultural uses. This explains the net increase in natural areas.

These results are the effect of the transformation dynamics that occurred in the period 2000–2018 in terms of (i) persistences, (ii) agricultural conversions, and (iii) net transformations. These dynamics are reported in Table 4 which accounts for the surfaces affected by each specific change category in the two reference years: the surface that, in 2018, was affected by the specific change is reported both in absolute terms (expressed in hectares) and in relative terms compared to the extension that, in 2000, was classified in that same category [1]. A graphical representation of the data reported in Table 2 is provided by the map shown in Figure 1.

Table 4. Land use transformation in the period 2000–2018.

Cod.	Change Category	Macro Category	Surface (ha) in 2018	% of Surface in 2000
PU	Artificial Surfaces	Persistence	4064.70	97%
RA	Arable Soil	Persistence	28,890.96	91%
PC	Permanent Crops	Persistence	20,578.08	93%
PP	Pastures	Persistence	109.70	5%
PH	Heterogeneous Areas	Persistence	6881.67	46%
	Total Agricultural Areas	*Persistence*	*62,541.55*	*88%*
PN	Natural Areas	Persistence	106,188.86	96%
HE	Arable to Heterogeneous Areas	Agricultural Conversions	1492.60	5%
ER	Permanent Crop to Arable Areas	Agricultural Conversions	337.42	2%
AP	Pasture to Other Agricultural Use	Agricultural Conversions	262.36	11%
TC	Other Agricultural Use to Permanent Crops	Agricultural Conversions	536.18	1%
OC	Heterogeneous Areas to Permanent Crops	Agricultural Conversions	3140.65	21%
CP	Other Agricultural Use to Pastures	Agricultural Conversions	311.95	0%
UR	Urbanization of Natural or Agricultural Areas	Net Transformations	1043.35	1%
DC	Deforestation for Cropping	Net Transformations	4801.94	4%
DP	Deforestation for Pastures	Net Transformations	47.76	0%
DU	De-urbanization and De-impermeabilization of Soils	Net Transformations	118.51	3%
NA	Recolonization by Natural Areas	Net Transformations	7638.43	11%

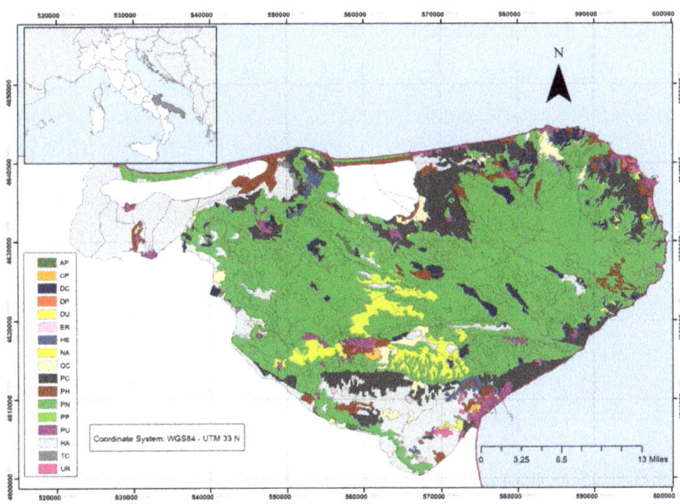

Legend: *AP: Pasture to Other Agricultural Use; CP: Other Agricultural Use to Pastures; DC: Deforestation for Cropping; DP: Deforestation for Pastures; DU: De-urbanization and de-impermeabilization of Soils; ER: Permanent Crop to Arable Areas; HE: Arable to Heterogeneous Areas; NA: Recolonization by Natural Areas; OC: Heterogeneous Areas to Permanent Crops; PC: Permanent Crops; PH: Heterogeneous Areas; PN: Natural Areas; PP: Pastures; PU: Artificial Surfaces; RA: Arable Soil; TC: Other Agricultural Use to Permanent Crops; UR: Urbanization of Natural or Agricultural Areas*

Figure 1. Representation of land use transformation in the period 2000–2018.

Regarding the *Persistences*, the analysis reveals that, for the most part, the various land uses have been retained substantially to the same extent in the analyzed period. This is certainly true for the artificial surfaces (PU), which have a persistence of 97%. Figure 1 shows that the persistence of natural areas (PN) is undoubtedly the most significant in terms of absolute surface quantities, seeing that green, the color indicating the persistence of natural areas, is the dominant color in the map, representing more than 106 thousand ha. Arable soil and permanent crops have also remained quite stable. In contrast, the heterogeneous areas and especially the pastures have undergone a significant transformation, with a low or very low persistence of 46% and 5%, respectively.

Partly, this is due to the *Agricultural Conversions*: 262 hectares of pastures (about 11% of the land extension in 2000) have been converted to other agricultural uses. Heterogeneous areas have also changed, yielding a substantial 3 thousand hectares (21%) to permanent crops and receiving 1.5 thousand hectares from arable surfaces.

As for the *Net Transformations*, the most significant one involves the conversion of areas previously destined for other purposes that have been reclaimed by nature. This amounts to over 7.6 thousand hectares subtracted almost exclusively from agricultural uses, comprising 55% (4.2 thousand ha) from heterogeneous areas, 25% (1.9 thousand ha) from pastures, and 15% (1.17 thousand ha) from arable surfaces. The analysis of net transformations highlights another major change represented by deforestation for cropping. In 2000, there were 4.8 thousand hectares of forestland that became agricultural cultivations in 2018. Finally, the urbanization of natural or agricultural areas covers an area of more than 1000 hectares, the majority of which (approximately 90%) were previously used for agriculture.

3.2. Assessment of Land Use Drivers

The information collected from the experts is summarized in Tables 5 and 6. The former is dedicated to the analysis of the drivers' ability in generating the LULCC. The data reported derive from questions *Q2* and *Q3* described above. The possible responses to question *Q2* were transformed into numbers using the following scoring: *not at all = 0, they were a collateral cause of marginal importance = 1, they were a relevant contributing factor = 2, they were the main cause = 3, they were almost the sole cause = 4*. The "Degree of Causality" column in Table 5 reports, for each driver, the average of the scores it received from each respondent. As this variable is based on the question concerning the extent to which drivers contributed to the observed changes, this value can be regarded as a measure of the drivers' propensity to generate one or more changes. The columns "# of Changes Caused" and "# of Experts Agreeing on Average" synthesize the responses to *Q3*. The first reports the number of different LULCCs caused by each driver according to the experts' views and is a sort of measure of the effectiveness of the drivers; the second indicates, for each driver, the average number of experts who identified it as a possible cause of the affected LULCC and is a measure of the intensity, perceived by the experts, of the drivers' action. The details of the information collected with Q3 are reported in Appendix A.

Among the physical and anthropogenic drivers, the one perceived as the most responsible (Degree of Causality 1.81) for producing LULCC is *D13 Topography*. It refers to the physical characteristics of the terrain and the morphology of the landscape, which, in the Gargano area, exhibit a significantly diversified territorial pattern with pronounced altimetric gradients, a distinctive mountainous coastline, and extensive forest cover. These elements are evidently perceived as particularly impactful on decisions regarding territorial organization by the interviewed experts. This driver has been identified as the cause of 13 LULCCs with a high level of agreement among the interviewed experts.

Within this group of drivers, *D5 Agricultural Extensification* (expansion of cultivated areas with lower input intensity) also plays a prominent role, being perceived as potentially responsible for 16 LULCCs, although with a lower causality than *D13* but with an even higher level of agreement among the experts. *D3*, which pertains to the modernization of agriculture, is also a significant driver, gathering the highest convergence among the interviewed experts (3.38) and influencing 13 changes.

Table 5. Causality of drivers of land use changes.

	Drivers	Degree of Causality	# of Changes Caused	# of Experts Agreeing on Average
Physical and Anthropogenic	D1 Population growth rural	0.18	10	1.4
	D2 Population growth urban	1.36	7	3.14
	D3 Agricultural intensification: modern	1.63	13	3.38
	D4 Agricultural intensification: traditional	1.18	15	2.2
	D5 Agricultural extensification	1.63	16	2.81
	D6 Irrigation (dams, rivers)	0.81	7	1.57
	D7 Improved crop varieties	1.36	9	3
	D8 Mining	0.27	3	1.33
	D9 Bushfire	1.72	11	1.9
	D10 Livestock	1.09	5	2.2
	D11 Road network	0.63	4	2.25
	D12 Soil type and fertility	1.54	8	2.25
	D13 Topography	1.81	13	2.23
	D14 Rainfall variability	1.27	13	1.69
	D15 Temperature variability	1.27	14	1.57
	D16 Wind intensity	0.54	5	1.2
Socio-economic	D17 Migration	1.81	12	3.41
	D18 Labor shortage	2.09	14	3
	D19 Cultural change in religious patterns	0.27	9	1.33
	D20 Level of education	1.36	11	1.63
	D21 Customary land tenure system	1.36	9	3
	D22 Economic rising living standard	1.72	11	2.18
	D23 Financial capital of rural farmers (poverty)	1.63	15	1.73
	D24 Foreign agricultural medium-scale investments	0.45	9	1.55
	D25 International funding/development aid	1.36	9	2.22
	D26 Credits by family, bank, government, or NGO	1.09	10	1.6
	D27 Science and research	1.18	9	1.88
	D28 Service offered by extension officers	1.09	6	2.5
	D29 Governmental laws	1.45	14	1.71
	D30 National agricultural programs	1.63	14	2.28
	D31 Fertilizer subsidies	1.63	8	3

Table 6. Land use change effects on local sustainable development.

Cod.	Change Category	Macro Category	Effect (sum)	Surface (ha) in 2018
PU	Artificial Surfaces	Persistence	1	4064.70
RA	Arable Soil	Persistence	10	28,890.96
PC	Permanent Crops	Persistence	10	20,578.08
PP	Pastures	Persistence	2	109.70
PH	Heterogeneous Areas	Persistence	−4	6881.67
PN	Natural Areas	Persistence	21	106,188.86
HE	Arable to Heterogeneous Areas	Agricultural Conversions	3	1492.60
ER	Permanent Crop to Arable Areas	Agricultural Conversions	−12	337.42
AP	Pasture to Other Agricultural Use	Agricultural Conversions	−11	262.36
TC	Other Agricultural Use to Permanent Crops	Agricultural Conversions	9	536.18
OC	Heterogeneous Areas to Permanent Crops	Agricultural Conversions	0	3140.65
CP	Other Agricultural Use to Pastures	Agricultural Conversions	11	311.95
UR	Urbanization of Natural or Agricultural Areas	Net Transformations	−20	1043.35
DC	Deforestation for Cropping	Net Transformations	−24	4801.94
DP	Deforestation for Pastures	Net Transformations	−11	47.76
DU	De-urbanization and De-impermeabilization of Soils	Net Transformations	15	118.51
NA	Recolonization by Natural Areas	Net Transformations	5	7638.43

Finally, in this category, a mention should also be made of *D9 Bushfire, D12 Soil Type and Fertility, D2 Population Growth Urban*, and *D7 Improved Crop Varieties* as being among the major driving forces behind LULCC in Gargano.

Among the socio-economic drivers, there is the one exerting the highest degree of causality overall: *D18 Labor Shortage*, with an index of 2.09. It is perceived to be the potential cause of 14 changes by an average of three experts. Other indicative drivers associated with a negative socio-economic trend are among the major contributors to land use changes in the Gargano area. These include *D17 Migration* and *D23 Poverty of Rural Farmers*, evidently seen as obstacles to maintaining adequately remunerative agricultural activities.

Within the socio-economic factors, there are also drivers representative of the effects of agricultural policy, such as *D30 National Agricultural Programs* and *D31 Fertilizer Subsidies* which are conceived as potentially very effective in terms of land use change by the experts.

Table 6 is based on the information from *Q1*. For this question, the responses have been encoded into numbers as follows: *Very negative* = −3; *Negative* = −2; *Slightly negative* = −1; *None* = 0; *Slightly positive* = 1; *Positive* = 2; *Very positive* = 3. For each LULCC, the table reports the sum of the scores assigned by respondents. This method of aggregation facilitates the interpretation of results, as conflicting connections with opposite signs will result in a decrease in the effect, while agreement reinforces the effect, producing an overall collective picture.

From this table, it emerges that *Persistences* are seen as substantially positive. *PN Natural Areas*, which also exhibit the maximum extension, shows the highest score. In other words, among the land cover categories, it is most strongly associated with the sustainable development of the area according to the experts' perception. This is consistent both with the concept of sustainability itself, which sees the conservation of resources, especially natural ones, as its main operation, and with the purpose of the area: since it is mostly covered by a National Park, it is institutionally a protected area. At the same time, the maintenance of agricultural areas, denoted by *RA Arable Soil* and *PC Permanent Crops*, is considered another essential element of the area's development path.

On the contrary, *Net Transformations* are generally perceived as negative in terms of sustainable development. *DC Deforestation for Cropping, DC Deforestation for Cropping*, and *DP Deforestation for Pastures* are considered warning signs to monitor closely. An exception is *DU De-urbanization and De-impermeabilization of Soils*, which, with a score of 15, is considered a positive element from the perspective of sustainable development for the area.

Finally, from the analysis of *Agricultural Conversions*, a clear message emerges: heterogeneous areas and pastures are considered important indicators of the socio-economic and ecological balance of the territory. Their reduction in favor of other agricultural arrangements (*ER Permanent Crop to Arable Areas* and *AP Pasture to Other Agricultural Use*) is seen as a potential threat, as evidenced by the strongly negative scores they accumulate. Conversely, their expansion (*HE Arable to Heterogeneous Areas* and *CP Other Agricultural Use to Pastures*) is viewed favorably in the context of sustainable development.

4. Discussion

The Table A1 (Appendix A) cross-references the drivers with the LULCC, allowing for the identification of the drivers that may have a greater influence on the sustainable development of the territory. These are, in particular, the factors causing the most impactful changes according to the experts' views. As explained, among these changes, there are *DC: Deforestation for Cropping* and *UR: Urbanization of Natural or Agricultural Areas*. Concerning the former, the experts identified *D3 Agricultural Intensification: Modern* and *D7 Improved Crop Varieties* with an agreement of the 63% and 45%, respectively, as possible causes. Other relevant drivers for this change potentially harmful to sustainable development, each identified by 27% of experts, are *D21 Customary Land Tenure System, D26 Credits by Family, Bank, Government or NGO, D29 Governmental Laws*, and *D31 Fertilizer Subsidies*. At the same time, *UR: Urbanization of Natural or Agricultural Areas* is influenced by *D2 Population Growth Urban* (72%) and *D22 Economic Rising Living Standard.*

On the other hand, *D13 Topography, D17 Migration, D18 Labor Shortage*, and *D20 Level of Education*, identified by 36%, 54%, 36%, and 36% of experts, respectively, are considered factors influencing the persistence of *PN Natural Areas. D17, D18*, along with *D5 Agricultural extensification*, are important drivers also for another sensitive change, *DU De-urbanization and De-impermeabilization of Soils.*

Given the above, we debate the potential implications that these land use changes may have on fragile ecological balances of the territory and community livelihoods. As for the former, ecological risks take the form of soil erosion, dry lands, water shortage [53] and biodiversity loss [54]. These imbalances would, in turn, affect the benefits provided by ecosystem services on nearby communities [55–57]. As for the latter, land use changes would alter food and livelihood production and security [58,59] and make smallholder farmers more vulnerable to climate variability or income sources so as to alter the original agricultural production into modern and 'touristic-oriented' farmhouses [60].

Nonetheless, these latter transformations ultimately attract new job opportunities [61], reshape social inclusiveness and social innovation in rural and peri-urban areas [62,63], emphasize existing cultural traditions [64,65], reduce outbound youth migration [65,66], strengthen cooperation among farmers and capitalize on local competitive advantages [67,68].

Therefore, land transformations, as a result of human activity, could also have severe impacts on the supply of land and generate conflicts in the local community spanning from private interests and income generation to conservation and restoration goals [69,70], thus resulting in diverse land configurations of the territory. Also, agricultural intensification and expansion over time has led to degraded ecosystems, thus affecting soil depletion, overexploitation and loss of the resilience of agro-systems [71].

Identifying the underlying drivers of the above dynamics is crucial for determining the actions to be taken to guide the LULCC process along desirable development trajectories. In this regard, the present work provides an insightful framework, enabling the distinction of drivers that require attention depending on whether the goal is to (i) control the most concerning LULCC or (ii) promote those that are more desirable.

Factors such as agricultural intensification and other agricultural support policies are among the first group, as they are perceived as potential sources of pressure on the natural component of the territory. On the other hand, these drivers (e.g., improved crop varieties, services provided by extension officers, national agricultural programs, fertilizer subsidies) counteract abandonment, which is promoted by the customary land tenure system and its land ownership fragmentation effects. These drivers play, therefore, a crucial role in the balanced development of the territory.

In the second group, there are drivers like topography, which makes certain areas inaccessible and remain naturally untouched, and other factors such as the level of education and the agricultural extensification process.

Some policy implications that favor a more desirable land cover directly stem from these facts:

- Supporting agriculture through targeted measures and actions conditioned by compliance with environmental constraints;
- Addressing the fragmentation of land ownership with initiatives aimed at promoting or rewarding the concentration of agricultural lands;
- Supporting agricultural extensification by providing forms of diversification and income support for farmers;
- Promoting environmental awareness aimed at developing an understanding of the natural environment and the local archaeological, naturalistic, historical, and cultural heritage value.

In addition, to combat agricultural abandonment and the loss of important ecosystem services, such as food production or soil fertility, the promotion of ad hoc 'proximity agriculture' and 0-km products and cuisine would reduce market asymmetries and sustain small farmers in the long run to ensure the economic viability of local agricultural production [72]. Further policy measures such as offering payments to farmers' initiatives (such as crop rotations, biodegradable soil covers) for the preservation of land and natural resources would help promote sustainable practices of the local agriculture.

The Regional Authority [73] has recently welcomed the latest European Union 'Farm to Fork' strategy, which rewards farmers for promoting adequate sustainable practices, creating further opportunities for the ecological transition of agri-businesses that are beneficial to local communities. [74,75].

Another value added of the present study is the reproducibility of the employed methodology to similar areas—e.g., other Mediterranean or EU countries—such that the insights deriving from this study can be beneficial in other contexts, other than providing homogeneity of comparisons across regions.

Some limitations of this study are the limited number of experts interviewed; the potential errors occurred in the GIS analysis, especially in areas prone to difficult identification, such as heterogenous or borderline areas; the use of a questionnaire to collect the information, as this facilitated quick data collection, on one hand, but limited the depth of explanations on the other; and the effort required of respondents hindered the collection of additional but relevant information, such as a specific quantification of the causal relationship between drivers and changes.

The next steps include the following: to increase the number of experts to bring additional perspectives; to broaden the temporal analysis by incorporating additional data points along the timeline to achieve a more precise trend assessment; to extend the analysis to quantify the drivers, as many of them can be operationally defined as variables using the latest data from the most recent census; to advance beyond expert perception and establish quantitative causal relationships; to conduct a comparative analysis of the perceptions of different stakeholders to capture diverse viewpoints on the LULCCs that have occurred and on future trends; and to extend the survey to include the perception toward LULCC of the farmer population of the area to quantitatively assess the causal relation between drivers and LULCC.

5. Conclusions

This study delved into the understanding of crucial aspects of land use and land cover change by identifying the major transformations that occurred in the area of Gargano in the period 2000–2018 and exploring the influential factors that played a pivotal role in driving these transformations. This work represents a preliminary analysis of the Gargano region's evolving land use as a foundational step in its exploration.

The analysis revealed that the area has experienced micro-changes, with a predominant portion characterized by persistences. These persistences, particularly in natural areas and agricultural activities, are perceived positively, signifying their potential contribution to the area's future sustainable development. More precisely, natural areas have, in general, experienced an increase in the overall surface, despite relinquishing nearly 5 thousand hectares to agriculture and another thousand to urbanization in the period analyzed, as they reclaimed, in the same period, more than 7.5 thousand hectares. It is worth highlighting the importance experts gave to pastures and heterogeneous areas which emerged as significant indicators of the socio-economic and ecological equilibrium of the area: their variations are considered relevant with reductions viewed as potentially detrimental and increases perceived as advantageous in terms of sustainable development. At the same time, urbanization and deforestation emerged as focal points of major concern, reflecting the heightened sensitivity surrounding these transformative processes.

Finally, with an eye to the broader field of rural land use change, the understanding of micro-transformations that occurred in the Gargano area, addressed in the present study, help the decision-making authority to reduce long-term asymmetries in view of adopting ad hoc strategies for land use changes and meeting the green transition more rapidly in the immediate future.

Author Contributions: Conceptualization, N.E., P.P. and A.L.; methodology, N.E. and P.B.; software, P.B.; investigation, N.E. and A.L.; data curation, N.E.; writing—original draft preparation, N.E.; writing—review and editing, A.L. and C.D.L.; supervision, P.P. and C.D.L. All authors have read and agreed to the published version of the manuscript.

Funding: This research was funded through PON "Research and Innovation" 2014–2020—Action IV.4, "Doctorates and research contracts on innovation issues", and Action IV.5, "Doctorates on green issues".

Data Availability Statement: No new data were created or analyzed in this study. Data sharing is not applicable to this article.

Conflicts of Interest: The authors declare no conflicts of interest.

Appendix A

Table A1. Causality of drivers of land use changes—no. (%) of experts (n = 11) who mentioned the driver (in row) as a possible cause of the change (in column).

				Persistences							Agricultural Conversions						Net Transformations		
	Drivers	*PU*	*RA*	*PC*	*PP*	*PH*	*PN*	*HE*	*ER*	*AP*	*TC*	*OC*	*CP*	*UR*	*DC*	*DP*	*DU*	*NA*
Physical and Anthropogenic	D1	0 (0)	1 (0.09)	2 (0.18)	2 (0.18)	0 (0)	1 (0.09)	0 (0)	1 (0.09)	1 (0.09)	0 (0)	2 (0.18)	0 (0)	1 (0.09)	2 (0.18)	0 (0)	1 (0.09)	0 (0)
	D2	7 (0.63)	0 (0)	0 (0)	0 (0)	0 (0)	1 (0.09)	2 (0.18)	0 (0)	0 (0)	2 (0.18)	1 (0.09)	0 (0)	8 (0.72)	1 (0.09)	0 (0)	0 (0)	0 (0)
	D3	0 (0)	6 (0.54)	5 (0.45)	0 (0)	0 (0)	1 (0.09)	1 (0.09)	5 (0.45)	5 (0.45)	4 (0.36)	5 (0.45)	1 (0.09)	0 (0)	7 (0.63)	1 (0.09)	1 (0.09)	2 (0.18)
	D4	0 (0)	5 (0.45)	3 (0.27)	1 (0.09)	2 (0.18)	2 (0.18)	2 (0.18)	2 (0.18)	2 (0.18)	3 (0.27)	3 (0.27)	2 (0.18)	0 (0)	2 (0.18)	2 (0.18)	1 (0.09)	1 (0.09)
	D5	0 (0)	5 (0.45)	5 (0.45)	2 (0.18)	1 (0.09)	1 (0.09)	4 (0.36)	3 (0.27)	2 (0.18)	2 (0.18)	4 (0.36)	4 (0.36)	1 (0.09)	3 (0.27)	2 (0.18)	3 (0.27)	3 (0.27)
	D6	0 (0)	3 (0.27)	1 (0.09)	0 (0)	0 (0)	0 (0)	0 (0)	2 (0.18)	1 (0.09)	1 (0.09)	0 (0)	0 (0)	0 (0)	2 (0.18)	0 (0)	1 (0.09)	0 (0)
	D7	0 (0)	6 (0.54)	2 (0.18)	0 (0)	1 (0.09)	0 (0)	0 (0)	5 (0.45)	3 (0.27)	2 (0.18)	2 (0.18)	0 (0)	0 (0)	5 (0.45)	0 (0)	1 (0.09)	0 (0)
	D8	0 (0)	0 (0)	0 (0)	0 (0)	0 (0)	2 (0.18)	1 (0.09)	0 (0)	0 (0)	0 (0)	0 (0)	0 (0)	0 (0)	0 (0)	0 (0)	1 (0.09)	0 (0)
	D9	1 (0.09)	2 (0.18)	0 (0)	2 (0.18)	1 (0.09)	1 (0.09)	0 (0)	0 (0)	1 (0.09)	0 (0)	1 (0.09)	2 (0.18)	2 (0.18)	3 (0.27)	0 (0)	0 (0)	5 (0.45)
	D10	0 (0)	0 (0)	2 (0.18)	4 (0.36)	0 (0)	1 (0.09)	0 (0)	0 (0)	0 (0)	0 (0)	0 (0)	2 (0.18)	0 (0)	0 (0)	2 (0.18)	0 (0)	0 (0)
	D11	5 (0.45)	0 (0)	0 (0)	0 (0)	0 (0)	0 (0)	1 (0.09)	0 (0)	0 (0)	0 (0)	0 (0)	0 (0)	2 (0.18)	0 (0)	0 (0)	1 (0.09)	0 (0)
	D12	0 (0)	4 (0.36)	2 (0.18)	0 (0)	0 (0)	1 (0.09)	0 (0)	3 (0.27)	2 (0.18)	0 (0)	1 (0.09)	1 (0.09)	0 (0)	4 (0.36)	0 (0)	0 (0)	0 (0)
	D13	0 (0)	1 (0.09)	1 (0.09)	2 (0.18)	1 (0.09)	4 (0.36)	3 (0.27)	1 (0.09)	0 (0)	2 (0.18)	2 (0.18)	4 (0.36)	0 (0)	0 (0)	1 (0.09)	2 (0.18)	5 (0.45)
	D14	0 (0)	2 (0.18)	1 (0.09)	0 (0)	1 (0.09)	1 (0.09)	3 (0.27)	1 (0.09)	2 (0.18)	1 (0.09)	2 (0.18)	2 (0.18)	0 (0)	2 (0.18)	0 (0)	2 (0.18)	2 (0.18)
	D15	0 (0)	2 (0.18)	1 (0.09)	0 (0)	1 (0.09)	1 (0.09)	3 (0.27)	1 (0.09)	2 (0.18)	1 (0.09)	2 (0.18)	2 (0.18)	0 (0)	1 (0.09)	1 (0.09)	2 (0.18)	2 (0.18)
	D16	0 (0)	0 (0)	1 (0.09)	0 (0)	1 (0.09)	0 (0)	2 (0.18)	0 (0)	0 (0)	0 (0)	0 (0)	0 (0)	0 (0)	0 (0)	1 (0.09)	0 (0)	1 (0.09)
Socio-economic	D17	1 (0.09)	0 (0)	0 (0)	3 (0.27)	4 (0.36)	6 (0.54)	6 (0.54)	0 (0)	1 (0.09)	1 (0.09)	1 (0.09)	6 (0.54)	2 (0.18)	0 (0)	0 (0)	5 (0.45)	5 (0.45)
	D18	0 (0)	1 (0.09)	1 (0.09)	5 (0.45)	6 (0.54)	4 (0.36)	6 (0.54)	0 (0)	1 (0.09)	1 (0.09)	1 (0.09)	5 (0.45)	1 (0.09)	0 (0)	1 (0.09)	3 (0.27)	6 (0.54)
	D19	1 (0.09)	0 (0)	0 (0)	0 (0)	1 (0.09)	1 (0.09)	3 (0.27)	0 (0)	0 (0)	1 (0.09)	1 (0.09)	1 (0.09)	0 (0)	0 (0)	1 (0.09)	0 (0)	2 (0.18)
	D20	1 (0.09)	0 (0)	1 (0.09)	0 (0)	0 (0)	4 (0.36)	2 (0.18)	0 (0)	0 (0)	1 (0.09)	2 (0.18)	2 (0.18)	1 (0.09)	2 (0.18)	0 (0)	1 (0.09)	1 (0.09)
	D21	0 (0)	3 (0.27)	0 (0)	4 (0.36)	4 (0.36)	0 (0)	2 (0.18)	2 (0.18)	0 (0)	2 (0.18)	0 (0)	3 (0.27)	0 (0)	3 (0.27)	0 (0)	0 (0)	4 (0.36)
	D22	6 (0.54)	2 (0.18)	0 (0)	0 (0)	0 (0)	1 (0.09)	0 (0)	3 (0.27)	1 (0.09)	2 (0.18)	0 (0)	0 (0)	5 (0.45)	1 (0.09)	1 (0.09)	1 (0.09)	1 (0.09)
	D23	1 (0.09)	1 (0.09)	1 (0.09)	3 (0.27)	1 (0.09)	1 (0.09)	3 (0.27)	1 (0.09)	1 (0.09)	1 (0.09)	0 (0)	6 (0.54)	0 (0)	1 (0.09)	1 (0.09)	2 (0.18)	2 (0.18)
	D24	0 (0)	2 (0.18)	2 (0.18)	0 (0)	0 (0)	0 (0)	1 (0.09)	1 (0.09)	1 (0.09)	3 (0.27)	0 (0)	0 (0)	0 (0)	2 (0.18)	1 (0.09)	1 (0.09)	0 (0)
	D25	0 (0)	2 (0.18)	3 (0.27)	0 (0)	0 (0)	0 (0)	0 (0)	4 (0.36)	1 (0.09)	4 (0.36)	2 (0.18)	0 (0)	0 (0)	2 (0.18)	1 (0.09)	1 (0.09)	0 (0)
	D26	0 (0)	0 (0)	1 (0.09)	0 (0)	0 (0)	3 (0.27)	0 (0)	1 (0.09)	1 (0.09)	3 (0.27)	1 (0.09)	1 (0.09)	0 (0)	3 (0.27)	1 (0.09)	1 (0.09)	0 (0)
	D27	0 (0)	3 (0.27)	3 (0.27)	0 (0)	0 (0)	1 (0.09)	0 (0)	3 (0.27)	1 (0.09)	2 (0.18)	2 (0.18)	0 (0)	0 (0)	1 (0.09)	0 (0)	1 (0.09)	0 (0)
	D28	0 (0)	3 (0.27)	4 (0.36)	0 (0)	0 (0)	0 (0)	0 (0)	3 (0.27)	1 (0.09)	3 (0.27)	1 (0.09)	0 (0)	0 (0)	0 (0)	0 (0)	0 (0)	0 (0)
	D29	0 (0)	2 (0.18)	1 (0.09)	1 (0.09)	1 (0.09)	3 (0.27)	0 (0)	2 (0.18)	1 (0.09)	3 (0.27)	1 (0.09)	3 (0.27)	1 (0.09)	3 (0.27)	1 (0.09)	0 (0)	1 (0.09)
	D30	0 (0)	4 (0.36)	5 (0.45)	1 (0.09)	1 (0.09)	0 (0)	1 (0.09)	3 (0.27)	1 (0.09)	5 (0.45)	3 (0.27)	1 (0.09)	1 (0.09)	4 (0.36)	1 (0.09)	0 (0)	1 (0.09)
	D31	0 (0)	7 (0.63)	3 (0.27)	0 (0)	0 (0)	0 (0)	0 (0)	5 (0.45)	1 (0.09)	3 (0.27)	1 (0.09)	0 (0)	0 (0)	3 (0.27)	1 (0.09)	0 (0)	0 (0)

Notes

[1] For the stake of clarity, here, we provide an example of how the first row of the Table 4, referring to the Change Category *Artificial Surfaces* (PU), should be understood: it indicates that 4064.70 hectares of the total area categorized as *Artificial Surfaces* in 2018 were also categorized in the same way in 2000. This area constitutes 97% of the total surface that was categorized as *Artificial Surfaces* in 2000. In other words, 97% of the artificial surfaces in 2000 remained unchanged in 2018, indicating a persistence rate of 97%.

References

1. Sisay, G.; Gessesse, B.; Fürst, C.; Kassie, M.; Kebede, B. Modeling of Land Use/Land Cover Dynamics Using Artificial Neural Network and Cellular Automata Markov Chain Algorithms in Goang Watershed, Ethiopia. *Heliyon* **2023**, *9*, e20088. [CrossRef]
2. Song, Y.; Yeung, G.; Zhu, D.; Xu, Y.; Zhang, L. Efficiency of Urban Land Use in China's Resource-Based Cities, 2000–2018. *Land Use Policy* **2022**, *115*, 106009. [CrossRef]
3. Aquilué, N.; De Cáceres, M.; Fortin, M.-J.; Fall, A.; Brotons, L. A Spatial Allocation Procedure to Model Land-Use/Land-Cover Changes: Accounting for Occurrence and Spread Processes. *Ecol. Model.* **2017**, *344*, 73–86. [CrossRef]
4. Verburg, P.H.; Overmars, K.P. Combining Top-down and Bottom-up Dynamics in Land Use Modeling: Exploring the Future of Abandoned Farmlands in Europe with the Dyna-CLUE Model. *Landsc. Ecol.* **2009**, *24*, 1167–1181. [CrossRef]
5. Ostrom, E. A General Framework for Analyzing Sustainability of Social-Ecological Systems. *Science* **2009**, *325*, 419–422. [CrossRef]
6. Caniglia, B.S.; Mayer, B. Socio-Ecological Systems. In *Handbook of Environmental Sociology*; Schaefer Caniglia, B., Jorgenson, A., Malin, S.A., Peek, L., Pellow, D.N., Huang, X., Eds.; Handbooks of Sociology and Social Research; Springer International Publishing: Cham, Switzerland, 2021; pp. 517–536. ISBN 978-3-030-77712-8.
7. McPhearson, T.; Cook, E.M.; Berbés-Blázquez, M.; Cheng, C.; Grimm, N.B.; Andersson, E.; Barbosa, O.; Chandler, D.G.; Chang, H.; Chester, M.V.; et al. A Social-Ecological-Technological Systems Framework for Urban Ecosystem Services. *One Earth* **2022**, *5*, 505–518. [CrossRef]
8. Shalaby, A.; Tateishi, R. Remote Sensing and GIS for Mapping and Monitoring Land Cover and Land-Use Changes in the Northwestern Coastal Zone of Egypt, Appl. *Appl. Geogr.* **2007**, *27*, 28–41. [CrossRef]
9. Kleemann, J.; Inkoom, J.N.; Thiel, M.; Shankar, S.; Lautenbach, S.; Fürst, C. Peri-Urban Land Use Pattern and Its Relation to Land Use Planning in Ghana, West Africa. *Landsc. Urban Plan.* **2017**, *165*, 280–294. [CrossRef]
10. Larigauderie, A.; Mooney, H.A. The Intergovernmental Science-Policy Platform on Biodiversity and Ecosystem Services: Moving a Step Closer to an IPCC-like Mechanism for Biodiversity. *Curr. Opin. Environ. Sustain.* **2010**, *2*, 9–14. [CrossRef]
11. Mienmany, B. *Analysis Land Use and Land Cover Changes and the Driving Forces: A Case Study in Kaysone Phomvihan District, Laos*; University of Porto: Porto, Portugal, 2018.
12. Radel, C.; Jokisch, B.; Schmook, B.; Carte, L.; Støen, M.; Hermans, K.; Zimmerer, K.; Aldrich, S.; Roy Chowdhury, R.; Munroe, D.; et al. Migration as a Feature of Land System Transitions Why Interactions between Land Systems and Migration Matter. *Curr. Opin. Environ. Sustain.* **2019**, *38*, 103–110. [CrossRef]
13. Calzada, L.; Meave, J.A.; Bonfil, C.; Figueroa, F. Lands at Risk: Land Use/Land Cover Change in Two Contrasting Tropical Dry Regions of Mexico. *Appl. Geogr.* **2018**, *99*, 22–30. [CrossRef]
14. Dibaba, W.T.; Demissie, T.A.; Miegel, K. Drivers and Implications of Land Use/Land Cover Dynamics in Finchaa Catchment, Northwestern Ethiopia. *Land* **2020**, *9*, 113. [CrossRef]
15. Uisso, A.; Tanrivermis, H. Driving Factors and Assessment of Changes in the Use of Arable Land in Tanzania. *Land Use Policy* **2021**, *104*, 105359. [CrossRef]
16. Muhammad, R.; Zhang, W.; Abbas, Z.; Guo, F.; Gwiazdzinski, L. Spatiotemporal Change Analysis and Prediction of Future Land Use and Land Cover Changes Using QGIS MOLUSCE Plugin and Remote Sensing Big Data: A Case Study of Linyi, China. *Land* **2022**, *11*, 419. [CrossRef]
17. Pulighe, G.; Baiocchi, V.; Lupia, F. Horizontal Accuracy Assessment of Very High Resolution Google Earth Images in the City of Rome, Italy. *Int. J. Digit. Earth* **2016**, *9*, 342–362. [CrossRef]
18. Senbeta, F. Community Perception of Land Use/Land Cover Change and Its Impacts on Biodiversity and Ecosystem Services in Northwestern Ethiopia. *J. Sustain. Dev. Afr.* **2018**, *20*, 108–126.
19. Chen, Z.; Huang, M.; Zhu, D.; Altan, O. Integrating Remote Sensing and a Markov-FLUS Model to Simulate Future Land Use Changes in Hokkaido, Japan. *Remote Sens.* **2021**, *13*, 2621. [CrossRef]
20. Liu, X.; Sun, R.; Yang, Q.; Su, G.; Qi, W. Simulating Urban Expansion Using an Improved SLEUTH Model. *J. Appl. Remote Sens.* **2012**, *6*, 061709. [CrossRef]
21. Pahlavani, P.; Askarian Omran, H.; Bigdeli, B. A Multiple Land Use Change Model Based on Artificial Neural Network, Markov Chain, and Multi Objective Land Allocation. *Earth Obs. Geomat. Eng.* **2017**, *1*, 82–99. [CrossRef]
22. Meyfroidt, P.; Lambin, E.F.; Erb, K.-H.; Hertel, T.W. Globalization of Land Use: Distant Drivers of Land Change and Geographic Displacement of Land Use. *Curr. Opin. Environ. Sustain.* **2013**, *5*, 438–444. [CrossRef]
23. Meyfroidt, P.; Roy Chowdhury, R.; de Bremond, A.; Ellis, E.C.; Erb, K.-H.; Filatova, T.; Garrett, R.D.; Grove, J.M.; Heinimann, A.; Kuemmerle, T.; et al. Middle-Range Theories of Land System Change. *Glob. Environ. Chang.* **2018**, *53*, 52–67. [CrossRef]
24. Lambin, E.F.; Turner, B.; Geist, H.J.; Agbola, S.; Angelsen, A.; Bruce, J.W.; Coomes, O.; Dirzo, R.; Fischer, G.; Folke, C.; et al. The Causes of Land-Use and Land-Cover Change: Moving beyond the Myths. *Glob. Environ. Chang.* **2001**, *11*, 261–269. [CrossRef]

25. Lambin, E.; Geist, H.; Lepers, E. Dynamics of Land-Use and Land-Cover Change in Tropical Regions. *Annu. Rev. Environ. Resour.* **2003**, *20*, 49205–49241. [CrossRef]
26. Mack, E.A.; Sauls, L.A.; Jokisch, B.D.; Nolte, K.; Schmook, B.; He, Y.; Radel, C.; Allington, G.R.H.; Kelley, L.C.; Scott, C.K.; et al. Remittances and Land Change: A Systematic Review. *World Dev.* **2023**, *168*, 106251. [CrossRef]
27. Tahmasebi, T.; Karami, E.; Keshavarz, M. Agricultural Land Use Change under Climate Variability and Change: Drivers and Impacts. *J. Arid Environ.* **2020**, *180*, 104202. [CrossRef]
28. Kanianska, R. *Agriculture and Its Impact on Land-Use, Environment, and Ecosystem Services;* Almusaed, A., Ed.; InTech: Vienna, Austria, 2016.
29. Reader, M.O.; Eppinga, M.B.; de Boer, H.J.; Damm, A.; Petchey, O.L.; Santos, M.J. Biodiversity Mediates Relationships between Anthropogenic Drivers and Ecosystem Services across Global Mountain, Island and Delta Systems. *Glob. Environ. Chang.* **2023**, *78*, 102612. [CrossRef]
30. Mastronardi, L.; Cavallo, A.; Romagnoli, L. A Novel Composite Environmental Fragility Index to Analyse Italian Ecoregions' Vulnerability. *Land Use Policy* **2022**, *122*, 106352. [CrossRef]
31. Nilsson, C.; Grelsson, G. The Fragility of Ecosystems: A Review. *J. Appl. Ecol.* **1995**, *32*, 677–692. [CrossRef]
32. European Committee of the Regions; Commission for Economic Policy; t33; Spatial Foresight; ÖIR. *A Territorial Approach for the Implementation of the SDGs in the EU: The Role of the European Committee of the Regions*; Publications Office: Luxembourg, 2019.
33. Demblans, A.; Martínez, M.P.; Lavalle, C. Chapter 19—Place-Based Solutions to Territorial Challenges: How Policy and Research Can Support Successful Ecosystems. In *Science for Policy Handbook*; Šucha, V., Sienkiewicz, M., Eds.; Elsevier: Amsterdam, The Netherlands, 2020; pp. 224–238. ISBN 978-0-12-822596-7.
34. Sustainable Development in the Mediterranean. Report 2020. Available online: https://flore.unifi.it/handle/2158/1249393 (accessed on 11 January 2024).
35. Egidi, G.; Zambon, I.; Tombolin, I.; Salvati, L.; Cividino, S.; Seifollahi-Aghmiuni, S.; Kalantari, Z. Unraveling Latent Aspects of Urban Expansion: Desertification Risk Reveals More. *Int. J. Environ. Res. Public Health* **2020**, *17*, 4001. [CrossRef]
36. Doblas-Miranda, E.; Alonso, R.; Arnan, X.; Bermejo, V.; Brotons, L.; de las Heras, J.; Estiarte, M.; Hódar, J.A.; Llorens, P.; Lloret, F.; et al. A Review of the Combination among Global Change Factors in Forests, Shrublands and Pastures of the Mediterranean Region: Beyond Drought Effects. *Glob. Planet. Chang.* **2017**, *148*, 42–54. [CrossRef]
37. Ruiz, I.; Sanz-Sanchez, M.-J. Effects of Historical Land-Use Change in the Mediterranean Environment. *Sci. Total Environ.* **2020**, *732*, 139315. [CrossRef] [PubMed]
38. Ferreira, C.S.S.; Seifollahi-Aghmiuni, S.; Destouni, G.; Ghajarnia, N.; Kalantari, Z. Soil Degradation in the European Mediterranean Region: Processes, Status and Consequences. *Sci. Total Environ.* **2022**, *805*, 150106. [CrossRef] [PubMed]
39. Giordano, S. Agrarian Landscapes: From Marginal Areas to Cultural Landscapes—Paths to Sustainable Tourism in Small Villages—The Case of Vico Del Gargano in the Club of the Borghi Più Belli d'Italia. *Qual. Quant. Int. J. Methodol.* **2020**, *54*, 1725–1744. [CrossRef]
40. Balena, P.; Pazienza, P.; De Lucia, C. Relazione Socio-Economica per l'adeguamento Del Piano Generale Di Bonifica 2021. Available online: https://www.google.com/url?sa=t&rct=j&q=&esrc=s&source=web&cd=&ved=2ahUKEwiAnIfllPiDAxVz8gIHHUA8DHMQFnoECBIQAQ&url=https://www.regione.puglia.it/documents/42866/5059784/Relazione+socio+economica.pdf/ea800472-264d-c34d-288f-b4decdd1421 (accessed on 3 December 2023).
41. Di Matteo, D.; Cavuta, G. Landscapes Protection and Eco-Development: The Case Study of Gargano National Park. *Geoj. Tour. Geosites* **2016**, *17*, 95–111.
42. Biscotti, N.; Bonsanto, D.; Laghetti, G. *Ethnobotanical Study on Traditional Use of Local Fruit Varieties in Gargano National Park (Apulia, Italy)*; Springer: Berlin/Heidelberg, Germany, 2021.
43. Xiuwan, C. Using Remote Sensing and GIS to Analyse Land Cover Change and Its Impacts on Regional Sustainable Development. *Int. J. Remote Sens.* **2002**, *23*, 107–124. [CrossRef]
44. Musakwa, W.; Niekerk, A.V. Implications of Land Use Change for the Sustainability of Urban Areas: A Case Study of Stellenbosch, South Africa. *Cities* **2013**, *32*, 143–156. [CrossRef]
45. McBratney, A.B.; Mendonça Santos, M.L.; Minasny, B. On Digital Soil Mapping. *Geoderma* **2003**, *117*, 3–52. [CrossRef]
46. CORINE Land Cover. Available online: https://land.copernicus.eu/en/products/corine-land-cover (accessed on 11 January 2024).
47. Amici, V.; Filibeck, G.; Rocchini, D.; Geri, F.; Landi, S.; Giorgini, D.; Scoppola, A.; Chiarucci, A. Are CORINE Land Cover Classes Reliable Proxies of Plant Species Assemblages? A Test in Mediterranean Forest Landscapes. *Plant Biosyst. Int. J. Deal. All Asp. Plant Biol.* **2018**, *152*, 994–1001. [CrossRef]
48. Ovejero-Campos, A.; Fernández, E.; Ramos, L.; Bento, R.; Martínez, G. Methodological Limitations of CLC to Assess Land Cover Changes in Coastal Environments. *J. Coast. Conserv.* **2019**, *23*, 657–673. [CrossRef]
49. Liu, L.; Yu, S.; Zhang, H.; Wang, Y.; Liang, C. Analysis of Land Use Change Drivers and Simulation of Different Future Scenarios: Taking Shanxi Province of China as an Example. *Int. J. Environ. Res. Public Health* **2023**, *20*, 1626. [CrossRef]
50. Zhou, Y.; Li, X.; Liu, Y. Land Use Change and Driving Factors in Rural China during the Period 1995–2015. *Land Use Policy* **2020**, *99*, 105048. [CrossRef]
51. Hailu, A.; Mammo, S.; Kidane, M. Dynamics of Land Use, Land Cover Change Trend and Its Drivers in Jimma Geneti District, Western Ethiopia. *Land Use Policy* **2020**, *99*, 105011. [CrossRef]

52. Burra, D.D.; Parker, L.; Than, N.T.; Phengsavanh, P.; Long, C.T.M.; Ritzema, R.S.; Sagemueller, F.; Douxchamps, S. Drivers of Land Use Complexity along an Agricultural Transition Gradient in Southeast Asia. *Ecol. Indic.* **2021**, *124*, 107402. [CrossRef]
53. Cao, C.; Song, W. Progress and Prospect of Ecological Risks of Land Use Change. *Front. Environ. Sci.* **2022**, *10*, 1077515. [CrossRef]
54. Semenchuk, P.; Plutzar, C.; Kastner, T.; Matej, S.; Bidoglio, G.; Erb, K.-H.; Essl, F.; Haberl, H.; Wessely, J.; Krausmann, F.; et al. Relative Effects of Land Conversion and Land-Use Intensity on Terrestrial Vertebrate Diversity. *Nat. Commun.* **2022**, *13*, 615. [CrossRef] [PubMed]
55. Liang, Y.; Song, W. Integrating Potential Ecosystem Services Losses into Ecological Risk Assessment of Land Use Changes: A Case Study on the Qinghai-Tibet Plateau. *J. Environ. Manag.* **2022**, *318*, 115607. [CrossRef] [PubMed]
56. Deng, G.; Jiang, H.; Zhu, S.; Wen, Y.; He, C.; Wang, X.; Sheng, L.; Guo, Y.; Cao, Y. Projecting the Response of Ecological Risk to Land Use/Land Cover Change in Ecologically Fragile Regions. *Sci. Total Environ.* **2024**, *914*, 169908. [CrossRef] [PubMed]
57. Gong, J.; Cao, E.; Xie, Y.; Xu, C.; Li, H.; Yan, L. Integrating Ecosystem Services and Landscape Ecological Risk into Adaptive Management: Insights from a Western Mountain-Basin Area, China. *J. Environ. Manag.* **2021**, *281*, 111817. [CrossRef] [PubMed]
58. Piergiovanni, A.R.; Margiotta, B. On Farm Survival of Apulian Legume and Cereal Landraces in Relation to Land Cover/Land Use Changes. A Case Study. *Ital. J. Agron.* **2021**, *16*, 1724. [CrossRef]
59. Peña-Rodríguez, F.-J.; Entrena-Durán, F.; Ivorra-Cano, A.; Llorca-Linde, A. Changes in Land Use and Food Security: The Case of the De La Vega Agrarian Shire in the Southern Spanish Province of Granada. *Land* **2023**, *12*, 747. [CrossRef]
60. Noni, I.D.; Ghidoni, A.; Menzel, F.; Bahrs, E.; Corsi, S. Exploring Drivers of Farmland Value and Growth in Italy and Germany at Regional Level. *Aestimum* **2019**, 77–99. [CrossRef]
61. Armanda, D.T.; Guinée, J.B.; Tukker, A. The Second Green Revolution: Innovative Urban Agriculture's Contribution to Food Security and Sustainability—A Review. *Glob. Food Secur.* **2019**, *22*, 13–24. [CrossRef]
62. Coles, R.; Costa, S. Food Growing in the City: Exploring the Productive Urban Landscape as a New Paradigm for Inclusive Approaches to the Design and Planning of Future Urban Open Spaces. *Landsc. Urban Plan.* **2018**, *170*, 1–5. [CrossRef]
63. Lombardi, M.; Lopolito, A.; Andriano, A.M.; Prosperi, M.; Stasi, A.; Iannuzzi, E. Network impact of social innovation initiatives in marginalised rural communities. *Soc. Netw.* **2020**, *63*, 11–20. [CrossRef]
64. Torquati, B.; Giacchè, G.; Venanzi, S. Economic Analysis of the Traditional Cultural Vineyard Landscapes in Italy. *J. Rural Stud.* **2015**, *39*, 122–132. [CrossRef]
65. Kondoh, K.; Kurata, H. *Agritourism, Unemployment, and Urban-Rural Migration*; Springer: Singapore, 2021; pp. 25–42, ISBN 9789811612312.
66. Dax, T.; Zhang, D.; Chen, Y. Agritourism Initiatives in the Context of Continuous Out-Migration: Comparative Perspectives for the Alps and Chinese Mountain Regions. *Sustainability* **2019**, *11*, 4418. [CrossRef]
67. Fagioli, F.; Diotallevi, F.; Ciani, A. Strengthening the Sustainability of Rural Areas: The Role of Rural Tourism and Agritourism. *Riv. Econ. Agrar.* **2014**, *69*, 155–169. [CrossRef]
68. The Equitable and Sustainable Well-Being Framework in Italy: An Action Plan for Its Use in Policy and Budget Decision Making | En | OECD. Available online: https://www.oecd.org/italy/the-equitable-and-sustainable-well-being-framework-in-italy-4f48c504-en.htm (accessed on 11 January 2024).
69. Quintas-Soriano, C.; Buerkert, A.; Plieninger, T. Effects of Land Abandonment on Nature Contributions to People and Good Quality of Life Components in the Mediterranean Region: A Review. *Land Use Policy* **2022**, *116*, 106053. [CrossRef]
70. Pérez-Ramírez, I.; Requena-Mullor, J.M.; Castro, A.J.; García-Llorente, M. Land Transformation Changes People's Values of Ecosystem Services in Las Vegas Agrarian Landscapes of Madrid Spain. *Land Use Policy* **2023**, *134*, 106921. [CrossRef]
71. Nyström, M.; Jouffray, J.-B.; Norström, A.V.; Crona, B.; Søgaard Jørgensen, P.; Carpenter, S.R.; Bodin, Ö.; Galaz, V.; Folke, C. Anatomy and Resilience of the Global Production Ecosystem. *Nature* **2019**, *575*, 98–108. [CrossRef]
72. Pascual, U.; Balvanera, P.; Díaz, S.; Pataki, G.; Roth, E.; Stenseke, M.; Watson, R.T.; Başak Dessane, E.; Islar, M.; Kelemen, E.; et al. Valuing Nature's Contributions to People: The IPBES Approach. *Curr. Opin. Environ. Sustain.* **2017**, *26–27*, 7–16. [CrossRef]
73. Strategia Farm to Fork: Operativa la Consultazione per la Sicurezza Alimentare in Epoca di Crisi. Available online: https://www.regione.puglia.it/web/agricoltura/-/strategia-farm-to-fork-operativa-la-consultazione-per-la-sicurezza-alimentare-in-epoca-di-crisi (accessed on 11 January 2024).
74. Frontiers | Agroecology in Large Scale Farming—A Research Agenda. Available online: https://www.frontiersin.org/articles/10.3389/fsufs.2020.584605/full (accessed on 11 January 2024).
75. The EU Pathway towards Sustainable Food Systems Transformation | Knowledge for Policy. Available online: https://knowledge4policy.ec.europa.eu/publication/eu-pathway-towards-sustainable-food-systems-transformation_en?gclid=Cj0KCQiAwP6sBhDAARIsAPfK_wYsbijWFk1vVnPCM_51bEtUMNq5hoOGpDbmt-lnFFBJ64BTis8tsQEaAh4EEALw_wcB (accessed on 11 January 2024).

land

MDPI

Article

Evaluating the Effectiveness of Land-Use Policies in Preventing the Risk of Coastal Flooding: Coastal Regions of Helsinki and Espoo

Faegheh Amani Fard [1,*], Kirsikka Riekkinen [1,2] and Havu Pellikka [1]

[1] Department of Built Environment, Aalto University, 02150 Espoo, Finland; kirsikka.riekkinen@aalto.fi (K.R.); havu.pellikka@aalto.fi (H.P.)
[2] National Land Survey of Finland, 00521 Helsinki, Finland
* Correspondence: faegheh.amanifard@gmail.com

Abstract: Effectively resolving environmental problems often involves not only technical solutions but also examining and potentially reforming the political and institutional frameworks that govern how societies relate to the environment. Coastal flooding due to rising sea levels, for example, poses a significant threat to waterfront areas. Land-use regulations represent an effective means of mitigating this risk. Despite the multitude of strategies developed for tackling similar problems, relatively few have concentrated on assessing the effectiveness of land-use policies in managing such issues. This study examines a framework for evaluating the effectiveness of land-use policies in preventing sea flood risks in the capital regions of Finland. While the focus of this research is on the coastal regions of the cities of Helsinki and Espoo, its implications extend internationally. By integrating Geographical Information System (GIS) and simulation tools, we simulate future land-use scenarios based on values that reflect the effects of land-use policies. This framework can be applied to other coastal regions worldwide facing similar challenges. Using land cover data and GeoSOS-FLUS software, land-use simulations of the target areas were generated. Land-use planning performance in the target areas exhibited positive changes, as fewer vulnerable land-use types were located within the sea flood risk zones in 2018 compared to 2000. This simulation also shows a strong similarity to actual land-use in 2018, confirming the framework's reliability. This paper presents a novel framework for evaluating the effectiveness of land-use policies in mitigating coastal flooding risks, focusing on the coastal regions of Helsinki and Espoo, Finland. By integrating GIS and simulation tools, the research demonstrates the utility of these tools in tracking land-use changes and analyzing policy impacts, enabling a nuanced assessment of policy effectiveness. Furthermore, the study highlights the importance of local knowledge and a localized approach to policy development, contributing to a deeper understanding of complex issues in urban planning and land-use management.

Keywords: land-use policy; land-use scenario; sea-level rise; coastal flood

1. Introduction

We live in a complex urban world that is undergoing emerging phenomena, intertwined development processes, and shifting connections [1]. It is inevitable for planners to face the evolving version of old problems [2–4]. In the context of the modern built environment, urban planning and land-use planning face challenges that encompass both wicked environmental problems and land-use issues. Addressing these intertwined challenges necessitates a simultaneous focus on institutional and political solutions, as technical remedies alone are insufficient for such problems [5]. Therefore, these problems can be addressed by making effective policies, and an effective policy usually covers a wide range of environmental concerns as well. Several wicked environmental problems can be tracked down to land-use cover change; therefore, land-use policies need to be evaluated in order to deliver the intended results. Moreover, the impacts and functions of land-use policies must be evaluated to pave the way for improvements.

This research focuses on sea-flood risks and their impact on waterfront urban areas. 'Flood risk' refers to the probability and potential adverse consequences of a flood event for human health, the environment, cultural heritage, and economic activity [6]. Coastal floods have been addressed by local governments with varying success, leading to new challenges. For example, in Vietnam, high dikes caused productivity issues in rice-producing deltas [7]. In Finland, cities such as Porvoo and Loviisa experienced sea floods with significant damage, such as the Loviisa sea wall failure in 1986 [8] and the Marjaniemi area flooding in Helsinki during the 2005 Gudrun storm [9].

Tracking the effects of land-use policies becomes essential as a guiding tool for improving land-use policies. Evaluating land-use policies ensures that land resources are utilized sustainably and responsibly. Furthermore, it enables the identification of unintended consequences or disparities that may arise from their implementation. This process aids in minimizing negative effects on communities, such as unequal access to resources or disproportionate environmental burdens. Evaluations provide valuable insights into the effectiveness and fairness of policies, enabling policymakers to address deficiencies or biases in the decision-making process. However, few scholars have attempted to evaluate land-use policies. Some efforts to evaluate land-use policies have been hampered by either the discontinuous effects of land-use policies in terms of planning and time or difficulties in detecting causal mechanisms between either the effects of land-use policy or other factors involved in the planning implementation process. Another impediment to policy evaluation is the likelihood of predicting possible futures rather than recognizing the effects of land-use regulations [10–12]. In fact, predicting possible futures rather than recognizing the direct effects of land-use regulations has hindered policy evaluation efforts. This study tries to fill this gap.

Thus, the present study aims to evaluate the effectiveness of Finnish land-use policies in preventing the risk of sea floods in the coastal regions of Helsinki and Espoo. This paper views policy effectiveness as a quality indicating a correspondence between the intended goals and the actual outcomes of the adopted policy, which can demonstrate whether the policy is working [11]. To accomplish this aim, a framework is examined for tracking the actual effects of land-use policies by simulating land-use covers in a time period expanding from 2000 to 2018. Using this framework, it will be possible to examine the actual impact of land-use policies without considering other factors or predicting possible future outcomes.

The paper effectively integrates GIS and simulation tools (GeoSOS-FLUS V2.0 software) into policy evaluation processes. By demonstrating the utility of such tools in tracking land-use changes and analyzing policy impacts, the research pushes the boundaries of conventional policy evaluation. Through the developed framework, land-use scenarios are simulated based on specific values, which reflect the effects of a policy set. This paper utilizes GIS and the GeoSOS-FLUS model in an analytic approach combined with reviewing policy documents and other relevant institutional publications to remove difficulties in detecting causal mechanisms as much as possible.

Moreover, this paper devises a system to classify land-use changes into desirable and undesirable categories based on their alignment with policy objectives. This innovative classification system enables a nuanced and direct assessment of policy effectiveness. It also emphasizes the importance of local knowledge in improving land-use policies, highlighting the need for a more localized approach to policy development and implementation. With these contributions, our research seeks to answer the crucial question of how effective municipal land-use policies are in mitigating the risk of coastal flooding in the waterfront areas of Helsinki and Espoo.

This paper studies waterfront areas of Helsinki and Espoo municipalities. Figure 1 represents the study area which consists of 57 city districts located in waterfront areas. This region is part of the Gulf of Finland. One of the greatest flood events happened in 2005. The Gudrun flood was a significant storm and subsequent flooding event that occurred in

January 2005 in several countries in Northern Europe, particularly Sweden and Finland. Storm Gudrun, also known as Cyclone Gudrun, was an intense extratropical cyclone that brought strong winds and heavy precipitation to the region [9].

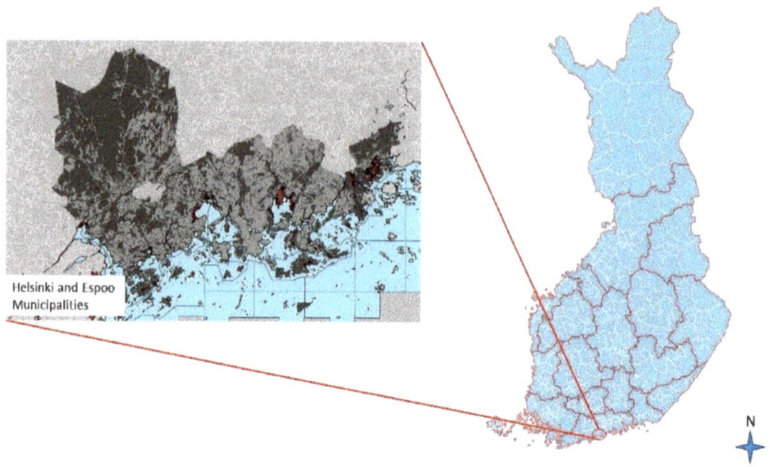

Figure 1. Coastal regions of Helsinki (60.1699° N, 24.9384° E) and Espoo (60.2055° N, 24.6559° E).

2. Theoretical Background

2.1. Restraining Wickedness through Effective Land-Use Policies

Institutional and political approaches are specifically useful when dealing with complex problems [5]. This is due to some shared characteristics among diverse types of wicked problems. In this context, wickedness is used to describe the nature of these problems as vicious, aggressive, untamable, and malignant, which are in contrast with beingness, being tamable, and solvable; in other words, wicked problems are aggressive like a lion, in contrast to the docility of a lamb [2,4,13]. The nature of an urban problem evolves with urban development and growth. As a result, the ubiquity of wicked problems in socio-ecological systems has become widely recognized in the literature, and it is evident in almost every pressing issue, such as global climate change, sustainability, land-use planning, resource management, and urbanization [13].

In addressing wicked problems, scholars recommend shifting from an optimistic view of science to embracing a post-normal science approach in policy analysis. Unlike technical solutions, institutional and political approaches are necessary for complex problems [5]. Post-normal science better suits the policy context's complexity, as traditional science may fail in this regard [14]. Although wicked problems are challenging to fully comprehend and tackle within centralized frameworks, scholars propose two main approaches. These involve using methodological approaches such as scenario planning and non-deterministic participatory methods to address uncertainties and employing instruments and tools to divide the problem into sub-problems or focus on specific goals [4,15].

Land-use is one of the three areas administered by the cadastral system, a system in which the whole information of a plot of land and its fiscal and legal purposes are recorded [16]. Policy contexts are highly dependent on the administrative system of a country. Thus, this study will focus on Finland's land-use planning system and land administration to avoid unnecessary complications. One of the ultimate goals of land-use policies is to mitigate the risks and issues that urban societies face. Policies are guided by ultimate goals such as sustainability, well-being, prosperity, and socio-economic balance. According to Howlett et al. [17], a policy framework consists of three main stages: policy design, policy implementation, and policy review. This study focuses on the last stage by assessing the effects of land-use policies. The main reason for restricting the scope of this article is the importance of making effective policies when dealing with wicked problems. One way of taking a cautious approach to wicked problems is to restrain the elements of wickedness by adopting effective policies [18]. Policy improvements can be made at any stage of the policy cycle. Still, the key message is that even small policy changes will redirect processes and maintain their effects for a long time [19]. The policy effectiveness concept is depicted in Figure 2.

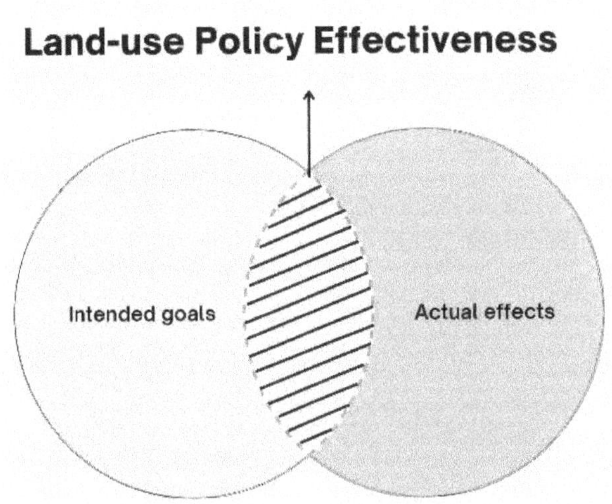

Figure 2. Land-use Policy Effectiveness.

Complex problems respond differently to political solutions depending on their attributes. Complex problems have been studied extensively over the past decades, and their characteristics have been modified by Conklin [20], Head et al. [3,20], Head [21], Ruhl and Salzman [22], Cajot et al. [23], Duckett et al. [4], and Alford and Head [24]. Among these, Ruhl and Salzman [22] investigated the attributes of massive problems based on their responsiveness to political solutions. This study considers massive problems equal to complex or wicked problems. Depending on the causal sources of the problem, causal mechanisms, and its cumulative effects, a policy can be highly effective in managing a massive problem or fail to have an influence on it. For a policy to be highly effective on a massive problem, the ideal conditions are listed in Table 1.

Table 1. The ideal conditions of causal sources, causal mechanisms, and cumulative effects for high policy effectiveness.

Source Attributes	Causal Attributes	Effect Attributes
Low number of sources	Limited scale (local and regional)	High level of detectability
Low diversity	High temporal limitation	High level of measurability
Clustered distribution		Proportional distribution over time
Large size on average (relative to effects)	Direct and proportional relationship	Proportional distribution over space
Alignment of incentives		High level of reversibility

Dealing with wicked problems is challenging for policymakers. However, since wicked problems respond to political solutions, aspects of a wicked problem can be managed by improving policies. Kirschke and Kosow [25] suggest using policy mixes for this purpose. Policy mixes are combinations of policy goals and the required instruments for achieving the intended goals [26]. According to the same study, some dimensions of wicked problems, such as complexity and uncertainty, can be addressed by improving specific aspects of policy mixes, such as adaptability, reversibility, robustness, comprehensiveness, and diversity of policy instruments.

2.2. Coastal Flooding

Sea floods are caused by a range of mostly weather-related phenomena, with climate change serving as the predominant driving force underlying the long-term change in flooding risks. It is useful to make a distinction between the long-term mean sea level and the sea level extremes, i.e., short-term variability around the mean. Both are affected by climate change. The rise in global mean sea level, caused by the thermal expansion of seawater and the melting of land-based ice, increases coastal flooding risks as the mean sea level rises. In addition, changes in sea level extremes relative to the mean may take place as the weather patterns change.

In the study area, the postglacial land uplift has so far compensated for the global sea level rise, resulting in a relatively stable mean sea level over the time period 2000–2018. Land uplift is a common process in the shoreline areas of Finland and Sweden which causes about 700 hectares of land to rise from the sea every year [27]. The rate of land uplift differs depending on the location; in Helsinki, it amounts to 3.7 mm/year [28], which is close to the current rate of global mean sea level rise [29]. In the future, global sea level rise will accelerate, overcoming the land uplift rate and causing mean sea level rise in the Helsinki region [30].

In the Baltic Sea, the astronomical tides are almost negligible, in the order of a few centimeters. Therefore, short-term sea level variations are governed by weather phenomena, mostly wind and air pressure variations. Sea levels in the Baltic Sea are relatively higher at times when westerly winds prevail because these strong wind patterns push more water into the almost enclosed basin of the Baltic Sea. In addition to water volume variations within the Baltic Sea basin, extreme sea level variations are caused by wind-induced internal redistribution of water within basins, air pressure variations, storm surges, and seiches in the various sub-basins of the sea [31]. Each of these phenomena can contribute to a sea level change of tens of centimeters on the Finnish coast: a compounding effect of several of these factors is required to produce a major coastal flood. Wind waves and, in rare cases, meteotsunamis [30] may also contribute to the height of the flood.

Climate change, on a global scale, has far-reaching and significant consequences. Research by Diez-Sierra et al. [32] indicates that the global average temperature could increase by at least 1.1 °C or up to a maximum of 6.4 °C by the year 2100. However, it is improbable that global warming will reach the most pessimistic scenario [32,33]. However, building a resilient urban environment remains essential. Urban flood events have been extensively discussed by scholars. Accordingly, ref. [34] suggests a tool for studying urban development's flooding consequences through a coupled urban growth-flood model that also considers policy, social, and physical factors influencing land cover patterns and flood impacts.

Climate change's primary impact in densely populated regions is the rise in sea levels and consequent damage from sea floods [35]. Other contributing factors include changes in salinity, ocean currents, and wind systems, influencing regional sea level fluctuations [36–39].

The causal sources and mechanisms of coastal flooding are represented in Figure 3. Accordingly, extreme sea level variation is caused by wind-induced internal redistribution of water within basins, air pressure variations, storm surges, meteotsunamis, and seiches in the various sub-basins of the sea [31].

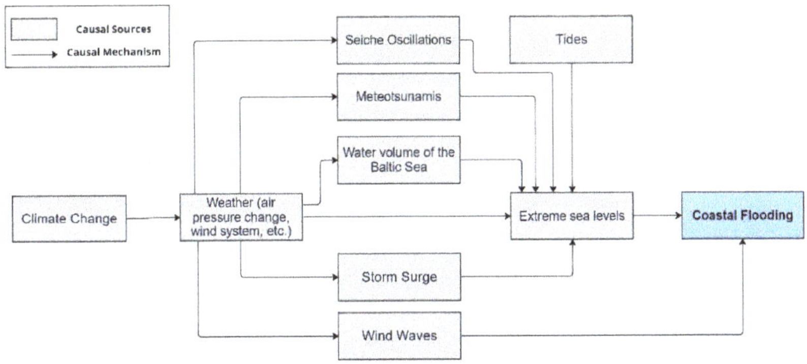

Figure 3. Causal sources and causal mechanisms of coastal flooding in the study area.

Coastal floods and wicked problems share common traits: multiple causal sources with complex relationships, ambiguous boundaries, and high uncertainty and complexity. Deep uncertainties, such as melting glaciers, further complicate sea level rise [40]. Managing sea flood risks emphasizes the significance of local knowledge [35], aligning with the "essentially unique" nature of wicked problems [2]. Sea floods are symptomatic of broader wicked problems such as ocean expansion, global warming, and climate change.

2.3. Coastal Regions of Helsinki and Espoo Municipalities

Helsinki and Espoo are in the southern part of Finland, with several densely populated districts. Since 2000, sea flood events have been more frequent in Finland as well as in other European countries. The range of sea level short-term fluctuations around the mean vary from 1.7 m in Degerby in the Archipelago Sea up to 3.1 m in Hamina in the eastern Gulf of Finland [41]. Consequently, several studies have been carried out in Helsinki and Espoo regarding the sea flood risk assessment to better understand and plan for future developments and to improve environmental protection. Additionally, some studies focused on estimating basin mass changes in order to develop effective water management strategies. A periodic rise and fall in Finnish terrestrial water storage occurs every 3.8 years, and snow plays a critical role in the annual cycle [42].

In Finland, 95% of the building codes set a minimum distance for residential or holiday buildings from the shoreline; the distance varies from 15 to 100 m and 89% of the municipalities have also defined the lowest construction height for buildings near the shoreline [9]. A minimum of 3 m above sea level was recommended for new constructions in Helsinki and Espoo [43]. The lowest recommended building height, above which structures affected by waterlogging cannot be built, is 0.5 m above the upper water line (HW50). The HW50 value is the flood height that occurs once every 50 years [44].

At a national level, the EU Floods Directive has had a great impact on the effort put into evaluating flood risks in coastal areas. The highest authority in flood risk management is the Ministry of Agriculture and Forestry and the subordinated Regional Environmental Centers. Moreover, municipal rescue authorities have an important role in flood risk management [9]. Climate adaptation policies in Helsinki are formulated by Helsinki's adaptation group (2016–2018), and these policies illustrate a climate-proof Helsinki in 2050 [45]. The "National Strategy for Adaptation to Climate Change" was adopted by the Finnish Parliament in November 2014 [46]. However, in 2006, there was no regional cooperation concerning sea flood prevention in Finland; therefore, the cooperation between city organizations only took place at the local level, especially in cases that had experienced sea flooding previously [9].

In 2007, Helsinki implemented a project (Rakennusvirasto 2007) to identify high flood-prone areas and at-risk buildings and suggested actions for risk mitigation. However, the project had limitations as it overlooked flood-prone islands and newly developing areas.

Additionally, monitoring flood-prone urban areas must be repeated regularly relative to the changes that happen in climate and mean sea levels. Currently, these aspects are covered by new measures taken, for instance, online map services (Tulvakarttapalvelu) provided by the Finnish Environment Institute (SYKE).

2.4. National and Municipal Land-Use Policies

A hierarchical system of three administrative levels organizes Finnish land-use planning [47,48]. The plans at the higher-level guide the plans at the lower level [1]. Nationally, the Ministry of the Environment and Finland's National Land-use Guidelines outline the future development of land-use. Finland's land-use and Building Act is the most important piece of legislation controlling land-use and spatial planning. The guidelines aim to ensure the appropriate actualization of national land-use solutions and to promote the implementation of international agreements and commitments [49].

The national land-use policies in Finland also consider various sector-specific policies and strategies. For example, the National Climate Change Adaptation Plan provides guidelines for addressing the impacts of climate change, including sea-level rise and coastal flooding. It emphasizes the need to integrate climate change considerations into land-use planning, risk assessments, and infrastructure development. Each municipality is responsible for preparing and maintaining a comprehensive land-use plan, which outlines the long-term vision and goals for land-use within its jurisdiction.

According to the legal consequences of land-use planning tiers in Finland, all the tiers of the land-use planning system are legally binding, and at the same time, they consider opportunities for non-binding and partially binding plans [50]. With the help of a legal-institutional role and the autonomy granted to the municipalities, land-use policies are mostly decided within municipalities. Municipalities use their autonomy in drafting local master plans. Local master plans are drafted under the supervision of municipal councils to provide long-term guidelines for land-use ordinances in the municipality [47]. However, some other factors might affect land-use controls on each level. For instance, environmental projects that monitor land-use changes by promoting climate change adaptation and environmental conservation values. FINADAPT was a project launched by the Finnish Environment Institute (SYKE) in 2003 to conduct a comprehensive research project that could help policymakers arrive at suitable policies [51].

3. Materials and Methods

3.1. Materials

To assess the effectiveness of land-use policies, two main datasets are used, which include the CORINE land-use cover dataset and the sea flood risk dataset provided by the Finnish Environment Institute (SYKE). The CORINE land cover 2000 (CLC2000) project provides information on land cover (LC) and its changes in 24 European countries, including Finland, between 1990 and 2000 [52]. This study uses CLC2000 and CLC2018 as reference material for the purpose of this study. However, the attributes of the CLC2018 dataset were improved in resolution and accuracy. CORINE land cover dataset classifies land-use data into four major classes, including artificial surfaces, agricultural areas, forests and semi-natural areas, and wetlands.

Using satellite imagery, the CLC2000 dataset was generated but has been deemed less reliable compared to newer technologies such as ESA Sentinel-2 dual date Landsat 8, which was used for CLC2018. The CLC2000 dataset exhibits low geometric accuracy (25 m) and a lower raster layer resolution. Furthermore, a generalization process was applied to aggregate small parcels into larger ones, affecting the dataset's accuracy. In response, the Finnish Environment Institute (SYKE) unified CORINE land-use cover data from 2000 to 2018 and enhanced dataset attributes to address these limitations. The improved outcomes include an increased resolution of 20 m for each layer, the unification of land-use type classifications, and the reversal of the generalization process. It is important to note that this project has not been publicly published, and access to it was granted to the researchers in response to an official request.

The Flood Risk Management Act (620/2010) of Finland demands that municipalities keep records of areas at risk of flooding. Among the different types of flood events, such as river floods, flash floods, urban floods, and floods from the sea in coastal areas, this study focuses on the last type. Due to the fact that sea level rise has accelerated the frequency of flood events in many regions, sea flooding is dependent on the mean sea level changes. This paper utilizes sea flood data provided by SYKE, which identifies the areas at risk of sea flooding and presents different probabilities of flood events. Due to the fact that flood risk zone data for 2000 is unavailable, there are limitations associated with flood risk data. The boundaries of risk zones are set based on the year 2018. The premier goal of the study was to investigate land-use changes that have happened within areas at risk of flooding. Evaluating these land-use changes leads us to recognize the effects of land-use policies that were applied to these regions.

In total, 27,366 km^2 of Helsinki and Espoo's coastal regions are at risk of sea flood events, but its frequency varies from once every two years to once every 1000 years. Table 2 represents shares of each land-use type that is located inside the sea flood risk zones. Based on this, land-use changes over a time period of 18 years have improved the allocation of artificial surfaces and agricultural lands. The greatest land-use change has happened in artificial surfaces, with a 373 km^2 reduction in the areas at risk of flooding events.

Table 2. Shares of each land-use type that is located within the flood risk zones.

Land-Use Type at Risk of Flooding	Area in 2000 (km^2)	Area in 2018 (km^2)	Performance (km^2)
Artificial surfaces	7232	6960.4	Decreased 271.4
Agricultural areas	3234.2	2861.2	Decreased 373
Forests and semi-natural areas	11,890.9	12,208.8	Increased 317.9
Wetlands	5008.9	5335.6	Increased 326.7

The present study utilizes international guidelines for land-use planning and regulations, national legislation, regional plans, local master plans, detailed plans, and associated projects in order to review the land-use policies that applied to the coastal areas of Helsinki and Espoo. Additionally, land-use cover changes in these urban areas are impacted by other policies, including land management policies, nature conservation goals, climate adaptation plans, and economic development incentives.

3.2. Methods

This study utilizes a back-casting method to detect the effects of land-use policies that are applied to the coastal regions of the study areas. The main distinctive feature of the back-casting method compared to other multi-scenario simulation methods that tend to predict possible futures of land-use cover is that this approach helps to find causes and solutions to a problem [53,54]. Using this approach, we can identify the factors that lead to a change under the influence of policies. Thus, the direct impact of policies on land-use changes becomes distinctive [55]. In this study, the envisioned future is defined based on the land-use cover in 2018. Land-use policy effects are detected according to the way this envisioned future has been achieved. The target scenario is referred to the envisioned land-use cover. The initial scenario is based on the land-use cover in 2000.

To operationalize this method, a land-use model was simulated based on the initial scenario using GeoSOS-Future Land Use Simulation (GeoSOS-FLUS) software. This software offers a multifaceted Cellular Automata (CA) allocation model. Recent advancements in CA modeling, such as LP-CA (Landscape-driven Patch-Based Cellular Automation), have enhanced its ability to incorporate information about landscape patterns [56]. It provides a better understanding of past and future urban developments by considering both landscape similarity and cell-by-cell agreement. The GeoSOS-FLUS also utilizes an artificial neural network (ANN) to consider multiple effects and establish intricate connections between land-use patterns and various driving forces. A comparison of the FLUS model with other most widely used urban growth models shows that while some models, such as the Cellular Automata-Markov model (CA-Markov), FLUS, Land Use Scenario Dynamics model (LUSD), and Land Change Modeler (LCM) met verification criteria, Conversion of Land Use and its Effects at Small extent model (CLUE-S) and Slope, land-use, excluded layer, urban extent, transportation, hillshade (SLEUTH) had limitations [57].

This paper follows the steps taken by Wang et al. [54]. The process of tracking the effects of land-use policies has three stages, including setting up initial and target scenarios, analyzing land-use policies, and simulating land-use scenarios. The initial land-use scenario consists of land-use type allocations within the boundaries of Helsinki and Espoo municipalities in 2000. Moreover, the target scenario includes the same information as in 2018. With a backcasting technique, a longer timeframe is helpful for capturing gradual trends, assessing the long-term effects of policies, and highlighting the significance of results. The initial and target land cover data enables the comparison of land-use patterns before and after the implementation of the policies.

The results of simulating land-use models can be interpreted to reveal those land-use policies that are responsible for the land-use changes. Since this study focuses on the significant areas at risk of sea floods, the result of this attempt can shed light on how effective land-use policies can help in sea flood risk prevention.

After listing the relevant land-use policies, a value was assigned to each land-use policy (see Table 3). The aim of this stage was to indicate the intensity and direction of land-use policy effects. Policy assignments are used to structure a conversion cost matrix which is the basis of simulation models. Policy assignment Q is dependent on the intensity and direction of the effects. The intensity of land-use policy effects (symbolized as P) is either Direct (D) or Indirect (I) and Strong (S) or Weak (W). The intensities are specified based on these conditions:

$$P = \begin{cases} 9, \ if \ the \ intensity \ is \ Direct \ and \ Strong \\ 5, \ if \ the \ intensity \ is \ Direct \ and \ Weak \ OR \ Indirect \ and \ Strong \\ 1, \ if \ the \ intensity \ is \ Indirect \ and \ Weak \end{cases}$$

The direction of the land-use policy function (symbolized as M) expresses whether the policy contributes to land-use change or prevents the change from happening. Thus, the direction would be positive (+) in the former condition or negative (−) in the latter condition. Policy assignment (Q) is calculated depending on P and M values according to the following equation.

$$Q = PM \tag{1}$$

Corresponding land-use policies with policy assignments are presented in Table 3. Table 3 also presents a short description of land-use policies.

A conversion cost matrix was created to discover a causal relationship between land-use policies and land-use changes. The value (V) of each cell in the conversion cost matrix is intended as:

$$Value = \frac{\sum k \in n_{ij} \, Q_k}{max\{\sum k \in n_{ij} \, Q_k\}} \tag{2}$$

where i represents the type of land-use transferred out; j represents the type of land-use transferred in; n_{ij} denotes a set of policy numbers for converting i-type land-use into j-type land-use; k denotes the policy number, and $1 \leq k \leq 17$; Q_k denotes the corresponding assignment of the policy number.

Table 3. Land-use policy collection and policy assignments.

Policy Name	Policy Action Mechanism Number		Policy Assignment
	Policy Number	Policy Intensity and Direction	
New residential areas are planned in Östersundom for which land elevation projects are needed to increase the height of land to prevent the risk of flooding (2012).	01-	D-S-+	9
Vegetation measures are planned in the coastal areas of Östersundom to reduce the speed of water runoff and the risk of flooding (2012).	02-	D-S-+	9
New developments include residential areas, business areas, green connections, and recreational areas, including recreational islands in Meri-Rastila-Vuosaari (2016).	03-	D-S-+	9
Residential areas and mixed land-uses are planned in Kruunuvuorenranta coastal areas (2016).	04-	D-S-+	9
Housing development in Pihlajisto with respect to natural and cultural values of Viikki old town bay and its surroundings (2015).	05-	I-S-+	5
Finnoonsatama will be a residential area; however, the recreational and conservation values of the area will be safeguarded (2015).	06-	I-S-+	5
Espoo's ecological network (EVN) is developed throughout almost all coastal areas to support biodiversity and sustainable use of cultural environments, ecological connectivity, and recreational networks (2008–2018).	07-	D-W-+	5
Developing a green network along the coastlines and naturally valuable areas, an environmental conservation plan, and supporting recreational uses for the city of Helsinki (2016).	08-	D-W-+	5
According to National land-use guidelines, municipalities must be prepaid for extreme weather events and new constructions must be located outside flood risk zone areas or otherwise ensure flood risk management (2008–2017).	09-	I-S-−	−5
Mapping flood-prone areas started in 2007 and giving instructions for flood preparation started by The European Parliament and The Councils on 23 October 2007.	10-	D-S-+	9
Construction in the shore zones is allowed only according to detailed local plans or a legally binding local master plan. This provision does not apply to (Land-use and Building Act, 2003):building required by agriculture and forestry or fishery	11-	D-S-+	9
building to serve the needs of national defense or frontier control;	12-	D-S-+	9
building required by navigation;	13-	D-S-+	9
building of an outbuilding within the curtilage of an existing residential building;	14-	D-S-+	9
repair of or limited extension of an existing residential building.	15-	D-S-+	9
Flood risks must be considered during planning and construction to ensure that new building developments are not located where they may be damaged by floods (Land-use and Building Act, 2003).	16-	D-S-+	9
Developments along the shores of the Baltic Sea should be located higher than the levels flood waters can be expected to reach once every 200 years on average, plus at least 30 cm to account for wave heights. This means the lowest recommended heights for the bases of buildings would be N60 + 2.60 m in Helsinki (2008).	17-	D-S-+	9

In this formula, $\sum k \in n_{ij}\, Q_k$ refers to the sum of the assigned values of all policies that stimulate or inhibit the conversion of land-use types. Moreover, *max* $\{\sum k \in n_{ij}\, Q_k\}$ is the maximum value of the sum of policy assignments in the table of all land-use conversions. The purpose of this step was to normalize Values to a number between 0 and 1. Since some policies were assigned negative values, it is inevitable to make $\sum k \in nij\ Qk$ negative. Therefore, the following must be applied when calculating V, so this value can take a positive number and represent different levels of land conversion costs at the same time.

$$\begin{aligned} if\ min\{Value\} \geq 0,\ V = Value\,; \\ if\ min\{Value\} < 0,\ V = Value + min\{Value\} \end{aligned} \tag{3}$$

If $V = 0$, land-use change is prohibited, and if $V = 1$, land-use change is encouraged. Other amounts of V between 0 and 1 represent low to high costs of land-use changes. This conversion cost matrix contains the initial parameters for land-use simulations which are carried out by Geo-SOS FLUS software. The spatial variables used in simulations were demographic concentration and proximity, route proximity, Aspect and Slope maps, and distances from the city center and sub-centers of the study region. An ANN-based estimation model was created using these driving forces (see Figure 4). This estimation model demonstrates the probability of occurrence for each land-use type.

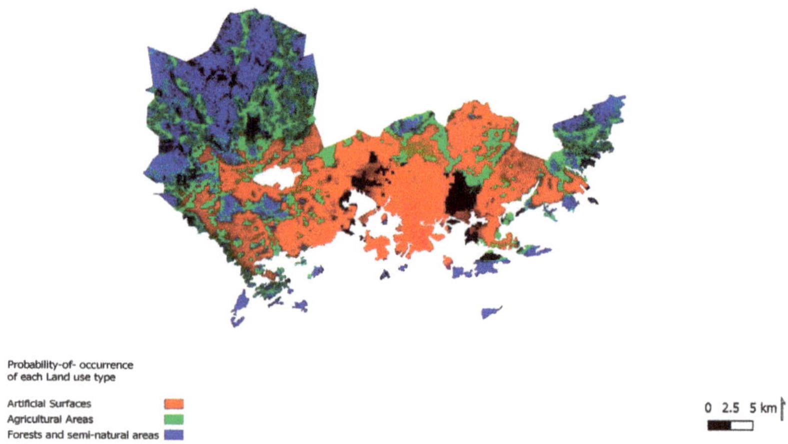

Figure 4. The ANN-based model.

To deliver a satisfying land-use scenario, multiple rounds of simulation were run, and the initial parameters of the conversion cost matrix were adjusted according to deviations in each round. Figure 5 depicts the first and last round of simulations. The adjustment criteria for land-use conversion from type A to type B, for instance, are as follows. Rule (1) states that if the simulated result of land-use type A was more than its actual amount and the simulation result of land-use type B was less than its actual amount, the corresponding value in the conversion cost matrix should be increased. Rule (2) says that if the simulated result of land-use type A is less than the actual amount and the simulation result of land-use type B is more than its actual amount, the corresponding value in the conversion cost matrix must be decreased.

GeoSOS-FLUS software enables users to check the similarity of their simulation to the actual land-use cover. The Kappa statistic tool was used to test the validity of the simulated land-use model, which indicates Kappa ecoefficiency and overall accuracy. The Kappa statistic tool measures the inter-rated reliability of variables, and the results of this test are only acceptable if they are greater than 0.4; otherwise, the model is deemed unreliable.

Figure 5. The first and the last round of land-use simulation.

Broadly, value A, based on expert knowledge, was assigned to each policy, indicating the intensity and direction of a land-use policy. The process continued with making a conversion cost matrix. This matrix represents the degree of change in each land-use conversion. The GeoSOS-FLUS tool considers this matrix as one of the main driving forces of land-use change. The next stage of the process consists of simulating land-use scenarios based on the conversion cost matrix and the demanded land-use areas in 2018. The initial scenario was used as a starting point. The areas of each land-use type in 2018 were inserted into the model as demanded land-use areas. Thus, the iteration rounds continued until the simulated areas reached the required areas for each land-use type. The outcome of the simulation was compared to the target scenario using the Kappa statistic tool. In case of a poor simulation result, the conversion cost matrix was adjusted. The simulation procedure was repeated until overall accuracy and Kappa ecoefficiency scores reached an acceptable rate.

4. Results

The present study aims at measuring the effects of land-use policies in the coastal regions of Helsinki and Espoo municipalities. The target area is further limited to areas at risk of sea flooding; however, land-use policies are applied to much more extensive regions and have diverse effects. Thus, selected land-use policies differ in the extent of their effects. The main results of this study are represented in Tables 4 and 5.

Table 4. The final conversion cost matrix.

Land-Use (from)	Land-Use (to)			
	Artificial Surfaces	**Agricultural Areas**	**Forests and Semi-Natural Areas**	**Wetlands**
Artificial Surfaces	1	0.1	0.22	0.26
Agricultural areas	0.9	1	0.32	0.4
Forests and semi-natural areas	0.99	0.16	1	0
Wetlands	0.2	0.21	0.22	1

The scale 0–1 reflects the degree of conversion. (0) indicates no conversion, and (1) complete conversion. Red indicates undesirable conversions, and green shows desirable conversions.

Table 5. Effective land-use policies in preventing the risk of coastal flooding.

Desirable Land-Use Conversion	Policy Numbers	Policy Names
Artificial Surfaces Agricultural Areas	11	(02)-Vegetation measures are planned in the coastal areas of Östersundom to reduce the speed of water runoff and the risk of flooding (2012) (03)-New developments include residential areas, business areas, green connections, and recreational areas, including recreational islands in Meri-Rastila and Rastila-Vuosaari. (2016) (05)-Housing development in Pihlajisto with respect to natural and cultural values of Viikki old town bay and its surrounding (2015) (07)-Espoo's ecological network (EVN) is developed throughout almost all coastal areas to support biodiversity and sustainable use of cultural environments, ecological connectivity, and recreational networks. (2008–2018) (08)- Developing a green network along the coastlines and naturally valuable areas, an environmental conservation plan, and supporting recreational uses for the city of Helsinki (2016) (09)-According to National land-use guidelines, municipalities must be prepaid for extreme weather events and new constructions must be located outside flood risk zone areas or otherwise ensure flood risk management (2008–2017) (10)-Mapping flood-prone areas started in 2007 and giving instructions for flood preparation started by The European Parliament and The Councils on 23 October 2007. (11)-Construction in the shore zones is allowed only according to local detailed plans or a legally binding local master plan. This provision does not apply to the building required by agriculture and forestry or fishery.
Artificial Surfaces Forests and semi-natural Areas	07, 08, 11	
Artificial Surfaces Wetlands	02, 07, 08, 09, 10	
Agricultural Areas Forests and semi-natural Areas	02, 07, 08, 11	
Agricultural Areas Wetlands	02, 03, 05,07, 08, 09, 10	

The value in each cell corresponds to land-use conversions. These values are influenced by land-use policies and their assignments. To translate these values into policy effectiveness, we need to classify land-use changes into desirable and undesirable groups of change. The desirability of a land-use change is detectable by reviewing the main objectives of land-use policies. Based on the national guidelines and the Land Use and Building Act (132/1999), the following objectives are relevant to the study areas of this study:

- Environmental protection and prevention of environmental hazards.
- Functionality of communities and good building.

To succeed in these objectives, it is better to position urban infrastructures and functional land-use types, such as residential areas and main transportation routes, at a safe distance from sea-flood risk zones. Additionally, natural areas and wetlands are helpful when dealing with natural hazards such as flooding events. Forests and semi-natural areas also support environmental protection aims, and non-motorized transportation systems are encouraged as green connection networks alongside the coastal areas of Helsinki and Espoo. Therefore, the land-use conversions are divided into two groups indicated by green and red colors in the final conversion cost matrix (Table 4). This classification reflects better choices of land-use types for areas at risk of coastal flooding. Green-colored land-use conversions are aimed at reducing coastal flood risk and safeguarding the functionality of communities. Other land-use changes colored in red do not affect or make little contribution to achieving the objectives listed in the Land Use and Building Act (132/1999).

To introduce effective land-use policies, the values in each green cell in Table 4 must be translated into a set of policy numbers. Accordingly, these policies listed in Table 5 are labeled as effective at preventing the risk of coastal floods.

In order to detect the degree of effectiveness, we should pay attention to the degree of conversion in Table 5. Among the desirable land-use conversions, land-use changes from agricultural to wetlands have the greatest degree of conversion. This means the policy combination involved in this change has made a wider impact compared to the policies that are involved in land-use changes from artificial to agricultural areas. On the other hand, among the undesirable land-use changes, the highest degree of change belongs to the land-use conversion from forests and semi-natural areas to artificial surfaces. Among the policy combinations listed in Table 5, policies related to the green network connection in Helsinki and Espoo's ecological network (EVN) are the most frequently used to promote desirable land-use changes. These policies could prevent coastal areas from the risk of coastal flooding, promote non-motorized ways of transport, and preserve environmental values and natural landscapes. Fiscal incentives, however, may be undermined by using the most demanded land in coastal areas to serve public needs instead of putting them into more profit-generating uses in order to generate tax income.

Results indicate that land-use policies applied to the study area were effective in reducing the risk of coastal flooding. The majority of vulnerable land-use types, such as artificial surfaces, have been planned to be outside the flood risk zones. However, land-use changes cannot be separated by the boundaries of flood risk zones. As a result, policies that appear effective in this study might not work for other purposes. Therefore, the overall effectiveness of the land-use policies applied to the study region is satisfying in terms of safeguarding the functionality of communities and preventing the risk of environmental hazards such as coastal flooding.

5. Discussion

This paper offers a framework for evaluating land-use policy effectiveness. Evaluation of policy is a complex task, and land-use planning poses wicked problems as well. More tellingly, urban planning generates wicked problems because it attempts to manage all the issues of urban living and the environment at once, just as what sustainability science is trying to accomplish [58].

Professionals have treated wicked problems differently over the past decades. The suggested mitigations advanced from 'taming', 'handling', and 'tackling', to 'working with', 'living with', and 'embracing' [14,18], showing an increasing acceptance of wicked problems as an integrated part of socio-economic systems [13]. When facing wicked problems, one reasonable way of handling them is to break the problem down into specific aspects. Land-use policies and coastal flooding are the aspects covered by this study.

Policy evaluation is demanded by several legislations to improve decision-making and policy outcomes. However, fewer scholars have attempted to evaluate land-use policies. However, land-use planners can benefit from the literature available on the evaluation of environmental policy instruments and waste management policies [10,47]. In both studies, statistical methods are suggested to measure the effectiveness of policies, while in land-use policy evaluation, location-based information and analysis can provide useful outcomes for planners and decision-makers. For improving the effectiveness of land-use policies, local knowledge is a crucial requirement, otherwise, the policy will fail either in implementation or in delivering the intended outcomes. Another tool used in related research is GIS which has been utilized to track land-use changes and analyze the effects of these changes on various aspects of the urban environment, such as preserving natural features [12].

The implementation of land-use policies has significant legal and administrative implications that shape the spatial and socio-economic dynamics of regions and communities. Understanding the difference between the legal and administrative effects of land-use policies is crucial for effective implementation and enforcement. Legal effects of land-use policies refer to the legally binding rules and regulations that govern land-use and development within a specific area. These rules are enforceable by law and can result in penalties or legal actions if not followed. While detailed plans have explicit legal obligations, master plans play a vital role in shaping land-use policies and decisions, and their adherence is typically expected, even if not strictly legally enforceable. On the other hand, administrative effects of land-use policies refer to the non-binding guidelines, recommendations, and procedures that inform the decision-making process related to land-use and development.

This paper refers to flood risk zones as a sample of local knowledge that enriches the application of land-use policies affecting coastal areas. Using software tools enables us to conduct explanatory research regarding policy effectiveness and actual outcomes caused by land-use policies. Recently, several researchers used GIS and simulation tools to detect land-use policy effects [54,55]. Although software tools can be beneficial, researchers must be aware of their possible limitations. For instance, it is better to compare the simulated areas of each land-use type with reality to ensure the data has not been altered through different processes.

The GeoSOS-FLUS software provides validity tests to compare the simulation model with actual land-use cover. In the case of this study, the model starts with simulating according to the land-use cover in 2000, and the simulation cycle will continue (according to variables set to the model) until areas of each land-use type reach the demanded amounts in 2018. As a result, the areas of each land-use type in the actual land-use map for 2018 and in the final simulation model are equal. However, the allocation of land-use types must be compared to reality. The Kappa statistic tool was used for this purpose. This test registered excellent results for the simulated land-use cover. The overall accuracy is 0.93, and the Kappa coefficient is 0.88, both of which are excellent.

Recognizing the inherent uncertainties in predicting future land-use changes and flood events, this study has taken several steps to ensure transparency and rigor in its approach. Comprehensive information about data sources and assumptions is provided, enabling policymakers to understand the basis of our predictions and the associated limitations. In addition, scenario analysis and sensitivity analysis have been conducted to explore different outcomes and assess the model's robustness. Moreover, this study actively engages policymakers, experts, and stakeholders throughout the research process to collaboratively address challenges and make well-informed decisions. By incorporating these strategies, we aim to facilitate meaningful discussions on policy implications and contribute to effective land-use planning in flood-prone coastal regions.

6. Conclusions

This paper aimed to evaluate the effectiveness of Finnish land-use policies in preventing the risk of coastal floods in the waterfront areas of Helsinki and Espoo. Based on the results of this study, Municipal land-use policies applied to the study area were highly effective in preventing the risk of coastal flooding.

Land-use ordinances are an effective means of starting new norms and values that can maintain their effects over time and scales. Therefore, land-use planning is a beneficial tool to mitigate climate change and promote resilient living environments [45]. Land-use policies are the backbone of the land-use planning system in Finland, and evaluating these policies is mandated by the European Union and other national legislations. This study develops a framework for evaluating the effectiveness of land-use policies in preventing sea flood risks. The target application of this framework is to track the actual effects of land-use policies. Through the developed framework, land-use scenarios are simulated based on specific values, which reflect the effects of a policy set. A multi-policy analysis can be done using this framework to assess the impact of multiple policies on different land-use conversions. The framework owes this functional capability to the GeoSOS-FLUS model, as it produces simulations using ANN-based models and on the basis of CA theory.

The process of evaluating land-use policies begins with setting up initial and target land-use scenarios. The former scenario is based on the land-use cover of the study region in 2000, and the latter represents the same information for the year 2018. There are four classes of land-use (Artificial Surfaces, Agricultural areas, Forests and semi-natural areas, and Wetlands) and twelve possibilities for land-use conversions. Furthermore, a list of land-use policies that are applied to the study region has been collected. However, the boundaries of sea flood risk zones were considered when narrowing down the list of relevant land-use policies.

As demonstrated by simulation results and validation scores, the developed framework outperforms existing methods for evaluating the effects of land-use policies. The process concludes by translating the values in the final simulation scenario into policy sets, establishing a causal relationship with land-use changes. These results specify the effects of land-use policies on changes between 2000 and 2018 in the coastal regions of Helsinki and Espoo. To assess the effectiveness of selected land-use policies in mitigating coastal flood risks, changes are classified into two groups: desirable land-use changes involving conversions from Artificial Surfaces to other types or from Agricultural Areas to other types excluding Artificial Surfaces. Natural areas and wetlands are deemed less vulnerable to flooding. Depending on the framework's purpose, land-use conversions' classification may vary, presenting a valuable functional capability.

The results show that land-use planning performed well in the coastal regions of Helsinki and Espoo. Based on the results, we can conclude that vulnerable land-use types were successfully relocated outside the boundaries of sea flood risk zones. Both the simulation results and the comparison of land-use types in 2000 and 2018 support this claim.

Using a novel framework, this study assesses the effectiveness of land-use policies in mitigating coastal flooding risks in Helsinki and Espoo, Finland. By integrating GIS and simulation tools, the research demonstrates the utility of tracking land-use changes and analyzing policy impacts. This approach pushes conventional policy evaluation boundaries. This paper emphasizes the importance of local knowledge, advocates for a more localized approach to policy development, and highlights the potential of combining policy initiatives and technology to address complex issues in urban planning and land-use management. Future studies may focus on a detailed analysis of land-use types for sustainable development of waterfront areas, focusing on small-scale study areas and expanding the number of land-use classes.

Author Contributions: F.A.F., K.R. and H.P.: Conceptualization; resources; writing—review and editing. F.A.F.: methodology; software; validation; formal analysis; investigation; data curation; writing—original draft preparation; visualization. K.R. and H.P.: supervision; project administration. K.R.: funding acquisition. All authors have read and agreed to the published version of the manuscript.

Funding: This research received no external funding.

Data Availability Statement: This paper utilizes CORINE dataset available at https://www.syke.fi/fi-FI/Avoin_tieto/Paikkatietoaineistot/Ladattavat_paikkatietoaineistot, accessed on 1 March 2023. The original dataset consists of incomparable datasets for the years 2000 and 2018. Therefore, upon researchers' request, an unpublished version of the land-use cover dataset, which was derived from the original one, was provided by the Finnish Environment Institute (SYKE) and used in this research.

Conflicts of Interest: The authors declare no conflict of interest.

References

1. Mäntysalo, R.; Kangasoja, J.K.; Kanninen, V. The paradox of strategic spatial planning: A theoretical outline with a view on Finland. *Plan. Theory Pract.* **2015**, *16*, 169–183. [CrossRef]
2. Rittel, H.W.J.; Webber, M.M. Dilemmas in a general theory of planning. *Policy Sci.* **1973**, *4*, 155–169. [CrossRef]
3. Head, B.W. Wicked Problems in Public Policy. *Public Policy* **2022**, *3*, 101–118. [CrossRef]
4. Duckett, D.; Feliciano, D.; Martin-Ortega, J.; Munoz-Rojas, J. Tackling wicked environmental problems: The discourse and its influence on praxis in Scotland. *Landsc. Urban Plan.* **2016**, *154*, 44–56. [CrossRef]
5. Perry, J. Climate change adaptation in the world's best places: A wicked problem in need of immediate attention. *Landsc. Urban Plan.* **2015**, *133*, 1–11. [CrossRef]
6. The European Parliament and the Council. *Assessment and Management of Flood Risks*; Directive 2007/60/EC. Jun. 2007; European Parlament: Brussels, Belgium, 2007.
7. Yuen, K.W.; Hanh, T.T.; Quynh, V.D.; Switzer, A.D.; Teng, P.; Lee, J.S.H. Interacting effects of land-use change and natural hazards on rice agriculture in the Mekong and Red River deltas in Vietnam. *Nat. Hazards Earth Syst. Sci.* **2021**, *21*, 1473–1493. [CrossRef]
8. Virkki, H.; Kallio, H.; Orenius, O. *Sea Level Rise and Flood Risk Assessment in Itä-Uusimaa*; GTK: Helsinki, Finland, 2006.
9. Juuti, P. Näin Aallot Lyövät Helsingin Rantaan Tulevaisuuden Myrskyissä—Vuonna 2005 Tulva Nosti Veden Kauppatorille, Mutta Paljon Pahempaan on Varauduttava. Available online: https://yle.fi/a/3-12121601 (accessed on 13 August 2023).
10. Mickwitz, P. A Framework for Evaluating Environmental Policy Instruments. *Evaluation* **2003**, *9*, 415–436. [CrossRef]
11. Scott, E.M. Setting, and evaluating the effectiveness of, environmental policy. *Environmetrics* **2007**, *18*, 333–343. [CrossRef]

12. Taylor, J.J.; Brown, D.G.; Larsen, L. Preserving natural features: A GIS-based evaluation of a local open-space ordinance. *Landsc. Urban Plan.* **2007**, *82*, 1–16. [CrossRef]
13. Xiang, W.-N. Working with wicked problems in socio-ecological systems: Awareness, acceptance, and adaptation. *Landsc. Urban Plan.* **2013**, *110*, 1–4. [CrossRef]
14. Batie, S.S. Wicked Problems and Applied Economics. *Am. J. Agric. Econ.* **2008**, *90*, 1176–1191. [CrossRef]
15. Lazarus, R.J. Super Wicked Problems and Climate Change: Restraining the Present to Liberate the Future. 2009. Available online: https://scholarship.law.georgetown.edu/facpub/159 (accessed on 27 June 2023).
16. Enemark, S. *A Cadastral Tale*; Análisis Geográficos; Instituto Geográfico Agustin Codazzi: Bogota, Columbia, 2006; pp. 147–159.
17. Howlett, M.; Perl, A.; Ramesh, M. *Studying Public Policy*; Oxford University Press: Oxford, UK, 2009.
18. Norton, B.G. The Ways of Wickedness: Analyzing Messiness with Messy Tools. *J. Agric. Environ. Ethic.* **2011**, *25*, 447–465. [CrossRef]
19. Adams-Schoen, S. Taming the Super-Wicked Problem of Waterfront Hazard Mitigation Planning: The Role of Municipal Communication Strategies. In *Contemporary Issues in Climate Change Law and Policy: Essays Inspired by the IPCC*; SSRN: Rochester, NY, USA, 2016.
20. Conklin, J. *Dialogue Mapping: Building Shared Understanding of Wicked Problems*; John Wiley & Sons, Inc.: Hoboken, NJ, USA, 2006.
21. Head, B.W. Forty years of wicked problems literature: Forging closer links to policy studies. *Policy Soc.* **2018**, *38*, 180–197. [CrossRef]
22. Ruhl, J.B.; Salzman, J. Climate Change, Dead Zones, and Massive Problems in the Administrative State: A Guide for Whittling Away. *Calif. Law Rev.* **2010**, *98*, 59–120. [CrossRef]
23. Cajot, S.; Peter, M.; Bahu, J.-M.; Koch, A.; Maréchal, F. Energy Planning in the Urban Context: Challenges and Perspectives. *Energy Procedia* **2015**, *78*, 3366–3371. [CrossRef]
24. Alford, J.; Head, B.W. Wicked and less wicked problems: A typology and a contingency framework. *Policy Soc.* **2017**, *36*, 397–413. [CrossRef]
25. Kirschke, S.; Kosow, H. Designing policy mixes for emerging wicked problems: The case of pharmaceutical residues in freshwaters. *J. Environ. Policy Plan.* **2021**, *24*, 486–497. [CrossRef]
26. Rogge, K.S.; Reichardt, K. Policy mixes for sustainability transitions: An extended concept and framework for analysis. *Res. Policy* **2016**, *45*, 1620–1635. [CrossRef]
27. Poutanen, M.; Steffen, H. Land Uplift at Kvarken Archipelago/High Coast UNESCO World Heritage area. *Geophysica* **2015**, *50*, 49–64.
28. Pellikka, H.; Johansson, M.M.; Nordman, M.; Ruosteenoja, K. Probabilistic projections and past trends of sea level rise in Finland. *Nat. Hazards Earth Syst. Sci.* **2023**, *23*, 1613–1630. [CrossRef]
29. Intergovernmental Panel on Climate Change. Ocean, Cryosphere and Sea Level Change. In *Climate Change 2021—The Physical Science Basis*; Cambridge University Press: Cambridge, UK, 2023; pp. 1211–1362. [CrossRef]
30. Pellikka, H.; Laurila, T.K.; Boman, H.; Karjalainen, A.; Björkqvist, J.-V.; Kahma, K.K. Meteotsunami occurrence in the Gulf of Finland over the past century. *Nat. Hazards Earth Syst. Sci.* **2020**, *20*, 2535–2546. [CrossRef]
31. Pellikka, H. *Dark-Blue Horizon: Sea Level Rise and Meteotsunamis on the Finnish Coast*; Finnish Meteorological Institute: Helsinki, Finland, 2020. [CrossRef]
32. Diez-Sierra, J.; Iturbide, M.; Gutiérrez, J.M.; Fernández, J.; Milovac, J.; Cofiño, A.S.; Cimadevilla, E.; Nikulin, G.; Levavasseur, G.; Kjellström, E.; et al. The Worldwide C3S CORDEX Grand Ensemble: A Major Contribution to Assess Regional Climate Change in the IPCC AR6 Atlas. *Bull. Am. Meteorol. Soc.* **2022**, *103*, E2804–E2826. [CrossRef]
33. Hausfather, Z.; Peters, G.P. Emissions—The business as usual story is misleading. *Nature* **2020**, *577*, 618–620. [CrossRef] [PubMed]
34. Pérez-Molina, E.; Sliuzas, R.; Flacke, J.; Jetten, V. Developing a cellular automata model of urban growth to inform spatial policy for flood mitigation: A case study in Kampala, Uganda. *Comput. Environ. Urban Syst.* **2017**, *65*, 53–65. [CrossRef]
35. Crane, R.; Landis, J. Introduction to the Special Issue. *J. Am. Plan. Assoc.* **2010**, *76*, 389–401. [CrossRef]
36. Alexandra, R. Climate Change Adaptation in Metro Vancouver: Examining the Role of Managed Retreat. Master's Thesis, University of Waterloo, Waterloo, ON, Canada, 2018.
37. Biber, E. The Sting of the Long Tail: Climate Change, Backlash and the Problem of Delayed Harm. *SSRN Electron. J.* **2008**, 1292529. [CrossRef]
38. Nicholls, R.J.; Cazenave, A. Sea-Level Rise and Its Impact on Coastal Zones. *Science* **2010**, *328*, 1517–1520. [CrossRef]
39. Rizzardi, K.W. Washington and Lee Journal of Energy, Climate, and the Environment. Article 4 3-2015 Part of the Energy and Utilities Law Commons, Environmental Law Commons, and the Natural Resources Law Commons. Seas, Receding Ethics? Why Real Estate Professionals Should Seek the Moral High Ground. 2015. Available online: https://scholarlycommons.law.wlu.edu/cgi/viewcontent.cgi?article=1093&context=jece (accessed on 27 June 2023).
40. Berkes, F. Implementing ecosystem-based management: Evolution or revolution? *Fish Fish.* **2011**, *13*, 465–476. [CrossRef]
41. Jaana, J.; Kristiina, N.; Paula, T. *Climate-Proof Living Environment-Methodologies, Tools and Practical Recommendations for Climate Change Adaptation in the Kymenlaakso and Uusimaa Regions and the Helsinki Metropolitan Area*; Gaia: London, UK, 2018.
42. Zhai, Y.; Yao, Y.; Guan, Q.; Liang, X.; Li, X.; Pan, Y.; Yue, H.; Yuan, Z.; Zhou, J. Simulating urban land use change by integrating a convolutional neural network with vector-based cellular automata. *Int. J. Geogr. Inf. Sci.* **2020**, *34*, 1475–1499. [CrossRef]
43. Uudenmaanliitto. *General Flood Map of Coastal Areas of Uusimaa*; Finish Government: Helsinki, Finland, 2008.

44. Land Use and Building Act (132/1999). Construction in the Waterfront Area. 1999. Available online: https://www.kuntaliitto.fi/opas-rakennusjarjestyksen-laatimiseen/6-opas-ja-mallimaarayksia/67-rakentaminen-ranta-alueella (accessed on 28 July 2023).
45. City of Helsinki. *Helsinki-Climate-Change-Adaptation-Policies-2019–2025*; Publications of the Central Administration, no. 43, 2019; City of Helsinki Administration: Helsinki, Finland, 2019.
46. Ministry of Agriculture and Forestry. *Finland's National Climate*; Ministry of Agriculture and Forestry: Helsinki, Finland, 2014.
47. Granqvist, K.; Humer, A.; Mäntysalo, R. Tensions in city-regional spatial planning: The challenge of interpreting layered institutional rules. *Reg. Stud.* **2020**, *55*, 844–856. [CrossRef]
48. Puustinen, S.; Mäntysalo, R.; Jarenko, K. The Varying Interpretations of Public Interest: Making Sense of Finnish Urban Planners' Conceptions. *Curr. Urban Stud.* **2017**, *05*, 82–96. [CrossRef]
49. Finish Government. *Government Decision on Finland's National Land Use Guidelines*; Finish Government: Helsinki, Finland, 2017.
50. Mäntysalo, R.; Kangasoja, J.K.; Kanninen, V. *Framing 'Evidence' and Scenario Stories in Strategic Spatial Planning*; Routledge: Abingdon, UK, 2016; Volume 16, pp. 364–377.
51. Carter, T.R. *Assessing the Adaptive Capacity of the Finnish Environment and Society Under a Changing Climate—Summary for Policy Makers*; European Commission: Brussels, Belgium, 2007.
52. Feranec, J.; Jaffrain, G.; Soukup, T.; Hazeu, G. Determining changes and flows in European landscapes 1990–2000 using CORINE land cover data. *Appl. Geogr.* **2010**, *30*, 19–35. [CrossRef]
53. Robinson, J.B. Futures under glass. *Futures* **1990**, *22*, 820–842. [CrossRef]
54. Wang, Y.; Shen, J.; Yan, W.; Chen, C. Backcasting approach with multi-scenario simulation for assessing effects of land use policy using GeoSOS-FLUS software. *MethodsX* **2019**, *6*, 1384–1397. [CrossRef] [PubMed]
55. Brunner, S.H.; Huber, R.; Grêt-Regamey, A. A backcasting approach for matching regional ecosystem services supply and demand. *Environ. Model. Softw.* **2016**, *75*, 439–458. [CrossRef]
56. Lin, J.; Li, X.; Wen, Y.; He, P. Modeling urban land-use changes using a landscape-driven patch-based cellular automaton (LP-CA). *Cities* **2023**, *132*, 103906. [CrossRef]
57. Zhang, Y.; Kwan, M.-P.; Yang, J. A user-friendly assessment of six commonly used urban growth models. *Comput. Environ. Urban Syst.* **2023**, *104*, 102004. [CrossRef]
58. Mancebo, F. Sustainability Science in the Light of Urban Planning. *Chall. Sustain.* **2017**, *5*, 26–34. [CrossRef]

land

Article

Do Land Use and Land Cover Scenarios Support More Integrated Land Use Management?

Roberta Rigo [1,2] and Thomas Houet [1,2,*]

1 CNRS UMR 6554 LETG, University de Rennes 2, Place du Recteur Henri le Moal, 35000 Rennes, France
2 LTSER Zone Atelier Armorique, 35042 Rennes, France
* Correspondence: thomas.houet@univ-rennes2.fr

Abstract: In agricultural landscape management, the conventional top-down approaches that primarily focus on market-led responses struggle to preserve the landscape elements essential for environmental sustainability. To address this deficiency, land use and land cover change (LUCC) scenarios promote an integrated understanding of landscape dynamics and highlight the inconsistency between the compartmentalisation of the public sector ("siloisation") and the necessity for management that reflects the interdependencies of socio-ecological systems. This study investigates the extent to which the creation and dissemination of LUCC scenarios lead to modifications in the values, attitudes, and behaviours of local actors engaged in land management, giving particular emphasis to the role of these scenarios in encouraging integrated management. To accomplish this objective, we interviewed local actors who actively participated in the co-construction of the scenario narratives or learned about the scenarios during dissemination workshops. We then analysed the data via a thematic and lexicometric analysis. The findings highlighted the dual function of these scenarios as a catalyst for pre-existing political will to promote integrated management and as a tool for raising awareness about major environmental challenges. At the group level, the outcomes encompassed aspects such as basing political decisions on the results of scenarios and fostering collaboration between institutions. These outcomes were observed among the actors involved in co-constructing scenarios or those with pre-existing motivations to pursue integrated management initiatives. Additional personal outcomes included an increased awareness of environmental challenges and the consolidation of non-formalised knowledge. We argue that combining co-construction and dissemination enhances the outcomes of scenarios considerably.

Keywords: co-construction; dissemination; impact pathways; integrated management; LUCC scenarios; outcomes of scenarios

1. Introduction

We live in an era in which scientific knowledge emphasising the urgent need for socio-economic paradigm shifts is broadly available [1–3]. Within the scientific community [4–6] and public discourse [7], terms such as "sustainability" and "low-carbon transition" are gaining increasing prominence. However, despite scientists highlighting the importance of valuing and protecting nature for its intrinsic value [8], capitalist systems primarily rely on a neoclassical approach [9,10] that disregards scientific knowledge and citizens' mobilisations against this approach. In the European Union's (EU) rural context, the environmental degradation caused by intensive agriculture and the limited ecological ambition of the recently approved Common Agricultural Policy (CAP) reflect the reticence to change the traditional paradigm [11,12]. Notably, the CAP remains the agricultural economy's fundamental pillar, even if it struggles to promote environmental sustainability and fails to conserve the landscape elements essential for preserving biodiversity, water, and soil [13].

In northwestern France, where this study was conducted, the bocage (hedgerow network) is iconic. Not only does it possess significant cultural and economic value, but it also exerts a considerable influence on biodiversity, field microclimates [14,15], soil–water transfers [16], and groundwater quality [17]. Nevertheless, despite laws and local initiatives promoting bocage maintenance, the length of hedgerows in Brittany continues to decline [18,19]. This reduction in hedgerow length benefits farmers who take advantage of selling wood and a simpler landscape mosaic that facilitates crop production by facilitating labour. This local example shows the failure of the current land management practices and specific top-down measures in translating scientific knowledge into practice. Instead of relying solely on market-led responses that target specific sectors, researchers have highlighted the urgent need to address sustainability issues with an integrated approach that reflects the interdependencies of the socio-ecological system [20,21].

Policy integration refers to "the collaboration of actors from two or more policy domains to integrate aims and concerns derived from one policy domain into another" [22], (p. 553). Initially applied to marine policy [23], integration is now gaining recognition in environmental management, which frequently confronts complex, cross-sectoral challenges [24]. However, a persistent barrier to achieving integrated policies lies in the division of public administration into "silos", composed of relatively stable groups of actors responsible for sub-sectoral policies, each with a specific perception of environmental issues [25]. This "pillarisation" or "siloisation" of the public sector complicates interagency collaboration and hinders a holistic approach to addressing interconnected problems [24,25]. In a framework proposed by Metz and Glaus [26], the authors highlighted how policy integration can be promoted not only by changing the legal framework, but also by favouring the involvement of actors in multiple policy-making sectors. The capacity of public institutions to deliver integrated policies has been examined at various spatial levels (e.g., see Candel et al. [27] for the EU level, Molenveld et al. [28] for the national level, and Metz and Glaus [26] for the catchment level) and across different environmental issues. Within the context of water management, Morgue et al. [29] showed that the little integration between water and land-use issues is one of the most significant challenges in achieving sustainable management.

While there are certain limitations to its application [30], developing sustainability science has been acknowledged as a promising approach to promoting policy integration and enhancing decision-making capacity [31,32]. Indeed, through the involvement of non-scientists in the generation of multidisciplinary, action-oriented research, sustainability science can create bridges between different actors, services, and projects [33].

LUCC scenarios have been recognised as valuable tools for achieving these objectives in land management by offering a spatially explicit representation of how land use and land cover could evolve over a pre-defined timeframe in response to socio-economic factors (such as demographic pressure and economic incentives favouring specific crops) and environmental variables (including climate change). They often combine narratives describing the drivers of LUCCs with mathematical models that translate these narratives into maps and vice versa [34].

This study aims to examine the outcomes, defined as changes in the values, attitudes, and behaviours of local actors [35], arising from creating and disseminating some catchment-level LUCC scenarios, including their role in promoting integrated management.

To meet the goal of this paper, we draw upon a range of previous research. Notably, Akkerman and Bakker [36] investigated the process of learning at boundaries and identified the fundamental mechanisms that facilitate this learning. These mechanisms include identification, which involves acquiring knowledge about the practices of others; coordination, which entails establishing stable and ongoing exchanges; reflection, which involves developing broader or alternative perspectives on practices; and transformation, which encompasses the co-construction of new practices. Besides investigations on learning in the presence of boundaries, a wide range of potential research outcomes has been studied (e.g., [37–39]) and categorised across various levels. For instance, Hamilton et al. [40]

identified project-level (i.e., research credibility, saliency, and legitimacy), individual-level (e.g., increased system understanding), group-level (e.g., the direct use of research results in technical documents, increased collaboration between actors), and system-level outcomes (e.g., a systematic change in attitudes). Moreover, potential outcomes specific to scenarios have also been described. Notably, scenarios have been recognised as a potential support to enhance the comprehension of a complex system by coupling multiple pieces of information into a coherent and systemic perspective of how landscapes may transform [41,42]. In parallel, they can widen the range of elements, drivers of change, and potential futures under consideration [42]. Additionally, scenarios facilitate communication and collaboration among participants with diverse perspectives, enabling them to address problem framing collectively [41,43]. This collective effort can promote an integrated management that accounts for interactions between landscape elements and socio-economic transformations [44], thereby overcoming the "siloisation" of the public sector. The occurrence of these various outcomes can be subject to different time lags, and the process employed to generate and communicate the research assumes a central role in shaping these outcomes. Muhonen et al. [45] summarised how the interactions between scientists and other actors can take various forms, resulting in different impact pathways. They identified thirteen pathways through which research can have societal impacts, ranging from the linear "classical pipeline pathway", where scientific knowledge permeates society through "knowledge creep", to more complex models of interaction. Although the impact pathways and societal outcomes of this research have been relatively well-documented, scientists may lack a comprehensive understanding of the land management system, including an overall vision of its local actors and the relations among them [46]. This limited understanding of social dynamics can lead to ineffective interaction strategies and hinder potential outcomes. In addition, establishing a coherent spatial extent of scenarios is essential for promoting desirable outcomes. Indeed, developing scenarios that align with the operating boundaries of local actors can foster a rise in these outcomes. Nevertheless, a trade-off exists between spatial resolution and spatial extent [47], i.e., if the goal is to highlight the interactions between landscape elements and socio-economic changes (such as the development of different agricultural systems or changes in population density), a high spatial resolution, down to the field level, is needed. However, scenarios with a higher spatial resolution often cover smaller areas, which may not correspond to the operating boundaries of certain land management actors. For instance, water and land managers are often responsible for separate administrative sectors with jurisdiction over different yet overlapping areas [48]. Therefore, evaluating how a coherent spatial extent between scenarios and operating boundaries influences these outcomes is essential to enhancing the outcomes of scenarios. In summary, when evaluating outcomes, it is crucial to consider the mechanisms of information exchange between researchers and local actors, the timing of the evaluation, and the spatial extent of scenarios as factors that play significant roles in shaping the outcomes of scenarios.

In this study, we examine the outcomes of the LUCC scenarios published by Houet et al. [19]. These scenarios were co-constructed with water and urban land managers and disseminated to a broader audience through a series of workshops. During these workshops, we conducted a first short-term assessment to evaluate the usability of the scenarios (for detailed information and results, refer to [49]). Then, between 7 to 11 months after the workshops, we interviewed some local actors who participated in the co-construction process or the dissemination workshops to identify personal and group outcomes in the long-term. This article focuses on a long-term assessment of outcomes.

2. Context and Materials

The assessed scenarios were developed as part of the EU INTERREG project ALICE[1] and focus on evaluating the impacts of agricultural and urban changes on water quality, quantity, and biodiversity (through the lens of landscape connectivity). The scenarios cover the Couesnon River catchment in northwestern France, which spans an area of 1130 km^2 [19], and consist of contrasting narratives illustrated by LULC maps.

Agricultural development could evolve in three directions: the "cerealisation", which favours producing cereals as biomass or biofuel; the "dairy intensification", which favours the intensification of local dairy/livestock production; and the "greening" direction, in which the CAP promotes dairy and livestock production based on grassland forage. Territorial development respects local urban management plans such as the BGINs policy, a local adaptation of the EU's Biodiversity Strategy that serves as a driver for changes in land use and land cover by limiting the possible expansion of urban areas. Territorial development is also driven by the social demand from inhabitants for certain types and locations of housing. Three different evolutions are proposed for this territorial development: in the first ("business-as-usual"), only the main BGINs are protected from urban sprawl, and the objective of zero urbanisation is not attained in 2040; in the second ("energy transition"), the urban sprawl is controlled, concentrated in the main urban areas, and the zero-urbanisation objective is met by 2045, strongly preserving BGINs; and in the third ("ecological citizen"), the zero urbanisation objective is not met. However, BGINs are strictly preserved, and local authorities favour ecological preservation and restoration.

Combining these 3 × 3 assumptions creates eight internally consistent scenarios (one combination is inconsistent). The five most contrasting scenarios (bold in Table 1) were selected to explore the widest diversity of possible future landscape changes.

Table 1. Combination of territorial and agricultural development assumptions. The five most contrasting scenarios (in bold) were selected to explore the widest diversity of possible futures. Source: [19].

		Territorial Development		
		Business-as-Usual	**Energy Transition**	**Ecological Citizen**
Agricultural development	Cerealzation	**DESERT OF CEREALS**	**ENERGY PERFORMANCE**	CONFLICTS
	Dairy intensification	**BUSINESS AS USUAL**	DIVERSIFIED BIOMASS	**DOUBLE PERFORMANCE**
	Greening	Ø	OPTIMISED BGINs	**GREEN ATTRACTIVENESS**

These assumptions were presented to, refined, and completed with stakeholders during a first participatory meeting. A second participatory meeting helped to specify them by considering local characteristics through narratives and participatory mapping. The researchers then refined and quantitatively illustrated the narratives and calculated the impacts of the simulated LUCCs on biodiversity and water resources.

The project was initiated by scientists who collaborated with water and urban land managers to develop the scenarios. The narratives were co-constructed with a panel of local actors and used as inputs for an LUCC model to simulate future landscape changes [50,51]. Then, the results of the scenarios were disseminated to a broad audience through 13 workshops (see Supplementary Material S1 and S2 and [49] for more details). During each workshop, the method and results of the scenarios were presented to an audience with the same professional activity. Following the Muhonen et al. [45] framework, we followed two pathways to create the research outcomes: the "collaboration pathway" for local actors who co-constructed scenarios and the "interactive dissemination pathway" for local actors who participated in the dissemination workshops. During the workshops, the participants engaged in discussions about the results of the scenarios and the implications they had for their respective professional activities. To assess the usability of the scenarios, an interactive polling system was used and the results indicated that local actors generally considered the scenarios to be legitimate, salient, and credible. Moreover, the scenarios were effective in increasing the awareness of future land management problems and understanding the interdependence of landscape elements [49]. The survey conducted during the workshops revealed that the respondents had high confidence in the research results, and this confidence did not differ significantly among the professional activity groups nor as a function of familiarity with the study area. In addition, representatives considered the projected water quality and quantity trends to be more relevant to their professional activity compared to the biodiversity trends, while technicians and members of civil society expressed no significant difference.

The scenarios envisioned LUCCs in the Couesnon catchment (Figure 1). The interest lies in the fact that farming activities have resulted in significant water quality issues, making residents particularly sensitive to water management tensions [52,53]. Additionally, water quantity tensions caused by climate change have recently emerged as an environmental problem.

Figure 1. The extent of the scenarios covers the SAGE Couesnon (light blue). The intermunicipalities of Couesnon Marche de Bretagne (orange) and the Fougères metropolitan area (green) share the same urbanisation policy (i.e., SCOT Pays de Fougères). The yellow dots represent the city of Fougères. The sage Couesnon covers 1130 km^2.

The Couesnon catchment represents well the multiscale decision system while maintaining a manageable level of complexity, and scenarios were developed at a consistent extent for local water managers' structure, the "SAGE[2] Couesnon", an administrative organisation in charge of managing water resources sustainably. Regarding urban land management, the intermunicipality level serves as its main organisational level. Intermunicipalities are groups of municipalities that jointly manage local public services and promote sustainable development in urban and rural areas. The management of urbanisation within the catchment is often carried out through a "territorial coherence scheme" (SCOT[3]), which involves multiple intermunicipalities jointly coordinating a unique urban development plan. In the Couesnon catchment, there are nine intermunicipalities in total, with two of them covering a significant portion of the territory and sharing the same SCOT (Figure 1).

3. Methods

3.1. *Interviews*

To ensure a comprehensive understanding of the outcomes of the scenarios, we selected one to two interviewees from each workshop for an in-depth interview. The selection criteria aimed to include a diverse range of perspectives and roles within land management and focused on actors who had solid interpersonal connections with the other participants and a high visibility in the initiatives and actions taken by their organisation or group. The workshop sessions involved key actors from water management, agriculture, and urban planning sectors including farmers, technicians, and elected representatives at different organisational levels (see Supplementary Material S2 for details). To limit acquiescence bias, the interviews were conducted by a researcher who participated in the dissemination of the scenarios, but not in their construction.

The interviews were conducted 7–11 months after the workshops. Indeed, we considered that this time lag allowed for group effects to emerge while ensuring that the interviewees could recall the workshop supports and discussions. All the interviewees gave orally informed consent to participate in the study.

In total, we conducted 20 interviews that lasted 27–88 min (mean: 52 min). The objective of these interviews was to gain insights into the network of local actors involved in land management at various organisational levels and to understand the outcomes of the LUCC scenarios for these actors. The interviews were structured into three parts: (1) In the first part, a semi-structured interview approach was employed to understand the interviewees' professional activity, their relations with other land management actors, and the main benefits they derived from participating in the workshop. If necessary, a follow-up question about the salient points of the workshop was asked. (2) The second part of the interview involved a reactivation technique, where the interviewees were encouraged to revisit and further discuss the workshop situation with the aid of images. The aim was to gain insights into additional outcomes of the scenarios. To facilitate this discussion, a booklet summarising the methodology and results of the scenarios was provided, along with the interactive poll responses collected during the workshops. (3) The final part of the interview focused on framing how the interviewee had experienced the workshop and what the interviewee had considered to be the outcomes of the scenarios. This allowed for a detailed exploration of specific outcomes and ensured alignment between the interviewer and interviewee's understanding of the situation.

3.2. Data Analysis: Thematic and Lexicometric Analysis

The interviews were transcribed manually using Sonal software [54] and examined using a thematic and lexicometric analysis. The former was conducted on the entire corpus, allowing for a more refined understanding of the diagram depicting the land management actors, their interconnections, and the different organisational levels at which they operated. The lexicometric analysis, on the other hand, was performed on a sub-corpus using the Iramuteq[4] open-access software. The sub-corpus consisted of interview excerpts specifically related to the main outcomes of the workshops. Each interview extract was coded as a text associated with relevant variables, such as the interviewee code, the professional activity, and the correspondence between the interviewee's working scale and the spatial extent of the scenarios. To prepare the sub-corpus for this analysis, all the interview excerpts were compiled into a single file and acronyms were standardised. Additionally, compound words were combined into single units following the procedure recommended by Carmago and Justo [55]. The sub-corpus was then automatically lemmatised, with words categorised into active forms (i.e., adjectives, adverbs, nouns, and verbs) used for the analysis and supplementary forms excluded from the analysis.

Following the preparation of the corpus, we employed the ALCESTE method [56] to perform a descending hierarchical classification (DHC). This method sorted the text into classes that had highly similar vocabulary within their classes but different vocabulary between them. To quantify the associative strength between the words and their respective class, Iramuteq used the χ^2 test. Only words with an χ^2 of $\geq$3.84 (indicating a probability of 0.05 of being randomly associated with the class) were assigned to a given class. The resulting classes and their relationships were visualised through a dendrogram and a factorial correspondence analysis (FCA) resulting from the DHC [55]. The FCA plotted the most statistically significant words for each class on a two-dimensional plane. Once the classes were defined, we qualitatively analysed the extracts that contained the classified words.

Interview citations were translated literally from the original language (French) to English. Figure 2 represents the overall workflow followed to develop the Couesnon scenarios and assess their outcomes.

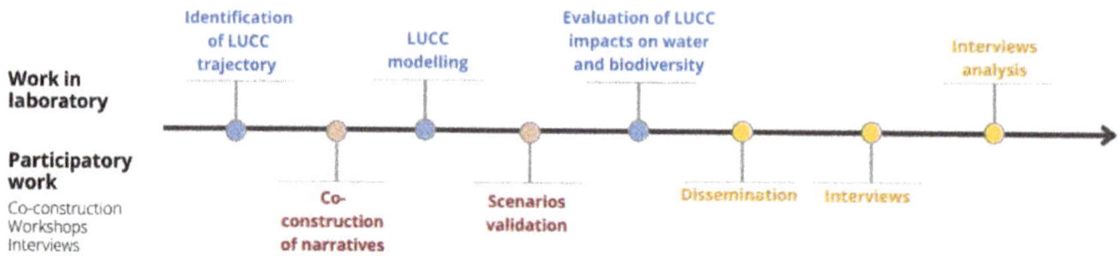

Figure 2. Overview of the method used to develop Couesnon scenarios (combining studies carried out by researchers only in laboratory—in blue—and steps involving local actors—in red) and assess their outcomes (in yellow). This study concerns results related to the steps marked in yellow.

4. Results

4.1. The Land-Management System

Figure 3 illustrates the influential documents, financial mechanisms, and local actors involved in the land management of the Couesnon catchment, including its water management, agriculture, and urban planning (additional information in Supplementary Material S3). The connections between levels and sectors were characterised mainly by administrative, financial, and technical aspects. Administrative connections involved the integration of documents into lower-level directives and legal documents. Financial connections encompassed funding projects and providing incentives. Technical connections provided support in defining policies and administrative documents. The actors were classified based on their sector (i.e., water management, agriculture, or urban planning), their organisation level (i.e., communal, sub-regional, catchment, regional, national, or Europe), and their type (i.e., administration, political, technical, or civil).

Water management primarily occurs at the catchment level, as reflected in the scenarios' spatial extent. The Local Water Commission (CLE[5]) and the Syndicat of the Couesnon catchment are, respectively, in charge of approving and implementing the Water Development and Management Plan (SAGE). The SAGE is connected to urban planning through legal documents and human interactions involving technical support.

On the other hand, the integration between water policies and agriculture relies on farmers who apply for funding mechanisms. These mechanisms primarily come from the EU and are supplemented by regional funding and sustainable farming projects, with the Breizh Bocage and Terre de Source[6] projects being prominent examples. At the local level, the development of an agricultural system that benefits water quality and quantity depends on the availability and willingness of farmers. In the current administrative system, the regional level appears to be effective for promoting integrated water and agricultural policies, as it encompasses political and administrative organizations that cut across different sectors.

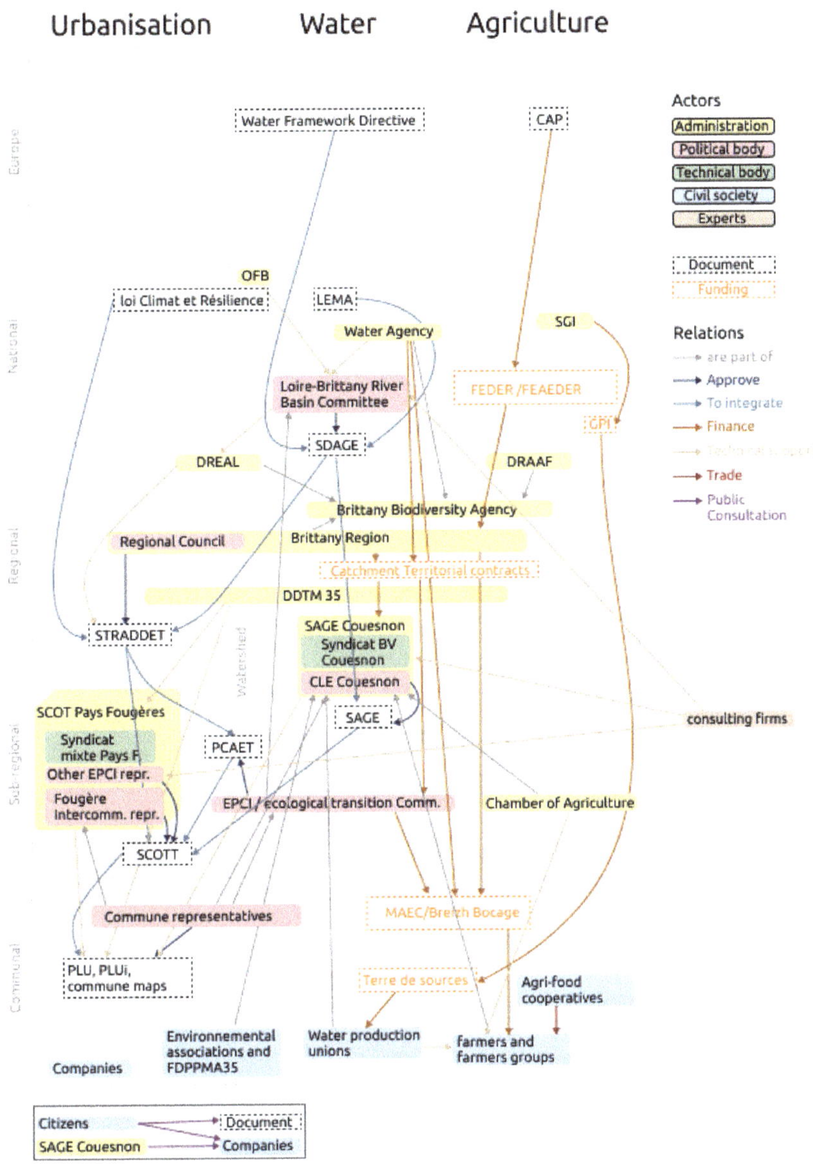

Figure 3. Actors involved in managing water in the Couesnon catchment and their relations with urban planning and agriculture actors. The diagram is not exhaustive and does not represent elements of agriculture or urban planning that are not related to water management.

4.2. Lexicometric Analysis

In order to gain insights into how the interviewees perceived the scenarios, we applied a lexicometric analysis (DHC) to the semi-structured section of the interviews, which addressed the primary outcomes of the workshops. The analysed corpus consisted of 6424 instances, including 486 hapaxes (7.6%), with an average of 338 occurrences per text (Table 2 reports the most recurrent words and their frequency).

Table 2. Most recurrent words of the semi-structured section of the interviews and their frequency.

Words	Frequency	Words	Frequency
Go	42	Small	14
Scenario	38	Couesnon	13
Questioning	30	Feel, tool, interesting	12
Water, thing	25	Project, become aware	11
See, territory	23	Remember, word	10
Level	20	Take, talk, moment, cereal, arrive	9
Representative, find, think	18	Return, put, river catchment	8
CLE	17	Energy, true, work, tendency, use, Scot, look at, mark, culture, covid, change, climate change, Brittany, opinion, agriculture	7
Allow, people	16		

Overall, 747 active forms and 98.4% of the text segments were included in the analysis, which is considered to be an effective use of elementary context units [55]. The application of DHC to all the interviews revealed that the discourse was organised around four principal semantic blocks or classes, as depicted in Figure 4.

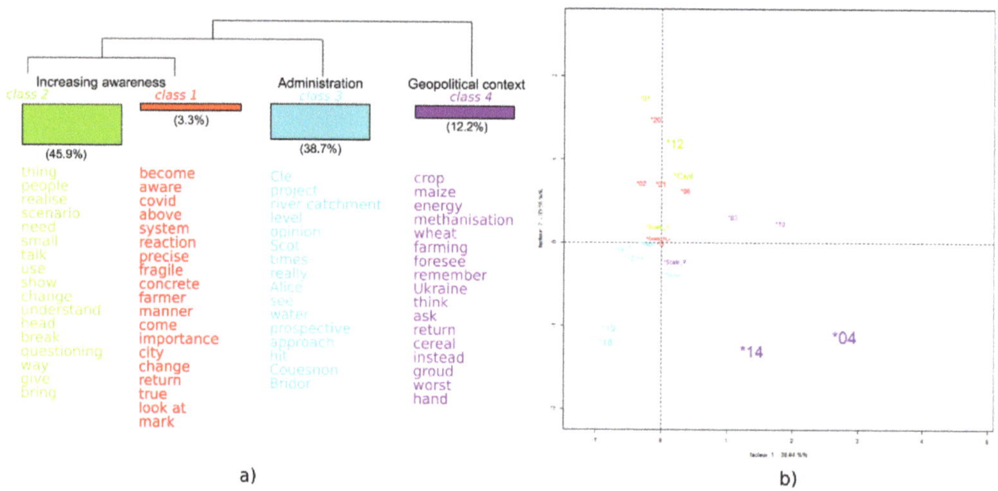

Figure 4. (**a**) Dendrogram of the four lexical classes obtained from descending hierarchical classification (DHC) of the active words, with the percentage of all classified words in each class, and (**b**) factorial correspondence analysis (FCA) of variables in each of the lexical classes obtained in the DHC. Variables are the interviewee code, the correspondence between the area covered by scenarios, and interviewee's organisation level and interviewees' professional activity. The spatial scale of the area covered by scenarios was considered to be completely consistent with that of the interviewee's organisation level (Scale_Y) when the latter was the Couesnon catchment (or entirely covered by it), not consistent (Scale_N) when the latter was another catchment or lay entirely outside the Couesnon catchment, or partially consistent (Scale_P) when the latter partially overlapped the Couesnon catchment (e.g., SCOT Pays de Fougères).

The first (the most significant) and second classes contained words associated with rising awareness. The third class focused on the practical application of scenarios within an administrative context, while the fourth class encompassed words related to the geopolitical context.

The technicians and elected representatives tended to employ more terms linked to instrumental applications, whereas the farmers and residents (coded as "civils") considered scenarios as a means to increasing awareness. When examining the scenario outcomes, most interviewees mentioned words associated with agriculture and water, while words related to urban planning were notably absent[7].

During this phase of the interviews, there was minimal mention of the technical aspects of the scenarios, including their spatial extent, resolution, underlying assumptions, or the specific protocol employed during their development.

Conversely, analysing the contextual use of the words within the fourth class, it became evident how the interviewees connected one of the scenarios, which described an increase in cereal production and increased cereal planting, as a consequence of the war in Ukraine and the subsequent escalation in cereal prices.

Through a qualitative analysis of the entire corpus, we observed that the interviewees discussed how the current geopolitical context disrupts the existing status quo, tilting the world toward one of the scenarios. Although we cannot assert a cause–effect relationship between participating in the workshops and an increased ability to identify connections among landscape elements, the interviewees demonstrated a notable level of transversality when discussing the current geopolitical context. For instance, the interviewees who mentioned the increase in cereal production also talked more broadly about the homogenisation of the landscape and its effects on water quality and the economy:

> *Three things struck me: the apocalyptic scenario [producing only] wheat, the fact that Brittany could get out of this dairy-based economy, and the fact that the most environmentally-friendly and alternative scenario was the least applicable. For me, it was a sort of: 'But why will this scenario, which is the most local, the one that respects the environment and people, not be the one ultimately implemented?'. And then we always come back to economic questions. And I wonder today, with the war in Ukraine, if we are not moving towards [producing] more cereals here, too? And thus, the whole landscape changes.*
>
> *(Interviewee 12, Resident)*

> *It allows us to ask questions from a land-use viewpoint, but also consider the energy aspect. What does it mean if a farmer produces energy, and what are the potential consequences on our soil?*
>
> *(Interviewee 13, Resident)*

4.3. *Identification of Outcomes*

The lexicometric analysis showed that the interviewees considered scenarios as a tool for increasing awareness and as a support for administrative procedures. The thematic analysis identified more specific outcomes of the workshop for each interviewee (Table 3). Following the framework of Hamilton et al. [40], we classified the outcomes as either individual or group.

4.3.1. Individual Outcomes

The most commonly mentioned outcomes of the scenarios were individual. These included an increased awareness of water tensions, a consolidated political viewpoint, and the acquisition of new technical knowledge or factual information. The interviewees expressed an increased awareness of the potential water and biodiversity challenges the region might encounter. They emphasised the significance of implementing appropriate measures to address these issues and mitigate their impact.

> *What will happen in 20 or 30 years if the area of maize increases by 30% to 40% as planned? That shocked me.*
>
> *(Interviewee 04—Elected representative)*

> *Above all, the awareness of the fragility and the importance of reacting... In the scenarios, you showed well that today's choices will have an effect on 2050.*
>
> *(Interviewee 15—Representative)*

Table 3. The presence (green cell) or absence (white cell) of outcomes for all interviewees identified by thematic analysis. Interviewees participated in scenario co-construction (CC) or in scenario dissemination workshops (W).

				Outcomes					
				Personal Level			Group Level		
Int. Code	Role	Participation	Consistent Spatial Scale?	Rising Awareness	Consolidate Viewpoints	Learning/Assessing Current Situation	Collaboration between Actors	Base for Future Work	Decision Making/ Influence on Political Will
03	Farmer	W	Y						
07	Farmer	W	P						
13	Resident	W	N						
12	Resident	W	Y						
20	Representative	W	Y						
04	Representative	W	Y						
05	Representative	W	Y						
15	Env. Representative	W	Y						
02	Env. Representative	W	Y						
01	Env. Representative	W	P						
06	Water Technician	W	N						
17	Water Technician	W	N						
18	Water Technician	W	N						
08	Water Technician	CC	Y						
14	Water Technician	CC	P						
11	Agric. Technician	W	Y						
16	Env. Technician	W	P						
19	Env. Technician	W	P						
09	Env. Technician	W	N						
10	Env. Technician	W	N						

According to the survey submitted during the workshops, the participants stated that the scenarios enhanced their understanding of territorial dynamics rather than the challenges specifically faced by the territory. Nevertheless, in the long term, the interviewees predominantly mentioned an augmented awareness of future water-related issues. Some interviewees also highlighted that the workshops served to consolidate their pre-existing, non-formalised awareness:

> *This study left a mark on me, and the fact that it reinforced my thoughts gives it strength. Because we tell ourselves: 'What I feel is confirmed by a serious study', and thus it allows us to be more confident in the debates that can take place elsewhere.*
>
> *(Interviewee 05—Representative)*

In the semi-structured interviews, the interviewees cited the specific details and results of the scenarios, indicating that they acquired specific concepts:

> *One-third of the Couesnon's flow taken by the city of Rennes... Some key figures like that. We were not aware of this. [...] If I recall, [the quantity of water] also shocked us. The volume did not change much, but it was the distribution. The water did not arrive at the right time for agriculture. And this is catastrophic.*
>
> *(Interviewee 03—Farmer)*

4.3.2. Group Outcomes

The outcomes reported by most interviewees were primarily at the individual level. Nevertheless, some group outcomes were also mentioned. According to the short-term survey, representatives and technicians were more inclined to use the results of the scenarios compared to members of civil society. Additionally, water resource data were perceived as more likely to be used than biodiversity data. The long-term interviews aligned with the survey results.

During the workshops, technicians and representatives acknowledged the effectiveness of multidisciplinary scenarios as a tool for promoting integrated management. In the interview conducted several months after the workshops, technicians of the Brittany Region and the SAGE Couesnon confirmed this observation, recognising an increased level of transversality in their activities. First, a technician of SAGE Couesnon recognised closer collaboration between the SAGE and SCOT, two entities with distinct areas of expertise (water management and land management, respectively):

> *There was interesting joint work between SCOT and SAGE when we worked together on the ALICE project. [...] The same list of people worked on their workshops. [...] Because of this exchange that was created, he [a technician from the SCOT] made me give several presentations to the SCOT environment commission about water-quantity problems. He saw that we were talking about [urbanisation] differently with the ALICE project, and he said to himself: 'This message via the SAGE and the ALICE project [...] allows representatives to see that tomorrow they will face water-quantity problems.'*
>
> *(Interviewee 08—Water technician)*

> *We wouldn't have participated if only BGINs had been considered; we wouldn't have been legitimate. That's also why [...] I said that I couldn't see myself working on this without the SCOT, which works on the land-use-planning part.*
>
> *(Interviewee 08—Water technician)*

The collaboration between SAGE and SCOT was strengthened throughout the entire co-construction process of the scenarios, enabling both organisations to solidify their exchanges and cooperation.

Another example comes from the regional administration, which is formed by specific departments and services in charge of agriculture, water management, and urban planning (see S3 for more details). In this case, technicians operating at the regional level participated only in a scenario dissemination workshop.

> *It was good for us to have this exchange between the Region's services because sometimes we have different mind-sets. [The workshop] allowed a real exchange in transversality, regardless of the position of the services. [Before the workshops], a small group already existed, but it was not easy to lead. It was even a bit stuck. Sometimes there were misunderstandings, and it was difficult reach agreements. I think that [the workshop] was the starting point for moving forward. In the weeks that followed, we succeeded in making a common proposal to coordinate our policies on the bocage, water, and BGINs. It's not yet the end of the road, but I think that the work between technical services advanced a little afterwards. [...] For the moment, we don't have political approval. There are a lot of steps to go through. But it has made us think. That's for sure.*
>
> *(Interviewee 19—Environmental technician)*

It is worth mentioning that it was the technicians of the Brittany region who initiated the contact to organise the workshop, indicating a pre-existing motivation to enhance the cross-disciplinary nature of their policies. In this context, scenarios must be considered as tools that accelerate a pre-existing dynamic. The disparity in the spatial extent between the scenarios and their organisational level was not a limitation.

Furthermore, the scenarios directly influenced a decision-making process of the CLE, which cited the results of the scenarios in a notice regarding the potential establishment of the Bridor agrifood company in the territory.

> *For the CLE, there was a before and after [the ALICE project]. We saw it in the decisions that followed. [...] For Bridor, this was very clear. [...] Before ALICE, the question of the influence of today's decisions on long-term projections was not necessarily taken into account. [...] When the day for the notice[8] arrived, we did not oppose the establishment, because it is not our job, but we strongly warned that there was a limit that must not be reached. In fact, it was necessary to avoid situations with too much water consumption. [...] With the Bridor case, for all the representatives of the CLE it was like a little thing that happened, like saying: 'Ouch, we have a problem there! It's our water. We already don't have much of it. We have to think about our next decisions.' [...] There was a real shift, and this was felt in the relations between the territory of the Couesnon catchment through the CLE and Rennes. For the first time, we dared to say: be careful. You have to leave us some water. [...] It is this 'be gentle' that the CLE had never dared to say. We said it.*
>
> *(Interviewee 15—Representative)*

In this case, the members of the CLE referred to the scenarios as a scientific source, specifically highlighting the significant decrease in water availability in the catchment area by 2050. Based on this information, the CLE requested the agrifood company to provide a long-term estimation of its water consumption as a condition for obtaining permission to establish itself in the territory. Although this notice represents a single decision rather than a systemic change, the interviewed member of the CLE also emphasised the need for a transformation in the decision-making system. Indeed, the interviewee expressed the need to conduct a water resource availability analysis preceding an economic one, particularly when providing opinions on the establishment of industries.

4.3.3. Other Outcomes

In the months following the workshops, the main scientific and administrative partners were invited to the annual conference of the Brittany Biodiversity Agency. During this event, they presented the Couesnon scenarios as an example of a research project designed to support decision making. The research team and SAGE Couesnon also prolonged their collaboration to further refine the quantitative results of the scenarios and include them in the development of agricultural water use scenarios. In addition, the ALICE scenarios were referenced in the working version of the roadmap for the BGIN management of the Regional Directorate for the Environment, Planning and Housing (DREAL[9]), as an as an illustrative example of how research can serve as both an awareness-raising tool and a resource for decision-making support.

5. Discussion

5.1. The Impacts of Scenarios on Integrated Management

Although the local actors generally acknowledged the need for more integrated management, joint initiatives among land management services were not the norm in our case study. Legal obligations ensure a basic integration between water management and urban planning, primarily at the local level (i.e., catchment and intermunicipality). However, this integration between water management and agriculture remains limited, reflecting the absence of significant mechanisms at the local/catchment level and the dominance of EU and regional policies [19].

In our study, the interviewees acknowledged the valuable role of scenarios in strengthening the collaboration between communities of practice, such as between the SAGE and SCOT, and between the different services of the Brittany region. Indeed, even though these services belong to the same organisation, they have different habits, missions, and tools. During the workshops that involved representatives from these organisations, the notion of a "boundary" was most explicit (i.e., the interviewees referred to different areas of expertise, mindsets, and tools). However, a shared vocabulary was already present in both workshops and the scenarios did not introduce entirely new knowledge. Drawing on the classification proposed by Akkerman and Bakker [36], learning occurred through a reflection process, in which technicians from different services or organisations simultaneously examined their practices and reflected on ways to coordinate them.

During the interview phase, technicians from the region made a joint proposal to coordinate their financial and administrative tools. However, political approval was still needed. While the scenarios contributed to strengthening collaborations, the exchanges that followed the workshops were not organised around scenarios, but rather focused on day-to-day working issues. In this sense, the scenarios did not directly foster integrated management, but instead shed light on existing organisational limitations and controversial policies, reactivating a social dynamic of inter-organisation/inter-sector collaboration.

5.2. The Influence of Spatial Scale and Impact Pathways on Scenario Outcomes

The spatial mismatch between the scenarios (Couesnon catchment) and administrative boundaries (Brittany) did not hinder cooperation efforts. To capitalise upon the pre-existing willingness to work transversally, regional technicians were primarily interested in the connections between landscape elements, which can be well represented by local and high-resolution scenarios compared to higher-level ones. Far from considering that the spatial extent and resolution of scenarios do not influence the promotion of integrated management, we argue that local scenarios were suitable in this case, despite not aligning with Brittany's boundaries. At the local level, the SAGE and SCOT occasionally collaborated to provide advice on local urban planning schemes. However, their technicians did not participate in scenario construction with the conscious objective of fostering collaboration between their organisations. Therefore, a consistent extent was crucial to generating interest among these organisations for a co-construction process. Ensuring a consistent spatial extent was also crucial for the instrumental outcome, as evident in the mention of the ALICE scenario results in the notice regarding Bridor's potential establishment. Indeed, the water quantity results stemmed from the scenarios accurately reflecting the local reality.

An adequate spatial extent also generated curiosity among the participants in the dissemination workshops who had no specific expectations, such as representatives of the intermunicipalities. Many of these representatives acknowledged that the scenarios helped them enhance their awareness of local water tensions. Beyond the individual nature of this outcome, an elected representative at the Pays de Fougères intermunicipality and at the CLE considered that this personal increased awareness facilitated collaboration among the representatives who participated in the workshop. Indeed, the scenarios facilitated the establishment of a common understanding of the situation, particularly regarding water quantity issues, which became a widely recognised problem. To understand the complementarity between the co-construction and dissemination, it is worth noting that the representative who assisted us in organising the workshop for the intermunicipalities belonged to an organisation that was actively involved in the co-construction process and possessed in-depth knowledge of the process that led to the scenario outcomes.

Although our assessment is based on a specific case study, we believe that the lessons learned can be applied more broadly. Indeed, the complementarity of the co-construction and dissemination pathways could be relevant for any research on sustainable study. Moreover, the dual role of the scenarios, serving as catalysts for pre-existing political will to promote integrated management and as a tool for raising awareness, can be transferred to other scenarios.

5.3. Limitations and Open Questions

The findings of this study indicate that scenarios are effective in facilitating group outcomes, including for organisations operating at higher organisational levels. However, although the workshops encouraged interdisciplinary exchanges among organisations working at the same level, the interviewees did not report inter-level outcomes. Creating ad hoc workshops mixing actors working at different organisational levels could help to explore the potential of scenarios in generating inter-level discussions. Additionally, considering that economic factors significantly influence farmers' decisions, incorporating economic indicators into scenarios (e.g., yield loss due to climate change and biodiversity decline) would promote the integration of agriculture and water management at the local level.

The interviewees recognised the contribution of the scenarios in reflecting on agriculture and water tensions. However, there was a notable lack of discussion regarding urbanisation and biodiversity, despite their significant role in the scenarios. Only two interviewees, who were either involved in the entire scenario construction process or consistently engaged in the dissemination phase, discussed urbanisation. These results may have been influenced by the selection of interviewees, which included several farmers and agricultural technicians, but only one urbanisation technician. However, even the non-specialist interviewees, such as the residents and representatives, predominantly focused on agriculture and water. This suggests that sustained and repeated interactions between scientists and society can effectively foster a multidisciplinary perspective when analysing complex phenomena like landscape changes. Conversely, when these interactions are brief, the increased awareness of complex dynamics and expected challenges is limited to the concepts previously considered important.

Determining the ideal timing for assessments presents numerous challenges. From a research perspective, project schedules impose some constraints, while on the administrative side, the calendar of official document revisions may be hard to access. Similar to the call for open science, there is a need for easier access to the updated schedules of administrative document decisions at the institutional level. In addition, long-term assessment results can evolve, and instrumental outcomes may emerge during the revision of local technical documents such as SCOT and SAGE. This raises questions about the appropriate time lag between the dissemination of scenarios and the assessment of their contribution to integrated land management, which ultimately depends on the administrative and political context. Therefore, the time lag chosen for our study could be considered to have yielded middle-term outcomes when considering the institutional perspective.

The instrumental use of the scenarios became possible due to their dissemination. By sharing the results with representatives, a common foundation for discussion was established, rooted in scientific evidence together with political and economic discourses. In this case study, representatives perceived the involvement of scientists as a "neutral" party, bringing scientific elements rather than political convictions, which was crucial to considering water tensions as factual. With a limited window of opportunity to address climate change [57], scientists are increasingly assuming political positions and actively advocating for climate action. Without questioning the appropriateness of scientists' advocacy [58], this raises the question of when scientists should maintain a "neutral" position and when they should speak out. Answering this question extends beyond the scope of this study, but it underscores the importance of expanding the existing literature to explore how and under what circumstances scientists should assume specific political positions.

Finally, while the scenario-based high resolution maps facilitated awareness and collaboration at the regional and sub-regional levels, caution should be exercised in generalising these results to the application of lower-resolution scenarios at the national or continental levels. Furthermore, we recommend the inclusion of self-assessment in future scenario-based scientific projects during the co-construction and dissemination phases, an additional step that is often overlooked but valuable for enhancing the effectiveness and impact of scenarios.

6. Conclusions

The primary objective of this study was to assess the outcomes of land use and land cover change (LUCC) scenarios, particularly their role in promoting integrated management. Based on our analysis of a local case study, we conclude that:

- Despite the general acknowledgment of the need for more integrated management by local actors, joint initiatives among land management services remain rare.
- While the scenarios did not provide the necessary knowledge in a suitable format to technically support water management, they were successful in enhancing the collaboration between communities of practice. This was particularly evident for actors who were involved in the co-construction and those who participated in the dissemination workshops, where the concept of "boundary" was most explicit.
- The scenario dissemination workshops played a crucial role in initiating integrated management initiatives. However, technical discussions on integrating land management services relied more on external sources rather than the scenarios themselves.
- Consistency between the spatial extent of the scenarios and operational boundaries was crucial for engaging local actors in the co-construction process and for instrumental outcomes to emerge. However, when there was already a political will to promote integrated management, the alignment with operational boundaries was not as essential.
- The co-construction and dissemination pathways demonstrated a strong complementarity in fostering social outcomes.
- Overall, these findings highlight the potential of LUCC scenarios to drive collaboration and integrated management, although further efforts are needed to optimise their knowledge transfer and application in technical decision-making processes.

Supplementary Materials: The following supporting information can be downloaded at: https://www.mdpi.com/article/10.3390/land12071414/s1, S1: Presentation of the Couesnon spatially explicit scenarios: S2: Details of scenario-dissemination workshops; S3: Main actors involved in managing water in the Couesnon catchment and relations between.

Author Contributions: Conceptualization, R.R. and T.H.; methodology, R.R.; formal analysis, R.R.; investigation, R.R.; resources, R.R. and T.H.; data curation, R.R.; writing—original draft preparation, R.R.; writing—review and editing, R.R. and T.H.; project administration, R.R.; funding acquisition, T.H. All authors have read and agreed to the published version of the manuscript.

Funding: This work was supported by the project TERRANOVA the European Landscape Learning Initiative, which has received funding from the European Union's Horizon 2020 research and innovation programme under the Marie Sklodowska-Curie [grant agreement No 813904]. The output reflects the views only of the authors, and the European Union cannot be held responsible for any use which may be made of the information contained therein. This research also benefitted from the ALICE project (EAPA_261/2016), funded by the Atlantic Area: European Regional Development Fund through the INTERREG Atlantic Area 2020 Transnational Cooperation Program (http://project-alice.com/).

Data Availability Statement: Anonymised data and the interview protocol are available on the Zenodo Repository (10.5281/zenodo.8124840).

Acknowledgments: The authors thank the local actors who actively participated in the scenario-dissemination workshops and interviews.

Conflicts of Interest: The authors declare no conflict of interest.

Notes

1. INTERREG ALICE project: a project that combined international partners from five countries of the Atlantic region. This article focuses on the outcomes of scenarios developed for France. https://project-alice.com/alice-project/ (accessed on 22 December 2022)
2. "Schéma d'aménagement et de gestion de l'eau" in French
3. "Schéma de cohérence territoriale" in French
4. Available at (accessed on 5 June 2023): http://iramuteq.org/telechargement
5. "Commission locale de l'eau" in French
6. Terre de source official webpage (accessed on 1 December 2022): https://terresdesources.fr/
7. The word "ville" ("city" in English) was included in the dendrogram. However, in the corpus, it was not used concerning urbanisation but to a duality between urban and rural residents
8. Extract from the notice translated from French: "The Bridor file establishes forecasts up to 2030. However, industrial activity is expected to continue beyond that date. The results of the ALICE project conducted by scientists in the Couesnon catchment have highlighted the risk of a decrease in the water flow by half during the low-water period by 2050. Global changes are announced by the IPCC for a horizon further away than 2030 (2050 and 2100) and can no longer be ignored. The Bridor file, like all ICPE and Water Law files, should be able to assess impacts of their activities beyond the date on which their investments are planned." Available at: https://www.liffre-cormier.fr/wp-content/uploads/2022/02/2021-09-02_Avis-CLE-COUESNON.pdf (accessed on 22 December 2022)
9. "Direction Régionale de l'Environnement, de l'Aménagement et du Logement de Bretagne" in French

References

1. Renn, J. The Evolution of Knowledge: Rethinking Science in the Anthropocene. *HoST J. Hist. Sci. Technol.* **2018**, *12*, 1–22. [CrossRef]
2. Glavovic, B.C.; Smith, T.F.; White, I.; Smith, T.F. The tragedy of climate change science. *Clim. Dev.* **2021**, *14*, 829–833. [CrossRef]
3. González-Márquez, I.; Toledo, V.M. Sustainability Science: A Paradigm in Crisis? *Sustain. Rev.* **2020**, *12*, 2802. [CrossRef]
4. Kajikawa, Y.; Tacoa, F.; Yamaguchi, K. Sustainability science: The changing landscape of sustainability. *Sustain. Sci.* **2014**, *9*, 431–438. [CrossRef]
5. Wimbadi, R.W.; Djalante, R. From decarbonization to low carbon development and transition: A systematic literature review of the conceptualization of moving toward net-zero carbon dioxide emission (1995–2019). *J. Clean. Prod.* **2020**, *256*, 120307. [CrossRef]
6. Fang, X.; Bingbing, Z.; Xingyue, T.; Qun, M.; Jianguo, W. "What Kind of a Science is Sustainability Science?" An Evidence-Based Reexamination. *Sustainability* **2018**, *10*, 1478. [CrossRef]
7. Yeh, M.J. Discourse on the idea of sustainability: With policy implications for health and welfare reform. *Med. Health Care Philos.* **2020**, *23*, 155–163. [CrossRef]
8. Pereira, L.; Ferrier, S. Developing multiscale and integrative nature—People scenarios using the Nature Futures Framework. *People Nat.* **2020**, *2*, 1172–1195. [CrossRef]
9. Brand, U.; Wissen, M. Crisis and continuity of capitalist society-nature relationships: The imperial mode of living and the limits to environmental governance. *Rev. Int. Political Econ.* **2013**, *20*, 687–711. [CrossRef]
10. Spash, C.L.; Hache, F. The Dasgupta Review deconstructed: An exposé of biodiversity economics. *Globalizations* **2022**, *19*, 653–676. [CrossRef]
11. Lasa Lopez, A. Neoliberal conditionality to the European agricultural system: Free trade agreements as a paradigm. In *Transforming Food Systems: Ethics, Innovation and Responsibility*; Wageningen Academic Publishers: Wageningen, The Netherlands, 2022; pp. 58–62.
12. Pe'er, G.; Bonn, A.; Bruelheide, H.; Dieker, P.; Eisenhauer, N.; Feindt, P.H.; Hagedorn, G.; Hansjürgens, B.; Herzon, I.; Lomba, Â. Action needed for the EU Common Agricultural Policy to address sustainability challenges. *People Nat.* **2020**, *2*, 305–316. [CrossRef] [PubMed]

13. Bodin, Ö.; Robins, G.; Mcallister, R.R.J.; Guerrero, A.M.; Crona, B.; Lubell, M. Theorizing benefits and constraints in collaborative environmental governance: A transdisciplinary social-ecological network approach for empirical investigations. *Ecol. Soc.* **2016**, *21*, 40. [CrossRef]
14. Montgomery, E.; Montgomery, I.; Caruso, T.; Reid, N. Hedgerows as Ecosystems: Service Delivery, Management, and Restoration. *Annu. Rev. Ecol. Evol. Syst.* **2020**, *51*, 81–102. [CrossRef]
15. Forman, R.T.T.; Baudry, J. Hedgerows and Hedgerow Networks in Landscape Ecology. *Environ. Manag.* **1984**, *8*, 495–510. [CrossRef]
16. Ghazavi, G.; Thomas, Z.; Hamon, Y.; Marie, J.C.; Corson, M.; Mertot, P. Hedgerow impacts on soil-water transfer due to rainfall interception and root-water uptake. *Hydrol. Process.* **2008**, *22*, 4723–4735. [CrossRef]
17. Thomas, Z.; Abbott, B.W. Hedgerows reduce nitrate fl ux at hillslope and catchment scales via root uptake and secondary eff ects. *J. Contam. Hydrol.* **2018**, *215*, 51–61. [CrossRef]
18. Chever, T.; Parant, S.; Gonçalves, A. Evaluation du Programme Breizh Bocage 2. Bilan de Mise en Œuvre, Premiers Impacts sur les Territoires et Pistes d' Amélioration pour la Future Programmation. 2021. Available online: https://www.bretagne.bzh/app/uploads/sites/5/2021/11/Rapport-final-Evaluation-du-programme-Breizh-Bocage-2-Lot-2.pdf (accessed on 4 June 2023).
19. Houet, T.; Palka, G.; Rigo, R.; Boussard, H.; Baudry, J.; Poux, X.; Narcy, J.; Alvarez Martinez, J.M.; Balbi, S.; Mony, C.; et al. Land Use Policy European blue and green infrastructure network strategy vs. the common agricultural policy. Insights from an integrated case study. *Land Use Policy* **2022**, *120*, 106277. [CrossRef]
20. Martínez-fernández, J.; Banos-gonzález, I.; Esteve-selma, M.Á. An integral approach to address socio-ecological systems sustainability and their uncertainties. *Sci. Total Environ.* **2021**, *762*, 144457. [CrossRef]
21. Bondin, Ö. Collaborative environmental governance: Achieving collective action in social-ecological systems. *Science* **2017**, *357*, eaan1114. [CrossRef]
22. Tosun, J.; Lang, A. Policy integration: Mapping the different concepts Policy integration: Mapping the different concepts. *Policy Stud. ISSN* **2017**, *38*, 553–570. [CrossRef]
23. Underdal, A. What? Why? How? *Integr. Mar. Policy* **1980**, *4*, 159–169. [CrossRef]
24. Duffy, R.J.; Cook, J.J. Overcoming bureaucratic silos? Environmental policy integration in the Obama administration policy integration in the Obama administration. *Environ. Polit.* **2019**, *28*, 1192–1213. [CrossRef]
25. Candel, J.J.L.; Biesbroek, R. Toward a processual understanding of policy integration. *Policy Sci.* **2016**, *49*, 211–231. [CrossRef]
26. Metz, F.; Glaus, A. Integrated Water Resources Management and Policy Integration: Lessons from 169 Years of Flood Policies in Switzerland. *Water* **2019**, *11*, 1173. [CrossRef]
27. Candel, J.J.L.; Breeman, G.E.; Termeer, C.J.A.M. The European Commission's ability to deal with wicked problems: An in-depth case study of the governance of food security. *J. Eur. Public Policy ISSN* **2016**, *23*, 789–813. [CrossRef]
28. Molenveld, A.; Verhoest, K.; Wynen, J. *Why Public Organizations Contribute to Crosscutting Policy Programs: The Role of Structure, Culture, and Ministerial*; Springer: New York, NY, USA, 2021; Volume 54.
29. Murgue, C.; Therond, O.; Leenhardt, D. Toward integrated water and agricultural land management: Participatory design of agricultural landscapes. *Land Use Policy* **2015**, *45*, 52–63. [CrossRef]
30. Lang, D.J.; Wiek, A.; Wehrden, H. Von Bridging divides in sustainability science. *Sustain. Sci.* **2017**, *12*, 875–879. [CrossRef]
31. Miller, T.R.; Wiek, A.; Sarewitz, D.; Robinson, J.; Olsson, L.; Kriebel, D.; Loorbach, D. The future of sustainability science: A solutions-oriented research agenda The future of sustainability science: A solutions-oriented research agenda. *Sustain. Sci.* **2014**, *9*, 239–246. [CrossRef]
32. Clark, W.C.; Dickson, N.M. Sustainability science The emerging research program.pdf. *Proc. Natl. Acad. Sci. USA* **2003**, *100*, 8059–8061. [CrossRef] [PubMed]
33. Posner, S.M.; Cvitanovic, C. Evaluating the impacts of boundary-spanning activities at the interface of environmental science and policy: A review of progress and future research needs. *Environ. Sci. Policy* **2019**, *92*, 141–151. [CrossRef]
34. Houet, T.; Aguejdad, R.; Doukari, O.; Battaia, G.; Clarke, K. Description and validation of a "non path-dependent" model for projecting contrasting urban growth futures. *CyberGeo* **2016**. [CrossRef]
35. Matthews, K.B.; Rivington, M.; Blackstock, K.L.; Mccrum, G.; Buchan, K.; Miller, D.G. Raising the bar? The challenges of evaluating the outcomes of environmental modelling and software. *Environ. Model. Softw.* **2011**, *26*, 247–257. [CrossRef]
36. Akkerman, S.F.; Bakker, A. Boundary Crossing and Boundary Objects. *Rev. Educ. Res.* **2011**, *81*, 132–169. [CrossRef]
37. Del Corso, J.P.; Kephaliacos, C.; Plumecocq, G. Legitimizing farmers ' new knowledge, learning and practices through communicative action: Application of an agro-environmental policy. *Ecol. Econ.* **2015**, *117*, 86–96. [CrossRef]
38. Council, N.R. *Informing Decisions in a Changing Climate*; The National Academies Press: Washington, DC, USA, 2009; ISBN 9780309137379.
39. Wenger, E. *Communities of Practice: Learning, Meaning, and Identity*; Cambridge University Press: Cambridge, MA, USA, 1998.
40. Hamilton, S.H.; Fu, B.; Guillaume, J.H.A.; Badham, J.; Elsawah, S.; Gober, P.; Hunt, R.J.; Iwanaga, T.; Jakeman, A.J.; Ames, D.P.; et al. A framework for characterising and evaluating the effectiveness of environmental modelling. *Environ. Model. Softw.* **2019**, *118*, 83–98. [CrossRef]
41. Priess, J.A.; Hauck, J. Integrative scenario development. *Ecol. Soc.* **2014**, *19*, 12. [CrossRef]
42. Xiang, W.N.; Clarke, K.C. The use of scenarios in land-use planning. *Environ. Plan. B Plan. Des.* **2003**, *30*, 885–909. [CrossRef]

43. Butler, J.R.A.; Wise, R.M.; Skewes, T.D.; Bohensky, E.L.; Peterson, N.; Suadnya, W.; Yanuartati, Y.; Handayani, T.; Habibi, P.; Puspadi, K.; et al. Integrating Top-Down and Bottom-Up Adaptation Planning to Build Adaptive Capacity: A Structured Learning Approach. *Coast. Manag.* **2015**, *43*, 346–364. [CrossRef]
44. Badham, J.; Elsawah, S.; Guillaume, J.H.A.; Hamilton, S.H.; Hunt, R.J.; Jakeman, A.J.; Pierce, S.A.; Snow, V.O.; Babbar-sebens, M.; Fu, B.; et al. Environmental Modelling & Software Effective modeling for Integrated Water Resource Management: A guide to contextual practices by phases and steps and future opportunities. *Environ. Model. Softw.* **2019**, *116*, 40–56. [CrossRef]
45. Muhonen, R.; Benneworth, P.; Olmos-pe, J. From productive interactions to impact pathways: Understanding the key dimensions in developing SSH research societal impact. *Res. Eval.* **2020**, *29*, 34–47. [CrossRef]
46. Bürgi, M.; Hersperger, A.M.; Schneeberger, N. Driving forces of landscape change—Current and new directions. *Landsc. Ecol.* **2004**, *19*, 857–868. [CrossRef]
47. Houet, T.; Verburg, P.H. Exploring Futures in Landscape Agronomy: Methodological issues and prospects of combining scenarios and spatially explicit tools. In *Landscape Agronomy: Advances and Challenges of a Territorial Approach to Agricultural Issues*; Benoit, M., Ed.; Springer: Cham, Switzerland, 2022; pp. 163–181.
48. Garrick, D.; Alvarado-Revilla, F.; Jorgensen, I. Markets and misfits in adaptive water governance: How agricultural markets shape water conflict and cooperation. *Ecol. Soc.* **2022**, *27*. [CrossRef]
49. Rigo, R.; Houet, T.; Gonthier, C. Dissiminating land use land cover change scenarios for improving the role of Blue and Green Infrastructure: Evaluation and prospects. *in review*.
50. Houet, T.; Grémont, M.; Vacquié, L.; Forget, Y.; Marriotti, A.; Puissant, A.; Bernardie, S.; Thiery, Y.; Vandromme, R.; Grandjean, G. Downscaling scenarios of future land use and land cover changes using a participatory approach: An application to mountain risk assessment in the Pyrenees (France). *Reg. Environ. Chang.* **2017**, *17*, 2293–2307. [CrossRef]
51. Palka, G.; Houet, T.; Vacquié, L.; Forget, Y. Simulating interactions and feedbacks between land use and land cover changes for prospective scenarios with FORESCEM. *submitted*.
52. Dupas, R.; Minaudo, C.; Gruau, G.; Ruiz, L.; Gascuel-Odoux, C. Multidecadal Trajectory of Riverine Nitrogen and Phosphorus Dynamics in Rural Catchments. *Water Resour. Res. Am. Geophys. Union* **2018**, *54*, 5327–5340. [CrossRef]
53. Eau du Bassin Rennais Qualité des Eaux Brutes Bassin Versant du Haut Couesnon. 2019. Available online: https://bassin-couesnon.fr/bassin-couesnon/les-enjeux/qualite-de-leau/ (accessed on 4 June 2023).
54. Alber, A. Presentation of the Software Program SONAL. *Bull. Sociol. Methodol. Méthodologie Sociol.* **2018**, *137–138*, 176–200. [CrossRef]
55. Camargo, B.V.; Justo, A.M. Iramuteq Tutorial R INTERFACE for Multidimensional Analysis of Texts and Questionnaires. 2013. Available online: http://iramuteq.org/documentation/fichiers/IRaMuTeQ%20Tutorial%20translated%20to%20English_17.03.2016.pdf (accessed on 4 June 2023).
56. Reinert, M. Les "mondes lexicaux" et leur' logique à travers l'analyse statistique d' un corpus de récits de cauchemars. *Lang. Société* **1993**, *66*, 5–39.
57. Masson-Delmotte, V.; Zhai, P.; Pirani, A.; Connors, S.L.; Péan, C.; Berger, S.; Caud, N.; Chen, Y.; Goldfarb, L.; Gomis, M.I.; et al. (Eds.) IPCC Summary for Policymakers. In *Climate Change 2021: The Physical Science Basis. Contribution of Working Group I to the Sixth Assessment Report of the Intergovernmental Panel on Climate Change*; Cambridge University Press: Cambridge, UK; New York, NY, USA, 2021; pp. 3–32.
58. Nelson, M.P.; Vucetich, J.A. On Advocacy by Environmental Scientists What, Whether, Why, and How. *Conserv. Biol.* **2009**, *23*, 1090–1101. [CrossRef]

land

MDPI

Article

Spatial Tools for Inclusive Landscape Governance: Negotiating Land Use, Land-Cover Change, and Future Landscape Scenarios in Two Multistakeholder Platforms in Zambia

Freddie Sayi Siangulube [1,2,*], Mirjam A. F. Ros-Tonen [1], James Reed [2,3], Eric Rega Christophe Bayala [1,2] and Terry Sunderland [2,4]

1 Amsterdam Institute for Social Science Research (AISSR), University of Amsterdam, Nieuwe Achtergracht 166, 1018 VW Amsterdam, The Netherlands
2 Centre for International Forestry Research (CIFOR), Bogor 16115, Indonesia
3 School of International Development, University of East Anglia, Norwich Research Park, Norwich NR4 7TJ, UK
4 Faculty of Forestry, University of British Columbia, 2424 Main Hall, Vancouver, BC V6T 124, Canada
* Correspondence: f.s.siangulube@uva.nl; Tel.: +260-966292955

Abstract: Landscape approaches are being promoted as a form of negotiated governance to help reconcile competing land uses and identify common concerns for planning envisioned future landscapes. Multistakeholder platforms play a key role in these efforts. This paper aims to contribute to an emerging scholarship that explores how spatial tools can be used in such platforms as boundary objects and if and how they can contribute to inclusive landscape negotiations and governance. We used spatial mapping to observe and document stakeholder perceptions about drivers of land-use and land-cover change and desired future scenarios that accommodate competing land uses. We found that land-cover maps derived from satellite images helped participants identify land-use change dynamics and drivers. The ensuing community mapping of desired landscape scenarios in both multistakeholder platforms (MSPs) triggered a process of identifying common concerns and defining actionable priorities. However, in one MSP, stakeholders ultimately reached a compromise on a draft land-use map that was widely regarded as an entry point for further negotiations in Local Area Plans, while the other lacked consensus due to deep-seated social-cultural issues, such as social-class-based disagreements. This paper illustrates, first, that instead of focusing on the end product (participatory maps), understanding negotiation processes helps uncover why spatial tools may fail to achieve the intended purpose of reconciling land uses. Second, spatial tools only work for landscape approaches if MSPs are inclusive and foster a collaborative process that considers the views of all participants. The authors recommend that those steering MSPs stimulate them to evolve from "mere consultation forums" to "innovative, participatory platforms", encouraging stakeholders to engage in genuine negotiation processes that allow negotiated and alternative outcomes. We contend that such an approach, supported by spatial tools, is likely to contribute to the implementation of landscape approaches. Policymakers and land users can use these spatial tools as boundary objects in user-focused strategies that engender inclusive stakeholder participation and ensure legitimate, acceptable, and sustainable outcomes.

Keywords: landscape governance; spatial mapping; participatory scenarios; negotiations; multistakeholder platforms; Kalomo District; Zambia

1. Introduction

The complexity of addressing landscape challenges requires novel approaches to effectively conserve biodiversity while simultaneously satisfying human socioeconomic needs and planning the future [1,2]. Landscapes in the tropics are becoming increasingly fragmented along land-use features due to a range of socioeconomic, development, and cultural drivers [3]. As such, forest remnants in human-habited tropical landscapes and other natural habitats are now ecologically stressed hotspots. These landscapes are losing habitat connectivity, and their capabilities to provide ecosystem services such as food, folder, air purification, water infiltration, climate change regulation, and carbon sequestration are diminishing [3,4]. To address these complexities, there is a need to visualise how landscapes evolve and establish drivers of landscape change, especially in contested landscapes where human and non-human factors interact [4].

The use of spatial tools proliferated in urban and rural development and planning literature as important policy instruments [5], traditionally focusing on space efficiency and identifying sustainable land-use synergies and conflict management [6,7]. Later, spatial tools have been adopted in different natural resource management contexts to support decision-making processes, helping local people visualise their landscapes and, to some extent, informing the 'good governance' criteria of participation and inclusivity [8]. Several studies addressing the values and challenges of spatial tools as decision-support tools attempt to contextualise the social-cultural and economic parameters as inputs in the design principles of land-use planning [9]. Uncovering local perceptions and their experiences through participatory spatial tools that help visualise people's desired future landscapes has recently received attention in the landscape governance literature [10,11]. However, these participatory spatial tools are often employed in isolation without combining them with remote sensing. Empirical data on how a mix of multiple spatial tools might be integrated to enhance stakeholder negotiation processes and enable inclusive engagement in landscape governance remains scarce [11]. Moreover, technological limitations and user subjectivity influence landscape-level negotiations involving various stakeholders [11,12]. Therefore, holistic and user-friendly spatial approaches are crucial, particularly in decision-making processes involving local communities with extensive knowledge of landscape dynamics.

Participatory mapping, remote sensing maps, three-dimensional (3D) modelling, and place-based future scenario-building processes are examples of spatial tools that enable land managers and users to analyse spatial and temporal transitions to inform landscape governance strategies and attain collective objectives [11]. These spatial tools, notably time series remote sensing maps, can be effective aids as boundary objects that reveal landscape dynamics [13] and set the stage for stakeholders to negotiate common concern entry points in future scenarios [14]. In addition, participatory mapping enables multi-sectoral collaborative action at the local level to address socio-ecological challenges in landscapes that would otherwise be difficult to achieve with a single-sector approach. However, development planners, conservationists, policymakers, and land users increasingly acknowledge the importance of engaging various stakeholders in managing landscapes to secure a sustainable future [15]. Despite the acknowledgement, landscape planning processes are constrained by a lack of insights into the views of stakeholders with competing interests [16]. Therefore, one of the critical steps in strengthening landscape governance in multiple land-use settings is to uncover stakeholders' nuanced perspectives about their landscape, what they think is changing, and why [12,16].

In this context, integrated landscape governance, which invites multiple stakeholders to negotiate their claims during the planning process, supports an equitable dialogue while aiming to reconcile competing land uses through adaptive management systems. [17, 18]. In practice, this is often not easy to achieve. This paper seeks to contribute to this knowledge-implementation gap by focusing on the roles of spatial tools as boundary objects for negotiated governance in community-based landscape planning in the Kalomo District of Southern Zambia. The paper aims to contribute to an emerging scholarship that explores how spatial tools (remote sensing and participatory mapping) can help to negotiate land uses and if and how they can contribute to inclusive landscape governance. We analysed two multistakeholder platform (MSP) case studies in which groups of local actors negotiated their desired future landscapes. Considering that our analysis has significance for rural land use planning policy and practice, the study design included MSPs in a communal land and protected area. This paper addresses three research questions: (i) How has land cover changed over time in the Kalomo District? (ii) How do land users with varying and often conflicting interests construct drivers of landscape changes? (iii) How do spatial tools help negotiate the land-use trade-offs and contribute towards mapping a desired future landscape?

2. Conceptual Framework: Spatial Tools as Boundary Objects for Inclusive Landscape Governance

Star and Griesemer [19] were the first to introduce the boundary object concept to denote objects or entities with a common identity but different meanings in "intersecting social worlds" [20]. Opdam et al. [20] (p. 1441) associate 'boundary objects' with local landscapes that communicate meaning "among disciplines and between science and local communities". Thus, boundary objects are framed as negotiation or management tools to help uncover participants' diverse thinking about landscape dynamics while searching for common concerns [13,19]. In doing so, we recognise participants' individuality, competing interests, perceptions, and preferences on issues of mutual importance. In this study, we conceptualise remote sensing maps generated independently of community members' preconceived opinions as crucial aids to elicit community views about the future of their landscapes. The outputs (maps) would constitute part of the material for negotiating Local Area Plans (LAPs), which include several other villages. Furthermore, the participatory maps might be used to negotiate desirable future scenarios with other land users, decision-makers and district-level planners when formulating Integrated Development Plans (IDPs) at the district level.

Governance is defined broadly as the process of collective decision-making that involves the determination and implementation of policy actions through collective engagements involving state and non-state actors in allocating resources, accepting responsibility, exercising powers, and sharing the benefits of potential outcomes [21,22]. From a landscape governance theory perspective [17], this scope can be summarised into two governance strands relevant to scenario building in spatial planning [23,24]. First, a normative perspective describes inclusive governance as an ideal model with 'good governance concepts' including stakeholder inclusion, equitable participation, legitimacy and accountability, effectiveness, consistency, proportionality, and subsidiarity [8,9,25–28]. Another strand views inclusive governance as a specific process outcome rooted in deliberative governance [9,17,29,30]. This strand has recently been influenced by social-ecological system science that seeks to improve the understanding of the relationships between human-to-human interactions, institutions, and nature [31].

In theories related to integrated landscape approaches, inclusive landscape governance is elaborated as both a normative and a process concept in specific contexts aiming to challenge the status quo of conflicting goals and interests by promoting negotiation of sustainable future landscapes in a 'win-more-lose-less' situation [32]. In this vein, this study aims to illustrate the benefit of using spatial tools in landscape approaches to negotiating context-specific trade-offs at the landscape scale, such as development, food security, and conservation [14].

3. Materials and Methods

3.1. Study Area and Context

This research was conducted in Kalomo District, southern Zambia, in Chief Chikanta's Chiefdom, which is approximately 1754 km^2 in size (Figure 1). The chiefdom is part of the area where the COLANDS initiative[1] seeks to operationalise integrated landscape approaches. The study site is one of the most important livestock and food-producing regions in southern Zambia [33] and is in the proximity of significant biodiversity hotspots, such as the Sichifulo Game Management Area and Nazhila Water Catchment. The area hosts a large part of the biggest forest reserve in Southern Province, the Kalomo Hills Forest Reserve No. P-13 (KHF13) [34]. However, since the 1980s, land-use conflicts have progressively escalated, especially between food production systems, development and conservation in customary land and informal settlements in KFR13. Claims to the conservation area are based on customary land ownership and user rights, power struggles between the traditional system and the state, and entitlements based on historical or ancestral land ownership [35]. As the area's population increases, settlements are expanding, and demand for fertile land has put significant pressure on the forest reserve.

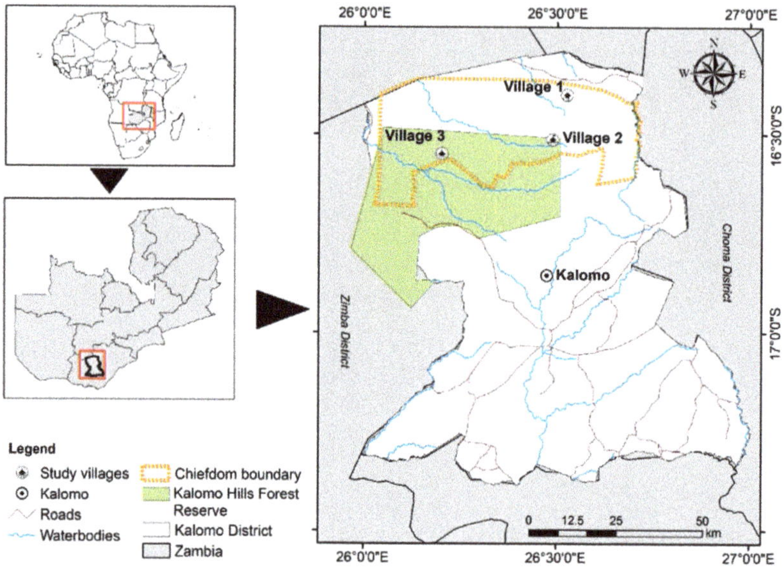

Figure 1. Location map showing Kalomo District, study villages 1 and 3 in the Chikanta Chiefdom (Source: modified from [36]).

In this regard, several stakeholders have proposed various land governance models that facilitate the harmonisation of various interests, including integrated landscape approaches [36]. National and local stakeholder meetings involving civil society, the state, traditional leaders, land users and researchers have been convened several times to try and remodel land-use governance in Kalomo District. For example, in August 2021, the Centre for International Forest Research (CIFOR) and the Zambian Forestry Department jointly organised a virtual national stakeholder dialogue attended by 39 participants from 24 organisations, including traditional leaders. In common with earlier participatory meetings, this high-level discussion concluded that there is a need to adopt inclusive governance approaches to restore landscape functions, improve livelihoods, and provide incentives and benefit-sharing mechanisms, a departure from the top-down approaches.

Against this background, village-level MSPs engaged in the landscape planning process, including a comprehensive analysis of landscape issues and identifying homegrown strategies to resolve land-use conflicts. Part of the planning process involved negotiating desired and sustainable future landscapes. The spatial tool employed in this study enhanced the negotiation processes in two local communities, Habulile (Village 1) and Siankwembo (Village 3) villages (Figure 1)[2].

3.2. Selection of Participants

In the COLANDS project, the communities of Habulile, Mudenda, Mubombo Ulilinyama, and Siankwembo, four MSPs known as Village Productivity Committees (VPCs), were purposely sampled. The VPCs are legal village-level MSPs recognised by the state under the Village Registration Act of 1994 [37]. MSPs were selected based on their land-use mix, demographic balance, period of existence, and frequency of meetings to deliberate on landscape issues. Ultimately, two MSPs were selected from Habulile and Siankwembo villages, respectively. The first MSP, multistakeholder platform Village 1 (further abbreviated as MSP1), is in Habulile village, near the chief's palace, which is considered the centre of decision-making power. The second MSP is in Siankwembo village, the oldest village on state land in a forest reserve. Despite being located on state land where statutory laws apply, the village manages local affairs via customary governance rules, demonstrating a typical space of governance contestations.

The village heads, locally known as *Sibbuku* (literally translated as 'bookkeeper'), coordinated the identification of eligible participants. The *Sibbuku* is the leader of the VPCs. The composition of VPCs includes electable village representatives and traditional leaders. In most cases, other participants outside the villages are invited, such as government officials, civil society organisations and the private sector [35]. Participation in the participatory mapping exercise was voluntary. However, a list of selected participants was cross-checked with those who had previously attended other village meetings. Suitable participants were assumed to have extensive knowledge about the landscape, having lived in the area for at least two years.

The age range for youth in Zambia is undefined and varies depending on activities. In the villages where the studies were conducted, a male or female under 21 is considered a youth. This socially accepted definition guided this study. MSP1 had 34 informants, including women, men, and youths, and 45 in the multistakeholder platform in Village 2 (further abbreviated as MSP2), with a similar composition (Table 1). Farmers' associations, women's groups, youth groups, community-based organisations, elderly people, civic leaders, local entrepreneurs, and ordinary village locals were all represented.

Table 1. Demographic details of participants in MSPs.

	MSP1	MSP2
Demography (no. of respondents)		
Gender		
Male adults	12	15
Female adults	13	15
Youths (mixed males and females)	10	15
Age in years		
Mean	39.5	40.3
Median	32	37
Level of education (%)		
Primary	60	53
Secondary	13	19
Tertiary	22	12
None (no formal education)	05	16
Land tenure and access to land (%)		
Land ownership		
<2 ha	19	51
2–10 ha	70	45
>10 ha	11	4
Wealth distribution (estimates of livestock)		
No. of cattle (total)	14,086	8985
No. of goats	22,900	1450
Others (sheep, donkeys, pigs)	321	634

Source: Authors' field data.

3.3. Data Collection and Materials

This section presents the methods and data types used in this study. Two sets of data were used: (i) we produced land-cover maps based on satellite images that were utilised as tools for MSP negotiations on drivers of deforestation and landscape degradation (Figure 2), and (ii) we analysed the participatory mapping processes to gain insights into negotiations for inclusive landscape governance in Kalomo District. Negotiation processes were analysed through participant observations and transcribing the proceedings (Section 3.3.3).

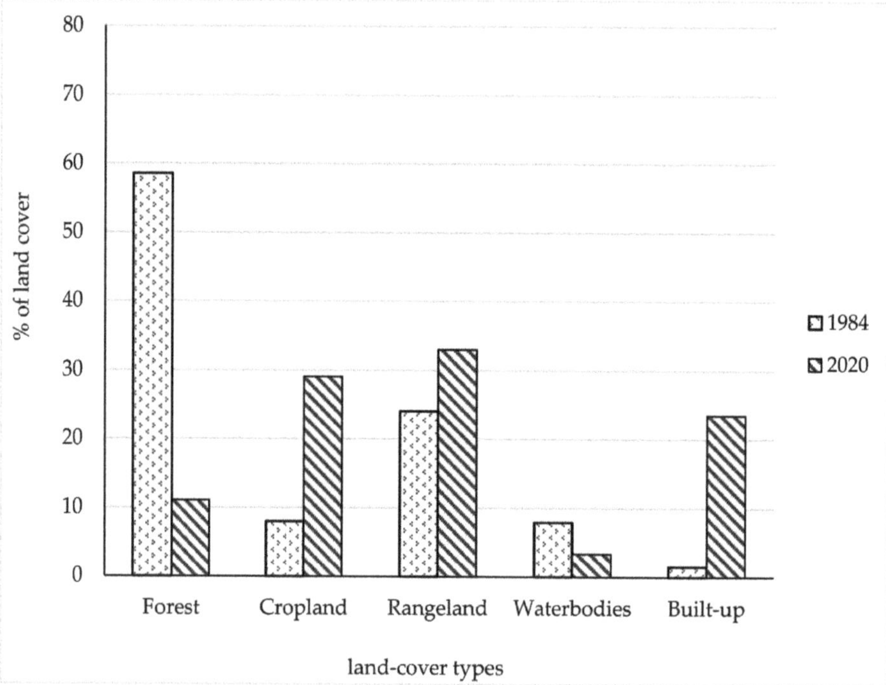

Figure 2. Land-cover types in 1984 and 2020 in the Chief Chikanta's area where the two villages are located.

3.3.1. Satellite Data Preparation and Analysis

Dataset

Two Landsat satellite images, the 1984 Landsat 5 Thematic Mapper (TM) and 2020 Landsat 8 Operational Land Imager and Thermal Infrared Sensor (OLI and TIRS), were used to assess land-cover change (see Appendix A Table A1). The two years were chosen for two reasons. First, the early 1980s coincided with land-use conflicts around Forest Reserves involving community members and state institutions, resulting in massive intra-village migrations that have influenced landscape dynamics. The second reason was the availability of cloud-free satellite images in both years [38].

Land-Cover Classes and Reference Data

The cloud-based Google Earth Engine was used to analyse the satellite images through Machine Learning. Following Asubonteng et al. [10], the classification scheme used was based on descriptions provided by key informants and ground-truthing (Table 2). The main land-cover types were discussed and agreed upon with key informants familiar with the landscape and followed the national classification [39]. This was important to ensure stakeholders could relate to their perceived landscape composition and configurations [10]. The land-cover types included forest, cropland, rangeland, waterbodies, and built-up/bare area.

Table 2. Land-cover classes used in the study.

Land-Cover Type	Description
Forest	Areas with a spatial extent of at least 0.5 hectares, woody vegetation with a canopy cover of more than 10%, and a tree height of more than 5 m.
Cropland	Land actively utilised to grow agricultural crops, such as annual and perennial crops that may be irrigated or rain-fed, for commercial, peasant, and small-scale farms.
Rangeland	Land comprising woodland, rangeland, grasslands, plains, shrubland and pans in river basins and water channels.
Waterbodies	Waterlogged areas formed naturally or artificially, including perennial and non-perennial streams and rivers, swamps, lakes, and dam lakes.
Built-up area/bare land	Characterised by low and high-intensity infrastructures, exposed soils, surface areas (roads), wastelands and rock outcrops, and all probable bare land.

Source: Based on the national land classification dataset [39] and stakeholder interviews.

Studies indicate that reference datasets should represent approximately 0.25% of the study area [40]. We used Google Earth Engine and JavaScript to process Landsat images and produce land-cover maps. Google Earth Engine is increasingly used in land-cover studies (e.g., [41,42]). In this study, we visually inspected high-resolution satellite imagery and historical data to extract training and validation samples. Guided by the classification scheme (Table 2), land-cover reference data for each land category in the current image (2020), cluster sampling was applied. Random sampling was undertaken in each cluster of given land-cover classes using high-resolution images on the online Google Earth Engine platform [43]. The allocation of sample points for each class was performed by estimating the average area of each land-cover class of the cluster [43,44] with sufficient samples across the different land types of interests to overcome the effects of sampling imbalances [45]. In the forest class, 139 points were sampled, 119 for cropland, 127 for rangeland, 120 for waterbodies, and 122 for built-up areas. For the 1984 map, 438 reference points were sampled from archived historical maps produced by the Forest Department and the District Council's physical planning department (between 80–90 points in each class). A total of 1065 land-cover sample points in various land-cover types in the study area were collected for 1984 and 2020. In geospatial data training and validation, the common practice is allocating more samples to training than validation [45]. For each of the images, the respective samples were divided into training (70%) and validation (30%) subsets [46].

Image Classification

Landsat images covering the entire area for 1984 and 2020 were accessed and processed in the Google Earth Engine environment using JavaScript processing chains to extract spectral–temporal metrics as predictors for land-cover classification. Google Earth Engine is a powerful cloud computing platform that provides advanced processing and visualisation tools while addressing time constraints for processing and data storage capacity issues [44] and hosts an extensive catalogue of datasets (see [44]). As a first phase, all images with less than 30% cloud cover were filtered using the mask function (FMASK) to have cloud and cloud shadow-free image collections as far as possible [47]. The results were visually evaluated and redefined throughout the masking process until the best result was reached [48]. The seasonal composites technique was used to extract spectral–temporal data, with the median reduction to generate cloud-free seasonal composites [49]. Southern Zambia's climate, which is composed of three seasons, namely a rainy-hot season (December–March), a cold-dry season (April–July) and a hot-dry season (August–October), guided the process of filtering satellite images. Images from the rainy and cold seasons were discarded because of the high amounts of clouds. For both 1984 and 2020, images between September and October were used in this study.

Using the above land-cover samples, the Random Forest Classifier in Google Earth Engine was used to train the model on the 1984 and 2020 Landsat images and produce land-cover maps. This study used 746 representative samples to train the Supervised Random Forest classifier to generate the land-cover maps for both years.

Accuracy Assessment and Change Detection

Assessing the accuracy of the land-cover map produced from the remotely sensed product is a widely accepted practice [44]. Accuracy assessment provides information about the reliability of the maps or any spatial information for use in decision-making processes. Classification accuracies of the 1984 and 2020 maps were evaluated using 319 reference data in Google Earth Engine.

Three error matrixes were computed based on the classification results. The accuracy assessments of the 1984 and 2020 land-cover maps were evaluated using the validation samples (30%) obtained from the visual assessment of raw images in Google Earth Engine. Classification accuracy for each map was obtained by matching the validation samples with thematic maps using a confusion matrix, as described by [43]. These comprised the producer accuracy (PA), which shows the likelihood that a pixel was properly classified in a particular class, and the Kappa coefficient (K), which is a measure of agreement between observed and expected values. The classified maps were exported later into QGIS 3.28.0 for change detection and analysis of land-cover transitions using the Semi-Automatic Classification Plugin [48].

3.3.2. Participatory Mapping Protocol

The participatory protocol presented in this section was used in all MSPs (Table 3). A two-day MSP meeting, segmented into three sessions, was convened, and the process lasted approximately 9–12 h. The first session introduced the research aims, deliberated on the remote sensing maps, and discussed the drivers of landscape change. During this session, the research team familiarised the participants (men, women, and youths) with the land-cover classification proposed by key informants, i.e., forest, cropland, rangeland, waterbodies, and built-up areas. Next, participants compared the land-cover map derived from the 2020 satellite image with their own knowledge of the area to discuss land-use issues and drivers of landscape change. These discussions formed a basis to reflect on and identify priorities for the future consistent with the local context, the problems perceived, land-use trends in the area, and envisaged resource demands. Finally, they broke up into focus groups of men, women, and youths to engage in participatory mapping of desired future landscapes—working in groups to create illustrative maps that identify the landscape features they hope to realise in the future. Taking 2020 as a base year in which the research project started, future landscapes were projected until 2040. Other studies [10,12] suggest a forecasting range between 10–30 years. However, community members assumed that in 20 years (2040), the current generations would still be 'available' to implement land-use strategies proposed in the project. The variables determining the projected future landscape were based on the participants' desired mix of land-cover types to accommodate their main agricultural activities (small-scale subsistence and cash-crop farming, livestock rearing, conservation farming based on intercropping trees and crops, and home gardening), water needs, settlement needs, and cultural assets. Variables underpinning the choice of a 20-year projection period included expected population growth and migration and settlement patterns, changing food systems, and climate change impacts. On the second day, participants reconvened to negotiate common concerns illustrated in respective participatory maps and integrate them into a single common map.

Table 3. Participatory mapping process in the two MSPs.

Stages	1. Preparatory Phase	2. Diagnosis of Drivers of Land Use	3. Participatory Land-Use Mapping in Focus Groups	4. Collective Participatory Map
Activities	Researchers prepare remote sensing maps. This follows study area verifications with key stakeholders [a] and ground-truthing of GPS coordinates.	Participants identify drivers of landscape change by interpreting land-cover maps produced in stage 1.	Sketch mapping in various focus groups (men, women, and youths).	Participatory sketching of desired future scenarios with inputs from all maps drafted in the focus groups.
Actors	Research team, GIS expert, and village leaders from each MSP.	MSP participants	Groups of men, women and youths involved in stage 2.	Chief's representative, village headmen and selected village participants, CBO reps, GIS expert, and Forestry Department official.
Tools	Topo sheet from the Forestry Department and GPS.	Topo sheet and land-cover maps, stationery for writing.	Sketch mapping of current land cover; identify issues and causes.	Participatory maps and discussion notes.
Measure of degree of inclusivity	Consultations with local leaders and some villagers conversant with the area.	Moderation during the meeting to allow equitable participation, engagement and transparency in the discussions.	Mediation to have equity and participation through participatory mapping in focus groups based on gender and age.	Mediation in MSPs to ensure balanced participation.

[a] Key stakeholders include personnel from the Physical Planning department at the Ministry of Local Government and Rural Development, a member of the Council of Elders and two village head persons familiar with the area of interest. Key: CBO = Community-based organisation, GIS = Geographical information system, GPS = Geographical positioning system, MSP = Multistakeholder platform. Source: Author compilation based on field data.

To avoid unintentional misinterpretations, several concepts such as 'negotiations', 'land cover', 'future landscape', and 'governance' were defined in the Tonga language (spoken by most participants, including the first author) at the outset of the meeting [49]. The participants were presented with two sets of maps for discussion. The first was a 1968 topographic sheet map acquired from the Forest Department's provincial office (reference number FDSP11/04/2018, scale 1:200,000). The topographic map helped participants to evaluate important biophysical features and appreciate the dynamics in the area of interest. After that, remote sensing maps (Section 3.3.1) illustrating landscape statuses in 1984 and 2020 were presented. These were utilised to initiate a discussion on drivers of landscape change and negotiate a common future landscape scenario. After lengthy discussions, the participants negotiated what they perceived to be the drivers of landscape change.

The three focus groups in the respective MSPs comprised 9–15 participants segregated by gender and age. Focus group discussions have a long tradition of being used as a data-gathering method in conservation and social science research [50]. Compared to interviews, focus groups provide more flexibility for data collection through interactions and co-production of meaning to uncover in-depth aspects that would be obscured in traditional interviews [51]. Given that the study was conducted in a gendered society with distinct roles for men and women [52], gender and age-segregated focus groups were necessary to ensure equitable participation. Focus group participants were encouraged to discuss and contribute to the list of landscape change drivers as they deemed suitable. They later sketched maps of imagined future landscapes. Grids were developed since the participatory maps were not drawn to scale, and the envisioned land-use classes were intuitively assigned ratios (converted into percentages) against the total land-cover area [10]. Participants in the main meeting analysed maps designed in the focus groups. All participants were involved in examining each map, and finally, collaboratively adjusted the focus group maps and reintegrated the negotiated maps into a new collective map expressing collective views.

3.3.3. Data Analysis

The proximate and underlying drivers of land-cover change, alongside the transcribed data of negotiation processes from all sessions, were qualitatively analysed using AT-

LAS.ti 22 using concept analysis and coding to determine the key variables and how they interacted to produce change.

4. Results

The results are presented in four parts: the first is associated with land-cover dynamics in Chief Chikanta's area using remote sensing to help participants visualise the context of the landscape dynamics between 1984 and 2020 (Section 4.1). The second part presents the participants' use of land-cover maps to identify drivers of change (Section 4.2) and relate those to village-level activities (Section 4.3). The last part presents the deliberations on the desired future landscapes (Section 4.4).

4.1. Land-Cover Dynamics in Chief Chikanta's Area

4.1.1. Accuracy Assessments

The 1984 land-cover map had an overall accuracy of 88.8% and a Kappa value of 0.83, whereas the 2020 map had an overall accuracy of 90.8% and a Kappa coefficient of 0.9. For 1984, Producer Accuracy for the forest was 99.3%, cropland 88.9%, rangeland 93.4%, and waterbodies and the built-up area were 100%, respectively. For 2020, the Producer Accuracy for the forest was 88%, for cropland 100%, for rangeland 92.5%, for waterbodies 86%, and for built-up/bare land 79.1%. The accuracy assessments are indicative that the classification is reliable for both years.

4.1.2. Land Cover in Chief Chikanta's Area in 1984 and 2020

In 1984, the entire landscape in Chief Chikanta's area comprised five predominant land-cover types of different proportions and spatial extent (Figure 3). The forest, which comprised Riparian, Miombo, Mopane, and Kalahari woodlands [39], had the highest proportion, covering 58.5% of the approximate 1754 km^2 of total land cover. Forest patches were concentrated along the rivers in the central area extending from the Nanzhila Water Catchment in the Kalomo Hills Forest reserve (KHR-13) to the north, with some forest patches on the south of the chiefdom. Cropland accounted for 8% of the total land and was extensively spread out in the southern part of the chiefdom and some patches dotted around the forest reserve. Rangeland (including seasonal wetlands) accounted for 24.1%, which, until now, served as grazing areas for livestock farmers due to the ability of the rangeland to retain moisture during dry seasons. Although open surface waterbodies were difficult to detect, the inundating evergreen deciduous forests were indicative of the dense network of the Nanzhila water system. Waterbodies, including rivers, *dambos* (permanent wetlands), ponds, and dam lakes, covered 7.9%. The Kalomo Hills Forest Reserve was established partly to protect the Nazhila Water Sub-catchment, one of the largest in the lower Kafue Basin. The built-up area, which includes bare land, village settlements, a road network, and other infrastructures, accounted for 1.5%.

4.1.3. Land-Cover Change in Chief Chikanta's Area

Between 1984 and 2020, the area's spatial distribution and extent of land-cover types changed, mainly due to village-to-village migrations from the south, central and northern areas. Small-scale farming activities (crops and livestock) are concentrated in the south and east of the area. In 1984, the forest took the largest part with 58.5%; cropland covered 8.0% of the area, rangeland 24.1%, waterbodies (*dambos*, wetlands, open water) covered 7.9%, and the built-up area accounted for 1.5%. In 2020, the major land-cover categories were cropland (29.1%), rangeland (32.9%), and built-up areas (23.5%), while waterbodies (3.3%) and forest (11.1%) had significantly dropped. Figure 2 compares the land cover in Chief Chikanta's area in 1984 with that in 2020, which will be further discussed below.

Table 4 lists the land-cover transitions in the Chikanta landscape between 1984 and 2020, accounting for 1428.8 km^2 (81.5%). This implies that only 18.5% (325.2 km^2) of the landscape—represented by the total of the bolded diagonal figures in Table 4—remained unchanged compared to the 1984 land-cover status. Net forest loss in this period amounted to 84.1%, predominantly due to conversion to rangeland (33.1%), cropland (28.5%), and built-up area (22.5%). Intermittent droughts led to some patches of farmland transitioning to other land uses (8.0%), as some farmers abandoned agriculture and opted for other activities such as charcoal production. Waterbodies decreased by 7.9%, which could be attributed to anthropogenic activities causing siltation, hence the marginal gains in forest cover (12.5%).

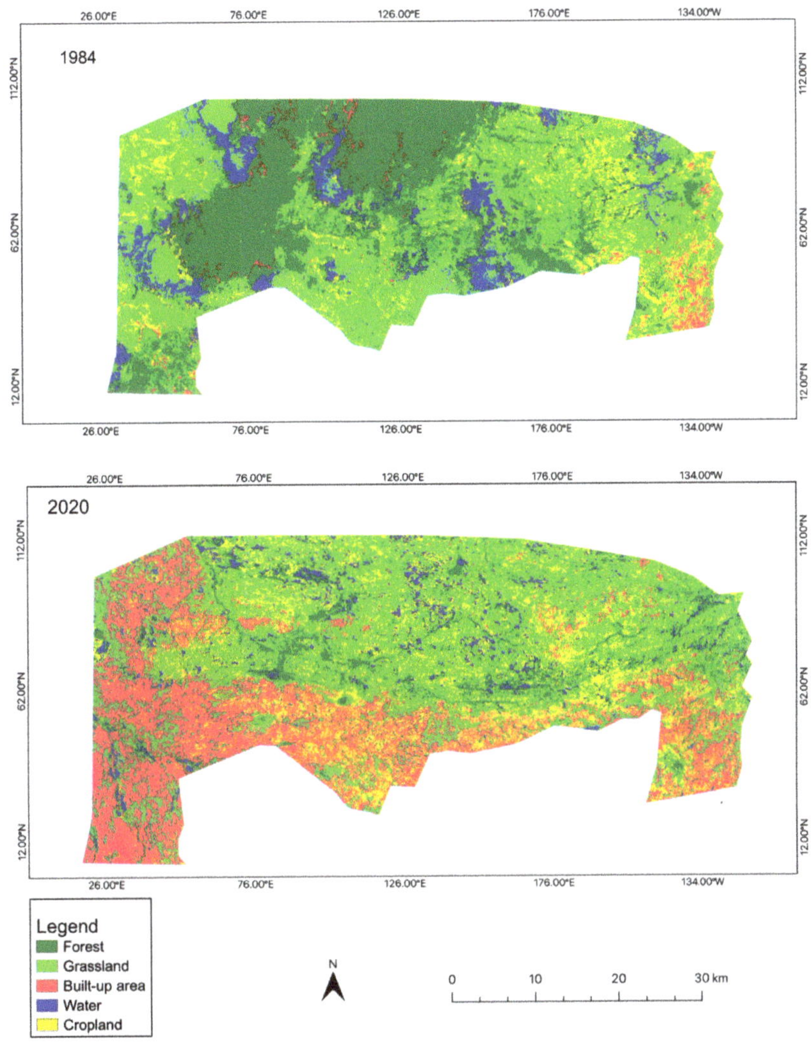

Figure 3. Land-cover maps for 1984 and 2020 for Chief Chikanta's area, Kalomo District, used to discuss drivers of landscape change in MSPs.

Between 1984 and 2020, built-up areas reached 412.5 km^2, an additional 23.5% compared to 1984. These results corroborate the Kalomo District report [53], showing that rural infrastructure expansion has significantly increased due to expanding smallholder farming. In the same period, cropland increased by 29.1% to 510 km^2. Similarly, rangeland gains from almost all other land uses increased by 33.0%, which could be attributed to various causes (Section 4.3), including livestock activities that convert forests to savannah rangeland. Both the expansion of agriculture and livestock resonate with the government's policy to accelerate rural economic growth through food production [54], albeit at the expense of natural habitats and their biodiversity.

Table 4. Landscape transition in km^2 and percentage (%) of the total land-use cover.

Land-Cover Types	Forest	Cropland	Rangeland	Water-Bodies	Built-Up/Bare	Total for 1984
Forest	**125.42**	292.41	339.46	37.74	231.23	1026.26
	12.22	*28.49*	*33.08*	*3.68*	*22.53*	*58.51*
Cropland	6.90	**46.04**	42.96	2.08	42.51	140.48
	4.91	*0.33*	*30.58*	*1.48*	*30.26*	*8.01*
Rangeland	42.63	133.89	**140.29**	13.06	92.32	422.20
	10.10	*31.71*	*33.23*	*3.09*	*21.87*	*24.07*
Waterbodies	17.33	30.61	47.65	**5.21**	38.18	138.99
	12.47	*22.02*	*34.29*	*3.75*	*27.47*	*7.92*
Built-up/bare	1.73	7.35	8.27	4.71	**8.26**	26.09
	6.64	*28.18*	*31.72*	*18.04*	*31.66*	*1.49*
Total for 2020	194.01	510.30	578.64	58.56	412.50	1754.01
	11.06	*29.09*	*32.99*	*3.34*	*23.52*	*100.0*

Note: The first row indicates land proportions in km^2 and italicised figures in percentages. Unchanged is the total of all diagonal figures (bold), which is 325 km^2 (18.5%), and the total transition is the total for 1984 and 2020 minus the unchanged area (i.e., 1754.01 − 325.5 is 1428.51 km^2).

4.2. Drivers of Landscape Change: Participants' Views

We elicited participants in each MSP to produce a list of drivers of landscape change that could explain the trends observed in the land-cover maps. Table 5 shows the resulting 25 drivers of landscape change, grouped into six categories, some exacerbating changes and others slowing them down. These include social and interpersonal, cultural, ecological, environmental, economic, climate change, and governance-related (organisation, policy, and legislation) variables.

Table 5. Drivers of land-cover/land-use change.

Aspects of Drivers of Land-Use Change (Direct and Proximate)	Perception Scores in MSP1			Perception Scores in MSP2		
	Men	Women	Youths	Men	Women	Youths
1. Social and interpersonal						
Lack of trust and legitimacy	++	+	++	+	++	++
Lack of accountability	+++	+++	++	+++	++	+++
Population increase	+		++			+++
Exclusive participation		+++			+++	
Gender equality		+++	+		++	+++
Migration		+++	+++			+++
Corruption in land administration	+	+++	+++	+++	+++	+++
2. Cultural aspects						
Lack of respect for traditional rules and norms	+++	++		+	+	
Degradation of *Malende* (sacred forests)	++					
3. Ecological/environmental						
Increased consumption of ecological services (water)	+	+	+++	++		+++
Deforestation	+++	+++	+++	+++	+++	+++
Animal diseases	+++		++	+++		+++
Climate change (associated with persistent droughts)	+++	+++	+++	+++	++	+++
Floods		+		++		
Cultivation in watersheds	++	++	++			++
Illegal logging			++	+++		+

Table 5. *Cont.*

Aspects of Drivers of Land-Use Change (Direct and Proximate)	Perception Scores in MSP1			Perception Scores in MSP2		
	Men	Women	Youths	Men	Women	Youths
4. Economic						
Increase in food demand	++	++	+++			
Expanding infrastructure	+++	+++	++	+++	+	+++
Demand for charcoal (biomass energy)	++	+	++		+	++
Agriculture	+++	+++	++	++	+++	+++
Overgrazing	++	+++	+++	+++	+++	+++
Lack of off-farm employment	+		+++	++		+++
Demand for non-timber forest products		+				
5. Governance (Organisation, Policy, and Legislation)						
Policy inconsistencies				+	++	+++
Unrealistic demands by the leader					+	
Settlements disputes	++	++	++	+++	+++	+++
Poor coordination among government departments				++	+	+++
Partisan politics/interference	++			++		+

Key: Important driver—+++, Moderately important driver—++, less important driver—+ (empty cell—not mentioned).

The findings show differences in perceptions of drivers of land-use change by gender, age, or MSP. Participants explained the drivers based on their memories of lived experiences related to economic, environmental, and social events from the 1980s to 2020. A lack of trust, legitimacy, and accountability in leaders and corruption were the primary societal drivers of land-use change, undermining public trust in land administration institutions. Villagers' perceived mistrust in the authorities and a lack of legitimacy in decision-making mechanisms contribute to inequities in access to land, especially for the marginalised. As a result, some people often disregard local rules governing usage and access to land. For example, they said cultivation or grazing in wetlands and riverbanks, sacred landscapes, and increased indiscriminate cutting of trees for charcoal production are forbidden. However, rule enforcement favours some and does not apply equally. The issue of migration as a driver of landscape change (between villages and from outside the chiefdom) was thorny, perhaps because most are migrants. However, the discussions hinted that migration instigates the demand for additional land for settlements and cropland. The group discussions also identified the effects of economic policies of the 1980s that routed considerable investment to support agriculture and boosted the demand for land.

Youths in all groups did not identify any cultural aspect, whereas men and women, especially in MSP2, rated cultural aspects as low. All groups concurred that in addition to economic factors such as the demand for food, which triggers a triple effect of overgrazing, demand for cropland, and biomass energy, ecological and environmental variables, including deforestation and climate change, are crucial in driving landscape change. Governance-related variables such as poor policy and institutional coordination in land management between agricultural officers and forest managers contribute to the failure to offer extension services to address unsustainable practices. Political interference by politicians was reported to influence the land distribution, thus escalating settlement disputes, for instance, with regard to the settlements in KHF13.

4.3. Deliberating Landscape Dynamics to Uncover Shared Challenges

4.3.1. MSP1: Habulile Village

In discussing various aspects of landscape transitions presented in the land-cover maps, participants related their experiences to some past and current social and environmental events. Initially, the discussions were amicable, with no disagreements emerging, even though males predominated the main meeting, as youths were conspicuously silent (probably for cultural reasons). Eventually, discussions became heated, especially in focus group discussions. Gender, age, knowledge of the area, and social status all influenced how the maps were interpreted.

The elderly participants (>50 years old) recounted the effects of some historical environmental events and the impacts on the landscape, notably the devastating three-year drought of 1992–1994. They attributed droughts to deforestation and lacking respect for sacred landscapes (locally called *Malende*, the rain gods). The environmental and socioeconomic impacts of droughts have reportedly put immense pressure on the forest reserve, rivers, and wetlands. For instance, due to low pasture availability and poor crop harvests, most households eventually resorted to harvesting non-timber forest products for food and charcoal production. Following such experiences, all participants identified pasture management as a common concern. Among some measures to help manage communal pastures, participants proposed strengthening local rules to regulate grazing regimes on communal pastureland.

Participants debated unplanned settlements and the need for a systematic land allocation approach. They also concurred with the youths' proposal that the higher hierarchy officials in the customary administration should be engaged in holding local village leaders accountable for land allocation. This referred to large parcels of land allocated to the so-called 'immigrants' for cropland and settlement. They agreed that settlements had grown exponentially in the last three decades, particularly near the chief's palace and towards the forest reserve. Similarly, between 1984 and 2020, participants echoed that the land resource had been under pressure, with cropland and pastureland increasing to almost their full carrying capacity. Accordingly, this was triggered by enhanced agricultural production methods (using animal draught power, tractors, and fertilisers). As a result, public and private institutions have gradually increased their investments in infrastructure development, such as schools, health facilities, telecommunications, energy transmission infrastructure, agricultural commodities storage facilities and road networks.

In discussing natural resource governance issues, female participants noted that previous efforts to negotiate land uses were characterised by insufficient stakeholder participation regarding land access and resource rights. Similarly, the male group voiced concerns that village stakeholders are rarely included in the formulation of local area plans (LAPs) that feed into integrated development plans (IDPs) at the district level [they referred to the IDP and chiefdom's master plan of the 1970s]. Youth and women's groups reiterated that while they had not come across any Local Area Plan, "a land-use map that includes our collective aspirations is essential" (MSP1, October 2021). In concluding the discussions, all groups identified the need for a community forest management plan to guide the sustainability of their future landscapes. However, men were quick to add that considering previous initiatives facilitated by conservation NGOs (e.g., the USAID Community-Based Natural Resources and Sustainable Agriculture (CONASA) project supporting local communities in resource governance in the 1990s), tree planting was a 'failed project' and preferred assisted natural regeneration (prescribed fallow period for abandoned crop fields). To restore wetland areas, participants reiterated imposing punitive measures to deter unsustainable grazing behaviour.

4.3.2. MSP2: Siankwembo Village

As in MSP1, the remote sensing maps were evaluated, and participants agreed that a realistic scenario of landscape transformations was presented. Protracted disagreements hampered the discussions (main meetings and focus groups) on practically every issue. This highlighted deep-seated class-based differences—poor vs. the rich, immigrants vs. locals, and politically connected vs. commoners. This was observed when discussing grazing land, boundary disputes, and land management issues. Participants engaged in counter-accusations concerning the reasons for landscape change, while others uncompromisingly highlighted the impacts of migrants on the social order in the village (mostly from outside the chiefdom). This referred to the people displaced because of the establishment of the Kariba Dam who migrated and settled in the area in the 1970s. This group of people is considered wealthy in terms of cattle husbandry. To enable successful engagements, the moderator (a local person selected at the beginning of the meeting) often intervened to mediate the debates on their perceived differences.

Nevertheless, all participants acknowledged that since 1984, large tracks of forest land had been converted into cropland and grazing land, viewed as the two most competing land uses in this village. The youths contended that the massive cropland expansion seen in the land-cover map for 2020 results from influential farmers who allocate and accumulate fertile land to themselves, e.g., by opening patches of small fields in various locations and claiming ownership. They also stated that settlements had expanded, especially along waterbodies, as moist grass in water areas is suitable for livestock rearing. In the main meeting, some participants (later identified as large livestock farmers) did not agree with the rest that livestock impact forest loss and siltation in wetlands. However, women argued that the so-called wealthy farmers monopolised access to grazing land by corruptly manipulating the poor to masquerade ownership of livestock to access communal grazing areas. Women and youths bemoaned that rangeland has replaced once-traditional water sources (wetlands and streams), and tributaries of Nanzhila River have changed flow courses downstream. The critical common concern in MSP2 regarded issues of clarifying access and rights to rangeland, waterbodies, and forest resources.

4.4. Participatory Mapping of Envisioned Landscapes in Focus Groups

In the two MSPs, participatory mapping was carried out in various groups of men, women, and youths (Figure 4). These maps were eventually debated, and a unified map was created, aligning all groups' concerns (Figure 5). The envisioned future was projected as 2040 based on a 20-year time frame. In both MSPs, participants debated on what constituted a 'future'. They agreed that a two-decade forecast was feasible based on rural life expectancy, rural migration dynamics, and coverage of the area development plans. Three possible scenarios emerged from this process: agriculture-dominated, forest-dominated, or grassland-dominated, with preferences differing according to gender and age, as elaborated below.

Each group envisioned a land-use balance in the future landscape based on their preferences. In MSP1, men preferred a balance of forests and rangeland (grazing land) and desired to allocate cropland intercropped with trees (conservation farming). Infrastructure was envisioned to reduce by restricting 'newcomers'. They proposed reassigning abandoned farms into forested spaces rather than maintaining built-up areas. However, their male counterparts in MSP2 failed to produce any map as they could not agree on most issues that required harmonising individual interests. For example, some men preferred a rangeland-dominated landscape to graze cattle, others favoured a forest-dominated landscape, and others considered cropland a preferred landscape.

Women's preferences in MSP1 were motivated by a desire to improve their land rights and access. They desired an agriculture-dominated future landscape and forests with increased space for gardens (to produce vegetables to improve household incomes) and less rangeland. They environed a future landscape with restored water sources and preferred to maintain the 2020 status quo of the built-up area. Women in MSP2 aspired to a similar scenario, although they emphasised a future with sufficient water through activities to restore wetlands and water flows in streams so that they would not have to travel long distances to search for water, especially in the dry season.

Youths in MSP1 were concerned about the haphazard planned village layout. They desired a future that responds to the expected population growth by equitably allocating sufficient agricultural land to all households. Youths in both MSPs want a future with a grazing regime that limits the number of cattle that may be maintained based on the carrying capacity of an individual's land, allowing some communal grazing area (rangeland) to be reclaimed for forest regrowth. They anticipated this would decrease cattle disease transmission, which has been an issue in the area for some time. MSP1 youths, like those in MSP2, envision a future with less rangeland and more forests by interplanting trees and crops in farmland. Both youth groups envision allotting land for a dam to harvest rainwater and serve as a water source to relieve the demand on the already overused natural water system.

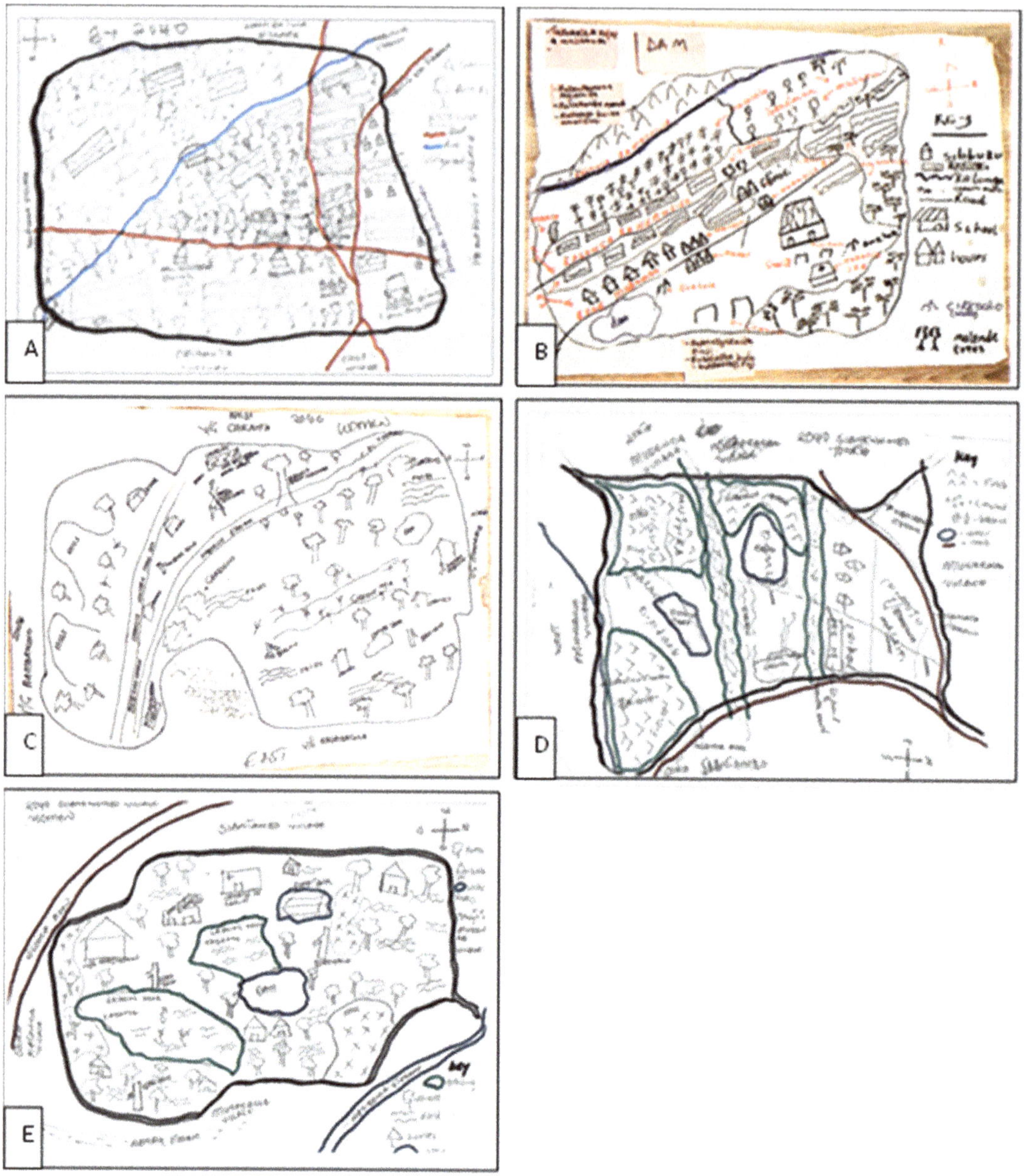

Figure 4. Participatory maps drawn in the two MSP workshops. MSP1—(**A**) male group, (**B**) youth group, (**C**) female group. MSP2—(**D**) female group, (**E**) youth group.

Negotiated Mapping of Desired Future Landscapes

Figure 5 depicts the final collaborative, participatory maps to capture common concerns, including issues of grazing land, deforestation, and access to water. In MSP1 (Figure 5A), all participants projected a forest-dominated landscape by implementing conservation agriculture systems that intercrop trees on the same piece of land to optimise the forest-cropland ratio. They also anticipate a future with restored natural water systems by strengthening local land-use rules and restrictions. By 2040, MSP2 participants (Figure 5B) envisioned a landscape characterised by a balance of cropland and rangeland. This was a compromise between two main land users, mainly male groups. They all agreed, however, to work with other stakeholders, including the Forest Department, to redesign the forest reserve's current land use to allow for a new model, such as a community-managed landscape, which would enable them to strengthen controls on any alleged illegalities. They claimed maintaining the landscape as a forest reserve is no longer attainable under the current circumstances as it perpetuates contestations between village stakeholders and state institutions (Forest Department).

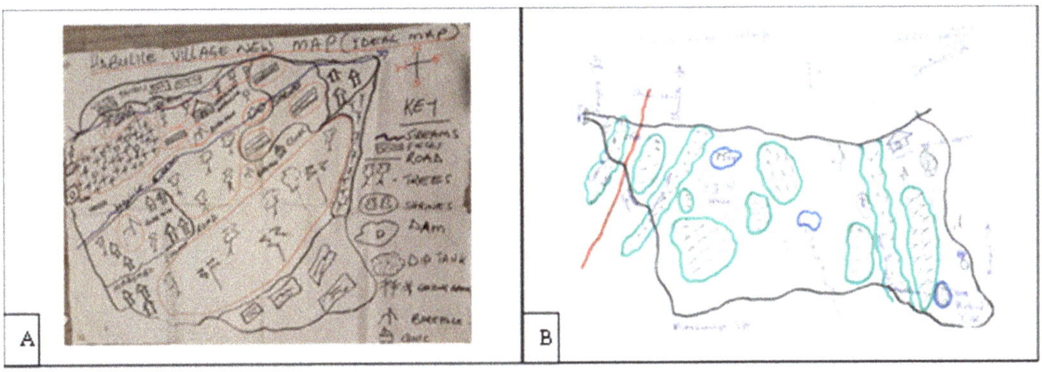

Figure 5. Combined participatory maps of the envisioned future landscapes in MSP1 (**A**) and MSP2 (**B**). (Note: different colours were used to delineate boundaries of various land-cover types).

5. Discussion

We explored the use of spatial tools as boundary objects in two MSPs in Zambia to identify common concerns in land-use negotiations. The study addressed three interrelated questions about how the landscape has changed between 1984 and 2020 and how diverse land users construct drivers of landscape change. Lastly, we sought to understand how stakeholders negotiate trade-offs that lead to mapping desired future landscape scenarios in the context of integrated landscape governance.

Participatory and remote sensing maps have a long history in spatial planning schemes in various contexts. The underlying assumptions are that spatial tools support inclusive landscape governance by inspiring transparent dialogue and addressing geodata needs in land-use planning processes [55]. With the renewed call for integrated landscape governance that focuses on multistakeholder processes [56], the question remains how to use these tools more inclusively by broadening the scope and representativity of stakeholders [23]. Further, inclusive governance paradigms must explore new technologies that make planning more accessible to illiterate, disabled, and marginalised people [11]. This study aimed to provide insights to address these concerns.

First, we demonstrated that allowing participants living in the landscape to interpret the remote sensing maps can enhance land managers' understanding of the local-level variables contributing to landscape dynamics that would otherwise be difficult to perceive in remote sensing data. A similar study in Zambia's rural context sought to understand the influence of socioeconomic factors on land-use change [57]. However, the present study adds empirical detail on the complexity of landscape dynamics, primarily related to broader socioeconomic issues occurring beyond the study area. The study's findings are consistent with the broader regional social and economic developments, revealing several drivers influencing landscape dynamics since the 1980s [58]. These drivers are related to the demand for biomass energy, agro-pastoral expansion, infrastructure developments, and poor governance at the sub-district level. Similarly, the period between 1984 and 2020 corresponds with continued deforestation exacerbated by ecological and anthropogenic factors such as drought and climate change, although at a much slower rate due to compensations from regrowth in abandoned agricultural land and a gradual decline in agro-pastoral activities owing to animal diseases and depressing macroeconomic activities, as well as improved law enforcement.

Second, the study indicates that remote sensing tools provided a good basis for local participants to engage in discussions on drivers of landscape change and find common ground while providing a feedback loop for researchers, resource managers and politicians. Combining different spatial tools ensured that different knowledge systems co-produced evidence-based problems and solutions. For instance, although remote sensing maps are complex, requiring specialised computer software, participatory mapping brings out lived experiences and knowledge of landscapes. Thus, this study confirms that combining the two approaches helped strengthen dialogue in the MSPs and contributed to the empowerment of those who could have previously been excluded from landscape governance. For example, we noted during the discussions that participants had fewer arguments about the state of the landscape as depicted in the spatial maps. In similar ways, land managers and policymakers may not doubt the remote sensing maps, but the story is incomplete. Differences in the interpretations and eventual mapping of envisioned future landscapes are understandable, given that local people hardly have the same priorities, perceptions, or future desires because their attachment to landscape differs [59].

Third, the application of participatory mapping in the two MSPs helped analyse the elements of inclusive governance. MSP1 has a long-standing governance practice in negotiating various local issues. However, inclusiveness in terms of accommodating diverse perspectives, especially those of women and youths (by men), is still influenced by local cultures, resulting in decisions lacking legitimacy. We noted that separating groups based on gender and age helped overcome cultural constraints that Mccall and Dunn [9] call 'participatory exclusion', i.e., being present in the meeting yet not fully able to engage due to (sociocultural) limitations. In MSP2, social differences, including wealth, gender, and the historical 'social divide' based on the area one hails from (migration issues), hindered effective engagement in the planning process. Although men could not achieve consensus in creating a map, their participation in debating the desired future scenario implies that the process was open to inclusive dialogue, which is the objective of using negotiation tools. The response from men who could not initially agree to disagree or agree on common concerns is encouraging. However, we cannot assume arbitrarily that their desire to engage in a final meeting reflects inclusivity.

Fourth, negotiation during the mapping of future scenarios can sometimes invoke personal emotions due to differences in attachments to landscapes that reflect personal feelings and interests [59]. Effective moderation is critical to successfully negotiating desired future landscapes involving several stakeholders, especially when power imbalances exist. The most powerful stakeholders sometimes sideline the views of marginalised groups regarding their desired future. Nevertheless, effective moderation and using spatial tools allowed the marginalised group to readily communicate their viewpoints by referring to the boundary object (maps) as the subject rather than focusing on their perspectives.

Based on this study, we can deduce that using spatial tools also poses some potential risks. To begin with, policymakers may be sceptical about the validity of participatory spatial tools and may not take them seriously [11,60]. For instance, participatory spatial tools may obscure power imbalances and hide some actors' ulterior objectives. This is because spatial tools are generally used in closed spaces of limited participants [11] whose selection process may deliberately exclude potential participants with divergent views. We also noted that sacred landscapes exist called *Malende*, but participatory mapping was challenged due to privacy and confidentially issues surrounding these sacred landscapes. This implies more attention is needed to the ethical considerations of how to deal with what is put on a map [61,62]. Finally, participatory spatial tools generally focus on local processes, keeping telecoupling effects out of sight, such as the ripple effects of global connectivity in the socioeconomic and environmental networks [63].

In the inclusive landscape governance literature, there is remarkable growth in the diversity of methods, tools, and applications of theories of spatial tools in land-use management (see [11]). These developments have methodological implications for this study. The most often used methods in land-use mapping literature are predicated on theories from sustainability science, collaborative planning, pragmatic planning and postmodern methods such as planning approaches that are "open to a diversity and plurality of styles and ideas" [64,65] (p. 474). Further, there is also a tendency to anchor spatial planning in geospatial and remote sensing technology to derive biophysical variables [66], while stakeholder perceptions in participatory mapping of desired future landscapes are at the margins of research [61,67]. Whereas this study has demonstrated the efficacy of combining the two approaches, further research is needed to address some limitations of this study. We focused on two small villages as case studies, implying that the findings should be interpreted in this specific context, but the methods can be applied more broadly. We recommend scaling up the combined use of remote sensing and participatory mapping to the district level to understand landscape dynamics and its governance implications beyond the village level and promote more inclusive landscape governance.

6. Conclusions

This paper provides insights into the question of how spatial tools can facilitate land-use negotiations and plan future sustainable landscapes in MSPs based on inclusive and equitable decision-making.

This paper shows that MSPs, as a form of negotiated governance, require practices that overcome inequalities and improve trust in participatory planning. We conclude that while remote sensing can provide a basis for discussing historical trends, participatory mapping of desired future scenarios potentially triggers a process to identify common concerns and co-produce actionable priorities. Prior to mapping, the process of identifying problems and agreeing on land-use priorities initiated a negotiation process of trade-offs and synergies involving pastureland, water, forestry, and agriculture. We showed that stakeholders in one MSP ultimately reached a compromise on a draft land-use map, which was widely regarded as a potential tool for further negotiations with the district land-use planners. Meanwhile, participants in the second MSP created a common map without consensus due to deep-seated social-cultural issues such as class-based differences of migration origin and land-use preferences. The disparity between the two illustrates that the potential of MSPs for negotiations must be understood in the context of the environmental histories of a place and within the constraints of place-based institutions rather than being generalised.

Finally, this paper illustrates two issues regarding spatial negotiations in MSPs: first, the same institutions designed to facilitate negotiations occasionally turn into spaces that engender resistance to successful negotiation outcomes from some participants, given differences in power positions and hidden agendas. This can be a source of illegitimacy and cause failures to implement the plans. Second, MSPs must be inclusive and foster legitimacy and a collaborative process.

The authors recommend that those steering MSPs stimulate them to evolve from "mere consultation forums" to "innovative, participatory platforms", encouraging stakeholders to engage in genuine negotiation processes that enable alternative outcomes. We contend that such an approach, supported by spatial tools, is likely to contribute to the implementation of integrated landscape approaches. Policymakers and land users can use these spatial tools as boundary objects in user-focused strategies that engender inclusive stakeholder participation and ensure legitimate, acceptable, and sustainable outcomes.

Author Contributions: F.S.S.: Conceptualization, Methodology, Data curation, Formal analysis, Visualisation, Writing—original draft. M.A.F.R.-T.: Conceptualization, Writing—reviewing and editing, Methodology, Validation, Supervision. J.R.: Investigation, reviewing and editing, Supervision, Funding acquisition. E.R.C.B.: Reviewing and editing. T.S.: Reviewing and editing, Supervision, Validation, Funding acquisition. All authors have read and agreed to the published version of the manuscript.

Funding: This research was funded through a grant from the International Climate Initiative (IKI) of the German Federal Ministry for the Environment, Nature Conservation, Building and Nuclear Safety (BMUB) (grant 18_ IV_084) for the Collaborating to Operationalize Landscape Approaches for Nature, Development and Sustainability (COLANDS) initiative carried out by the Centre for International Forestry Research (CIFOR) in collaboration with the University of British Columbia, the University of Amsterdam, and local partners in the countries of implementation.

Data Availability Statement: The data will be available from the corresponding author upon reasonable request.

Acknowledgments: We acknowledge the tremendous technical support in preparing the land-cover maps by Keddy Mbindo of the Forestry Department and Chenje Mtonga and Christopher Silengwa from the National Remote Sensing Centre staff in Zambia. We also thank Kwabena Asubonteng of the University for Development Studies in Ghana for offering his GIS and remote sensing expertise to check the geospatial procedures, analysis, and outcomes in this manuscript.

Conflicts of Interest: The authors declare no conflict of interest.

Appendix A

Table A1. Specification of satellite data used for the Chikanta landscape.

Year	Filter Dates	Image	Satellite	Band Combination
1984	1 September 1984–30 September 2020	LT05/CO1/T1_8 DAY_NDVI	Landsat	432
2020	1 September 2020–30 September 2020	LC08/C01/T1_8 DAY_TOA	Landsat	543

Notes

1 COLANDS (https://www.cifor-icraf.org/colands/, accessed on 16 October 2022) stands for Collaborating to Operationalise Landscape Approaches for Nature, Development and Sustainability and is an initiative led by the Centre for International Forestry Research (CIFOR) in cooperation with the University of British Columbia and the University of Amsterdam and local partners in Ghana, Zambia, and Indonesia (Reed et al. 2020).

2 Village 2 is part of the COLANDS initiative, of which this study is a part but was not included in this paper as it had no multistakeholder platform.

References

1. Bartolini, N.; DeSilvey, C. Landscape futures: Decision-making in uncertain times, a literature review. *Landsc. Res.* **2021**, *46*, 8–24. [CrossRef]
2. Reed, J.; Deakin, L.; Sunderland, T. What are "Integrated Landscape Approaches" and how effectively have they been implemented in the tropics: A systematic map protocol. *Environ. Evid.* **2015**, *4*, 2. [CrossRef]
3. Pelletier, J.; Paquette, A.; Mbindo, K.; Zimba, N.; Siampale, A.; Chendauka, B.; Siangulube, F.; Roberts, J.W. Carbon sink despite large deforestation in African tropical dry forests (miombo woodlands). *Environ. Res. Lett.* **2018**, *13*, 094017. [CrossRef]
4. Kusiima, S.K.; Egeru, A.; Namaalwa, J.; Byakagaba, P.; Mfitumukiza, D.; Mukwaya, P.; Mensah, S.; Asiimwe, R. Interconnectedness of ecosystem services potential with Land-Use/Land-Cover change dynamics in Western Uganda. *Land* **2022**, *11*, 2056. [CrossRef]
5. Do, T.H.; Vu, T.P.; Catacutan, D.; Nguyen, V.T. Governing landscapes for ecosystem services: A participatory land-use scenario development in the Northwest Montane Region of Vietnam. *Environ. Manag.* **2021**, *68*, 665–682. [CrossRef]
6. Lubida, A.; Veysipanah, M.; Pilesjo, P.; Mansourian, A. Land-use planning for sustainable urban development in Africa: A spatial and multi-objective optimization approach. *Geod. Cartogr.* **2019**, *45*, 1–15. [CrossRef]

7. Solly, A. Land use challenges, sustainability and the spatial planning balancing act: Insights from Sweden and Switzerland. *Eur. Plan. Stud.* **2021**, *29*, 637–653. [CrossRef]
8. McCall, M.K. Seeking good governance in participatory-GIS: A review of processes and governance dimensions in applying GIS to participatory spatial planning. *Habitat Int.* **2003**, *27*, 549–573. [CrossRef]
9. Mccall, M.K.; Dunn, C.E. Geo-information tools for participatory spatial planning: Fulfilling the criteria for "good" governance? *Geoforum* **2012**, *43*, 81–94. [CrossRef]
10. Asubonteng, K.O.; Ros-Tonen, M.A.F.; Baud, I.; Pfeffer, K. Envisioning the future of mosaic landscapes: Actor perceptions in a mixed cocoa/oil-palm area in Ghana. *Environ. Manag.* **2020**, *68*, 701–719. [CrossRef]
11. Ros-Tonen, M.A.F.; Willemen, L. Editorial: Spatial tools for integrated and inclusive landscape governance. *Environ. Manag.* **2021**, *68*, 605–610. [CrossRef] [PubMed]
12. Aggrey, J.J.; Ros-Tonen, M.A.F.; Asubonteng, K.O. Using participatory spatial tools to unravel community perceptions of land-use dynamics in a mine-expanding landscape in Ghana. *Environ. Manag.* **2021**, *68*, 720–737. [CrossRef] [PubMed]
13. Baggio, J.A.; Brown, K.; Hellebrandt, D. Boundary object or bridging concept? A citation network analysis of resilience. *Ecol. Soc.* **2015**, *20*, 2. [CrossRef]
14. Sayer, J.; Sunderland, T.; Ghazoul, J.; Pfund, J.-L.; Sheil, D.; Meijaard, E.; Venter, M.; Boedhihartono, A.K.; Day, M.; Garcia, C.; et al. Ten principles for a landscape approach to reconciling agriculture, conservation, and other competing land uses. *Proc. Natl. Acad. Sci. USA* **2013**, *110*, 8349–8356. [CrossRef]
15. Milder, J.C.; Hart, A.K.; Dobie, P.; Minai, J.; Zaleski, C. Integrated landscape initiatives for African agriculture, development, and conservation: A Region-wide assessment. *World Dev.* **2014**, *54*, 68–80. [CrossRef]
16. Beunen, R.; Opdam, P. When landscape planning becomes landscape governance, what happens to the science? *Landsc. Urban Plan.* **2011**, *100*, 324–326. [CrossRef]
17. Görg, C. Landscape governance. The "politics of scale" and the "natural" conditions of places. *Geoforum* **2007**, *38*, 954–966. [CrossRef]
18. Armitage, D.; Marschke, M.; Plummer, R. Adaptive co-management and the paradox of learning. *Glob. Environ. Chang.* **2008**, *18*, 86–98. [CrossRef]
19. Star, S.L.; Griesemer, J.R. Institutional Ecology, "Translations" and boundary objects: Amateurs and professionals in Berkeley's Museum of vertebrate zoology, 1907-39. *Soc. Stud. Sci.* **1989**, *19*, 387–420. [CrossRef]
20. Opdam, P.; Nassauer, J.I.; Wang, Z.; Albert, C.; Bentrup, G.; Castella, J.-C.; McAlpine, C.; Liu, J.; Sheppard, S.; Swaffield, S. Science for action at the local landscape scale. *Landsc. Ecol.* **2013**, *28*, 1439–1445. [CrossRef]
21. Ansell, C.; Gash, A. Collaborative governance in theory and practice. *J. Public Adm. Res. Theory* **2008**, *18*, 543–571. [CrossRef]
22. Mansourian, S.; Sgard, A. Diverse interpretations of governance and their relevance to forest landscape restoration. *Land Use Policy* **2019**, *104*, 104011. [CrossRef]
23. Ros-Tonen, M.A.F.; Willemen, L.; McCall, M.K. Spatial tools for integrated and inclusive landscape governance: Toward a new research agenda. *Environ. Manag.* **2021**, *68*, 611–618. [CrossRef] [PubMed]
24. van Oosten, C.; Gunarso, P.; Koesoetjahjo, I.; Wiersum, F. Governing forest landscape restoration: Cases from Indonesia. *Forests* **2014**, *5*, 1143–1162. [CrossRef]
25. Speer, J. Participatory Governance Reform: A good strategy for increasing government responsiveness and improving public services? *World Dev.* **2012**, *40*, 2379–2398. [CrossRef]
26. Gupta, J.; Vegelin, C. Sustainable development goals and inclusive development. *Int. Environ. Agreem. Politics Law Econ.* **2016**, *16*, 433–448. [CrossRef]
27. Pouw, N.; Gupta, J. Inclusive development: A multi-disciplinary approach. *Curr. Opin. Environ. Sustain.* **2017**, *24*, 104–108. [CrossRef]
28. Arts, K. Inclusive sustainable development: A human rights perspective. *Curr. Opin. Environ. Sustain.* **2017**, *24*, 58–62. [CrossRef]
29. Marcaletti, F.; Riniolo, V. A participatory governance model towards the inclusion of ethnic minorities. An action research experience in Italy. *Interv. Écon.* **2015**, *53*, 1–21. [CrossRef]
30. Dryzek, J.S. *Deliberative Global Politics: Discourse and Democracy in a Divided World*; Polity: Cambridge, UK, 2006.
31. Bäckstrand, K.; Khan, J.; Kronsell, A.; Lövbrand, E. (Eds.) The Promise of New Modes of Environmental Governance. In *Environmental Politics and Deliberative Democracy: Examining the New Promise of New Modes of Governance*; Edwards Elgar Publishing: Cheltenham, UK, 2010; pp. 3–27. [CrossRef]
32. Everard, M. Managing socio-ecological systems: Who, what and how much? The case of the Banas river, Rajasthan, India. *Curr. Opin. Environ. Sustain.* **2020**, *44*, 16–25. [CrossRef]
33. Ros-Tonen, M.A.F.; Reed, J.; Sunderland, T. From Synergy to Complexity: The trend toward integrated value chain and landscape governance. *Environ. Manag.* **2018**, *62*, 1–14. [CrossRef]
34. Upla, P.; Reed, J.; Moombe, K.B.; Kazule, B.J.; Mulenga, B.P.; Ros-Tonen, M.; Sunderland, T. Assessing the potential for private sector engagement in integrated landscape approaches: Insights from value-chain analyses in Southern Zambia. *Land* **2022**, *11*, 1549. [CrossRef]
35. Moombe, K.B.; Siangulube, F.S.; Mwaanga, B.M.; Mfuni, T.; Yanou, M.P.; Gumbo, D.J.; Mwansa, R.C.; Juunza, G. Understanding landscape dynamics A case study from Kalomo District. In *Operationalizing Integrated Landscape Approaches in the Tropics*; Reed, J., Ros-Tonen, M.A.F., Sunderland, T., Eds.; CIFOR: Bogor, Indonesia, 2020; pp. 148–175.

36. Siangulube, F.; Ros-Tonen, M.; Reed, J.; Djoud, H.; Gumbo, D.; Sunderland, T. Navigating power imbalances in landscape governance: A network and influence analysis in southern Zambia. *Reg. Environ. Chang.* **2022**, *23*, 41. [CrossRef]
37. Reed, J.; Ros-Tonen, M.; Sunderland Terry, C.H. *Operationalizing Integrated Landscape Approaches in the Tropics*; CIFOR: Bogor, Indonesia, 2020. [CrossRef]
38. Government of the Republic of Zambia. *The Registration and Development of Villages Act, Chapter 289 of the Laws of Zambia*; Government Printers: Lusaka, Zambia, 1994.
39. Mbanga, T.M.; Mulenga, M.C.; Membele, G. Monitoring forest cover change in Kalomo Hills local forest using remote sensing and GIS: 1984–2018. *J. Remote Sens. GIS* **2021**, *10*, 1–7.
40. Ministry of Lands and Natural Resources. *Integrated Land Use Assessment Phase II—Report for Zambia*; The Food and Agriculture Organization of the United Nations and the Forestry Department: Lusaka, Zambia, 2017.
41. Nasiri, V.; Deljouei, A.; Moradi, F.; Sadeghi, S.M.M.; Borz, S.A. Land use and land cover mapping using Sentinel-2, Landsat-8 Satellite Images, and Google Earth Engine: A comparison of two composition methods. *Remote Sens.* **2022**, *14*, 1977. [CrossRef]
42. Liu, L.; Xiao, X.; Qin, Y.; Wang, J.; Xu, X.; Hu, Y.; Qiao, Z. Mapping cropping intensity in China using time series Landsat and Sentinel-2 images and Google Earth Engine. *Remote Sens. Environ.* **2020**, *239*, 111624. [CrossRef]
43. Gorelick, N.; Hancher, M.; Dixon, M.; Ilyushchenko, S.; Thau, D.; Moore, R. Google Earth Engine: Planetary-scale geospatial analysis for everyone. *Remote Sens. Environ.* **2017**, *202*, 18–27. [CrossRef]
44. Phiri, D.; Morgenroth, J.; Xu, C. Four decades of land cover and forest connectivity study in Zambia—An object-based image analysis approach. *Int. J. Appl. Earth Obs. Geoinf.* **2019**, *79*, 97–109. [CrossRef]
45. Olofsson, P.; Foody, G.M.; Herold, M.; Stehman, S.V.; Woodcock, C.E.; Wulder, M.A. Good practices for estimating area and assessing accuracy of land change. *Remote Sens. Environ.* **2014**, *148*, 42–57. [CrossRef]
46. Naboureh, A.; Li, A.; Bian, J.; Lei, G.; Amani, M. A hybrid data balancing method for classification of imbalanced training data within google earth engine: Case studies from mountainous regions. *Remote Sens.* **2020**, *12*, 3301. [CrossRef]
47. Feizizadeh, B.; Omarzadeh, D.; Kazemi Garajeh, M.; Lakes, T.; Blaschke, T. Machine learning data-driven approaches for land use/cover mapping and trend analysis using Google Earth Engine. *J. Environ. Plan. Manag.* **2023**, *66*, 665–697. [CrossRef]
48. Congedo, L. Semi-Automatic Classification Plugin: A Python tool for the download and processing of remote sensing images in QGIS. *J. Open Source Softw.* **2021**, *6*, 3172. [CrossRef]
49. Schober, M.F.; Conrad, F.G.; Fricker, S.S. Misunderstanding standardized language in research interviews. *Appl. Cogn. Psychol.* **2004**, *18*, 169–188. [CrossRef]
50. Nyumba, O.T.; Wilson, K.; Derrick, C.J.; Mukherjee, N. The use of focus group discussion methodology: Insights from two decades of application in conservation. *Methods Ecol. Evol.* **2018**, *9*, 20–32. [CrossRef]
51. Morgan, D.L. Focus groups and social interaction. In *Interview Research the Complexity of the Craft*, 2nd ed.; Gubrium, J.F., Holstein, J.A., Marvasti, A.B., McKinney, K.D., Eds.; SAGE Publications Inc.: New York, NY, USA, 2012; pp. 161–176.
52. Chilala, C.; Jimaima, H. The Tonga art of Kulibanda: A literary and linguistic analysis. *J. Law Soc. Sci.* **2020**, *3*, 160–177. [CrossRef]
53. German Technical Cooperation (GTZ). *District Development Project-Southern: Strategic District Development Planning*; German Technical Cooperation (GTZ): Kalomo, Zambia, 2019.
54. Phiri, J.; Malec, K.; Majune, S.K.; Appiah-Kubi, S.N.K.; Gebeltová, Z.; Maitah, M.; Maitah, K.; Abdullahi, K.T. Agriculture as a determinant of Zambian economic sustainability. *Sustainability* **2020**, *12*, 4559. [CrossRef]
55. McCall, M.K.; Minang, P.A. Assessing participatory GIS for community-based natural resource management: Claiming community forests in Cameroon. *Geogr. J.* **2005**, *171*, 340–356. [CrossRef]
56. Kusters, K.; Buck, L.; de Graaf, M.; Minang, P.; van Oosten, C.; Zagt, R. Participatory planning, monitoring and evaluation of multi-stakeholder platforms in integrated landscape initiatives. *Environ. Manag.* **2018**, *62*, 170–181. [CrossRef]
57. Handavu, F.; Chirwa, P.W.C.; Syampungani, S. Socioeconomic factors influencing land-use and land-cover changes in the miombo woodlands of the Copperbelt Province in Zambia. *For. Policy Econ.* **2019**, *100*, 75–94. [CrossRef]
58. Mukwena, R. *Zambia at Fifty Years: What Went Right, What Went Wrong and Wither to? A Treatise of the Country's Socioeconomic and Political Developments since Independence*; Partridge Africa: Johannesburg, South Africa, 2016.
59. Ruggeri, D. The agency of place attachment in the contemporary co-production of community landscapes. In *Place Attachment*; Routledge: London, UK, 2020; pp. 243–260. [CrossRef]
60. Shantiko, B.; Liswanti, N.; Bourgeois, R.; Laumonier, Y. Land-use decisions in complex commons: Engaging multiple stakeholders through foresight and scenario building in Indonesia. *Environ. Manag.* **2021**, *68*, 642–664. [CrossRef] [PubMed]
61. McCall, M.K. Participatory Mapping and PGIS: Secerning facts and values, representation and representativity. *Int. J. E-Plan. Res.* **2021**, *10*, 105–123. [CrossRef]
62. Chambers, R. Participatory mapping and geographic information systems: Whose map? Who is empowered and who disempowered? Who gains and who loses? *Electron. J. Inf. Syst. Dev. Ctries* **2006**, *25*, 1–11. [CrossRef]
63. Friis, C.; Nielsen, J.Ø. On the system. Boundary choices, implications, and solutions in telecoupling land use change research. *Sustainability* **2017**, *9*, 974. [CrossRef]
64. Lagopoulos, A. Clarifying theoretical and applied land-use planning concepts. *Urban Sci.* **2018**, *2*, 17. [CrossRef]
65. Irving, A. The modern/postmodern divide and urban planning. *Univ. Tor. Q.* **1993**, *62*, 474–487. [CrossRef]

66. Mulenga, C.; Ngoma, A.; Sakala, J. Rural land use planning—The integration of shared resources mapping for improved communal tenure security: Experiences from Zambia. In *2019 Annual World Bank Conference on Land and Poverty*; World Bank: Washington, DC, USA, 2019; pp. 1–25.
67. Brown, G.; Kyttä, M. Key issues and priorities in participatory mapping: Toward integration or increased specialization? *Appl. Geogr.* **2018**, *95*, 1–8. [CrossRef]

land

MDPI

Article

Using Scenario Building and Participatory Mapping to Negotiate Conservation-Development Trade-Offs in Northern Ghana

Eric Rega Christophe Bayala [1,2,3,*], Kwabena Owusu Asubonteng [4], Mirjam Ros-Tonen [1,2], Houria Djoudi [3], Freddie Sayi Siangulube [2,3], James Reed [3,5] and Terry Sunderland [3,6]

1 Amsterdam Institute for Social Science Research (AISSR), University of Amsterdam, 1018 WV Amsterdam, The Netherlands
2 Department of Human Geography, Planning and International Development, University of Amsterdam, 1018 WV Amsterdam, The Netherlands
3 Center for International Forestry Research (CIFOR), Bogor 16115, Indonesia
4 Department of Natural Resources and Geo-Information Sciences, Faculty of Natural Resources and Environment, University for Development Studies (UDS), Nyankpala Campus, Tamale P.O. Box TL 1882, Ghana
5 School of International Development, University of East Anglia, Norwich Research Park, Norwich NR4 7TJ, UK
6 Faculty of Forestry, University of British Columbia (UBC), Vancouver, BC V6T 1Z4, Canada
* Correspondence: e.r.c.bayala@uva.nl or bayalaeric2@gmail.com; Tel.: +22-66-407-5088

Abstract: In multifunctional landscapes, expanding economic activities jeopardise the integrity of biodiverse ecosystems, generating conservation-development trade-offs that require multi-stakeholder dialogue and tools to negotiate conflicting objectives. Despite the rich literature on participatory mapping and other tools to reveal different stakeholder perspectives, there is limited evidence on the application of such tools in landscape-scale negotiations. This paper addresses this gap by analysing a participatory mapping process in Ghana's Western Wildlife Corridor, where a community-based landscape governance system called the community resource management area (CREMA) exists. Data from three participatory mapping workshops and focus group discussions with community and institutional actors reveal that increasing demand for food and natural resources and climate change impacts are drivers of landscape degradation, resulting in declining faunal and floral biodiversity and reduced ecosystem services. Meanwhile, community actors prioritise the expansion of farming land, while institutional actors prioritise forest conservation. However, scenario building and participatory mapping helped communicate each other's aims and reach a negotiated consensus. Finally, power relations, cultural and traditional rules, and differences in knowledge affected deliberations and decision-making. We conclude that scenario building and participatory mapping can contribute to an inclusive landscape approach, provided that well-functioning multi-stakeholder platforms are in place and facilitators adequately navigate power imbalances and recognise different kinds and degrees of knowledge.

Keywords: conservation-development trade-offs; consensus-building; participatory mapping; scenario building; integrated landscape approach; community resource management area (CREMA); northern Ghana

1. Introduction

Effective governance of multifunctional landscapes requires decision-making processes that consider divergent views, values, interests, and expectations. Such processes require balancing the influence and power of diverse stakeholders to formulate and maintain a common vision in the face of dynamic socio-ecological changes at the landscape scale [1]. A multitude of stakeholders have important roles in managing landscapes and natural resources, and as such a broad spectrum of views should be taken into account [2]. Therefore, there is a growing willingness and commitment to include previously marginalised groups, such as smallholders, women, Indigenous peoples, and local communities, in decisions concerning sustainable landscape governance and management [3]. Often, the people who have traditionally been excluded from decision-making are those who depend on resources the most.

Beyond the technical challenges of managing conflicting interests and finding compromises regarding land and resource use, landscape governance involves reconciling knowledge and considerations of key stakeholders, defining responsibilities, and concerted planning that expresses priorities in terms of objectives, needs, and challenges [4–6]. This typically requires a multi-stakeholder dialogue process [3,5,7]. Against this background, integrated landscape approaches that promote multi-stakeholder dialogue for land-use decision-making and practice are increasingly endorsed and widely implemented [8,9]. Compared to sectoral approaches, integrated landscape approaches place a greater emphasis on making land-use trade-offs explicit and subject to negotiation in multi-stakeholder platforms [5,8,10,11]. However, multi-stakeholder platforms and forums are often idealised as frameworks for collaboration and consultation. Equality of voice and realising equitable outcomes within such processes are often challenging [3]. Tools for identifying different stakeholder perspectives are needed to achieve inclusive negotiation of trade-offs with all relevant parties. Therefore, place-based planning is increasingly supported by spatial tools such as participatory mapping and scenario building [10]. Indeed, the use of participatory spatial tools in natural resource governance has become common due to their potential to stimulate more inclusive processes that consider social, cultural, and professional differences as well as diverging interests [2,12]. The findings of studies conducted by Robinson et al. [13], Aggrey et al. [14], and Asubonteng et al. [15] demonstrated the potential of these tools to bring landscape stakeholders together to define common concern entry points and to encourage critical consultation and collective knowledge co-production among various landscape stakeholders.

Participatory mapping is the process whereby local communities visualise a landscape pictorially with the assistance of technical partners, such as government agencies, non-governmental organisations, research institutes, and other actors involved in the development and land-related planning [16]. The resulting participatory maps reflect the priorities and concerns of the stakeholders who produce them and can be used to shape new spatial realities. It is, therefore, important that all those who will be affected by maps of their landscape actively participate in their creation [17]. A well-designed and facilitated multi-stakeholder process can enhance the inclusion of all key stakeholders in the mapping process. Participatory scenario building aims to discuss plausible futures of complex and dynamic systems among stakeholders with diverging interests to enhance the inclusion of local voices in decision-making, knowledge exchange, and social learning [18–20].

In Ghana, the Wildlife Division of the Forestry Commission initiated Community Resource Management Areas (CREMAs) that devolve natural resource decision-making to the community level and emphasise consultation and collaboration between key landscape stakeholders [21]. The CREMA system stresses the need for synergy of actions and consensus in decision-making between government agencies, NGOs, the private sector, community leaders and traditional chiefs, district assemblies, and relevant technical and financial partners [21]. This requires the effective participation of these actors in landscape decision-making and negotiation of trade-offs. However, CREMA governance faces constraints, including the often low capacity of governing bodies to lead discussions among stakeholder groups, particularly intercultural and inter-professional processes with divergent interests [22,23]. To make these diverging interests explicit, this paper analyses how scenario building and participatory mapping can help uncover actor aspirations for landscape and livelihood change and support consensus-building on land and natural resource use. Furthermore, it seeks to understand what happens during the process of trade-off negotiations in participatory mapping. Most studies involving the production of participatory maps focus on the final product (the map) and its interpretations, documenting little of the discussions and interactions between participants [13–15], with the process of consensus building rarely described [24]. Furthermore, attempts to operationalise integrated landscape approaches can engage different actor groups separately without joint negotiations hence failing to address power dynamics and their role in landscape decision-making [15].

The main question guiding this study is: What is the potential of participatory scenario building and mapping to support multi-stakeholder negotiations of landscape change in the Western Wildlife Corridor (WWC) CREMA governance system? Sub-questions are:

- How do local actors perceive changes in the landscape and their impacts on conservation and livelihoods?
- How do local actors visualise their desired future landscape and the underlying conservation and livelihood objectives?
- What factors influence negotiations of trade-offs and consensus on a common desired landscape?
- What lessons can be learned from the mapping process in relation to the operationalisation of integrated landscape approaches in the WWC?

After outlining the research methodology, this paper elaborates on the trade-offs and consensus concepts. It then analyses the landscape changes perceived by local stakeholders, their impacts on conservation and livelihoods, how they impact landscape composition and configuration now and in stakeholders' desired future, and the factors involved in negotiating potential trade-offs. After discussing the implications of the findings, the concluding section presents lessons and recommendations emerging from the multi-stakeholder participatory mapping process.

2. Methodology

2.1. Study Area

This study is a component of the Collaborating for Operationalising Landscape Approaches for Nature, Development and Sustainability (COLANDS) initiative, coordinated by the Centre for Forestry Research (CIFOR) in collaboration with the universities of British Columbia and Amsterdam and local organisations in the countries where it is implemented, namely Ghana, Zambia, and Indonesia [25]. The study was carried out in the Western Wildlife Corridor (WWC) in northern Ghana, including six communities spread among three CREMAs selected from the six within the WWC landscape (Figure 1). Builsa Yenning (BY), Moagduri Wuntanluri Kuwomsaasi (MWK), and Sanyiga Kasena Gavara Kara (SKGK) CREMAs were selected because they already had their devolution certificates and functional governance structures in place at the outset of the study (2019) [26]. Furthermore, studying three different CREMAs allows for a diversity of views on the evolution of the landscape and provides a representative spatial sample of the landscape. The communities are Fumbisi and Kunyinsa in the Builsa Yenning CREMA, Yizesi and Zukpeni in the MWK CREMA, and Nakong and Kwapun in the SKGK CREMA. They were chosen for their proximity to forest reserves, their role in the CREMA as the seat of the CREMA Executive Committee (CEC), and accessibility.

Located in the northern part of Ghana, the WWC belongs to the savannah ecological zone. Despite significant pressure on natural resources, this area harbours a rich biodiversity of flora and fauna, important for local livelihoods [27]. However, northern Ghana is characterised by a contentious socio-economic context and a high vulnerability to climate change, compounded by the extreme poverty faced by the majority of the rural population [28]. Local communities strongly depend on natural resources, resulting in high pressure on natural ecosystems, leading to landscape degradation and fragmentation [29]. The main livelihood activity is agriculture, which is essentially rainfed and depends on the maintenance of soil fertility [30]. Alongside agriculture, local people earn a living through hunting, charcoal production, mining, livestock, the sale of forest products, and petty trade (food processing and weaving) [31]. Thus, savannah landscapes such as the WWC are multifunctional areas supporting the local economy and achieving biodiversity conservation goals [26]. Such tropical, multifunctional landscapes are characterised by a mosaic of fragmented forests, areas of agroforestry and monocultures, as well as human settlements, which offer the potential to combine livelihood development and biodiversity conservation [32]. In terms of landscape governance, the WWC CREMAs, which were created relatively recently (between 2016 and 2019), face difficulties related to the functioning of governance bodies, financial and technical resources, and conflicts between various resource users [26].

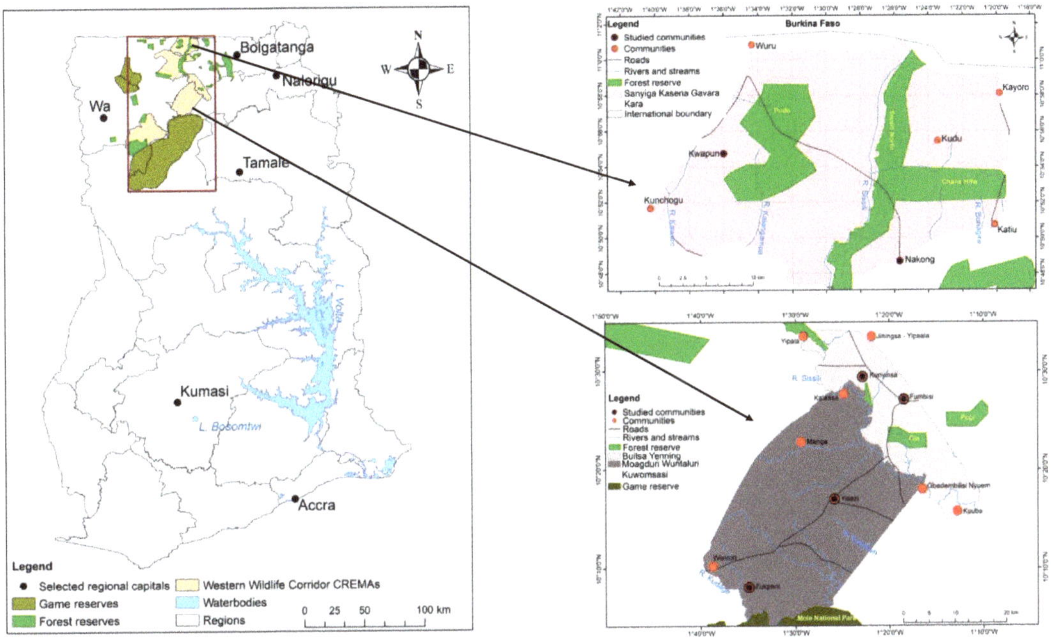

Figure 1. Location of the study area (Source: Authors).

2.2. Data Collection

The study employed focus group discussions and integrated scenario-building and participatory mapping in workshops for gathering data. A total of 34 focus group discussions were held from April to June 2021 with farmer groups, Fulani pastoralists, forest operators, women, youths (18–40 years)[1], and elders (60 years and over) in the three CREMAs. The aim was to capture the perceptions of each stakeholder group on the changes in the landscape, the drivers of change, and the resultant impacts on conservation and livelihoods. In collaboration with the Community Resource Management Committee (CRMC) leaders—community representatives in the CREMA governance structure—the participants (six per actor group) were chosen by convenience sampling based on availability and knowledge of the CREMA's functioning.

The integrated scenario-building and participatory mapping methodology used for this study adapted the approach developed by Asubonteng et al. [15]. It consisted of organising three participatory mapping and scenario-building workshops to discuss the past and anticipated changes, desired future landscapes, and stakeholders' underlying conservation and livelihood aims. This comprised scenario-building exercises and participatory mapping, first by each stakeholder group separately and then by all the stakeholders together (this last step was not conducted by Asubonteng et al. [15]). For the mapping, the Builsa Yenning (BY) and Moagduri Wuntanluri Kuwomsaasi (MWK) CREMAs were put together to form one landscape due to their proximity, while the Sanyiga Kasena Gavara Kara (SKGK) CREMA, which is more distant from the others, formed another landscape.

Basic maps (A1 map frames) of the respective CREMAs, showing landmarks such as forest reserve boundaries, rivers, roads, and well-known villages, were provided to facilitate the mapping process (see Figures A1 and A2 in the Appendix A). Then, symbols representing key land-cover types in the landscape were printed and cut out to allow participants to create their maps by sticking them onto the frame maps. These working materials, glue, and markers were made available to each group to map the composition and spatial configuration of the landscape in its current and desired future state. A time step of 20 years was considered for the scenarios and mapping with reference to the beginning of the CREMAs initiative. In addition, we considered that 20 years is sufficiently long to perceive changes and simulate future landscape dynamics (see [27,33–35]). The objective was, therefore, to understand the changes in the landscape from the beginning of the creation of the CREMAs (in the 2000s) to the present day (2021) and the perspectives of the stakeholders in terms of desired future landscapes in 20 years (the 2040s). Furthermore, the purpose was to identify stakeholders' land-use priorities and suggested scenarios to balance conservation and development objectives, which could contribute to the consensual and inclusive future planning of natural resource use in the area. Indeed, participatory scenario building allows for the inclusion of stakeholders' suggestions for pathways towards good landscape governance and integrated development [12,36].

Each workshop started with a PowerPoint presentation given by the research team (consisting of the first two authors) to harmonise understanding of the study's objectives and some concepts such as landscape, landscape composition, configuration, ecosystem services, landscape approaches, trade-offs, and consensus. This presentation was an opportunity to put the participants in the mood for the study. The moderator had a good sense of humour, sharing stories and jokes, and the explanations helped participants feel confident about the work and mapping to be done. It was also an opportunity to discuss the concept of CREMA and its objectives.

The first workshop was held with only local community representatives in Fumbisi on the 5th of December 2021 and attended by 12 participants (including representatives of farmers, youth, women, and CREMA leaders) from the three CREMAs targeted for the study. The participants comprised four representatives per CREMA (two per target community), selected by the CREMA leaders, ensuring a good spread across gender and age and good knowledge of the CREMA area. Unfortunately, attempts to include representatives of the Fulani pastoralists were unsuccessful due to tense relations between the population groups [37,38]. The second workshop involved only institutional actors and was held in Bolgatanga on the 6th of December 2021. It involved 12 participants, including representatives of public agencies (nine), the private sector (one), a research institute (one), and NGOs (one). The final workshop involved local communities and institutional actors together. This workshop, which took place in Navrongo on the 7th of December 2021, brought together 16 participants (10 from communities and six from public agencies and NGOs). In this last workshop, the representatives of the private sector and researchers were absent. Table 1 summarises the type and number of participants in the three workshops.

The three workshops enabled data collection on perceptions of the current state of the landscape; perceptions of desired landscapes by institutional actors and communities; and factors influencing trade-off negotiations such as power dynamics, knowledge, and culture. With participants' consent, photographs were taken, and the workshop processes were audio recorded, both during participatory mapping exercises and plenary discussions. Notes were also taken based on observations of participants' attitudes during the discussions.

Table 1. Type of participants in the workshops.

Type of Participant	Institution/Group	Number of Participants
	First Workshop	
Local community	Farmers	3
	Women	3
	Youths	3
	CREMA Executive Committee (CEC)	3
Total	-	12
	Second Workshop	
Public agencies	Builsa South District Assembly	1
	Forest Services Division (FSD) of the Forestry Commission	1
	Wildlife Division (WD) of the Forestry Commission	1
	Ministry of Food and Agriculture (MoFA) (Regional office)	4
	Environmental Protection Agency (EPA) (Regional office)	2
Private sector	Organisation for Indigenous Initiatives and Sustainability (ORGIIS)	1
NGO	Tree Aid	1
Research Institute	Forestry Research Institute of Ghana (FORIG) (Regional office)	1
Total	-	12
	Third Workshop	
Public agencies	Builsa South District Assembly	1
	Forest Services Division (FSD) of the Forestry Commission	1
	Wildlife Division (WD) of the Forestry Commission	1
	Ministry of Food and Agriculture (MoFA) (Regional office)	1
	Environmental Protection Agency (EPA) (Regional office)	1
NGO	Tree Aid	1
Local community	CREMA Executive Committee (CEC)	3
	Farmers	2
	Women	2
	Youths	3
Total	-	16

2.3. *Scenario Building and Mapping Process*

The scenario-building exercise was conducted following five main steps (see Figure 2). After the PowerPoint presentation in plenary, the workshop participants were asked first to describe the perceived changes in their landscapes since the creation of CREMAs; second, to identify the main drivers of these changes; third, to represent the current perceived key land-cover types (composition) of their landscapes; fourth, to give their perceptions on how their landscapes would evolve under a business-as-usual scenario; and fifth, to give the composition of their desired future landscape. Hence, the scenario-building exercise combined predictive forecasting (what will happen under a business-as-usual scenario) to make participants aware of how the landscape will likely look without interventions, and normative scenario building, focusing on participants' desired future landscapes [39]. The exercise was undertaken in plenary at each of the workshops, and each participant gave opinions regarding their respective landscapes (SKGK or BY-MWK for the community representatives and both landscapes for the institutional actors). It was organised in such a way as to focus on one landscape at a time. For example, while the SKGK representatives had discussions about their landscape at the local communities' workshop, the BY-MWK representatives silently observed and waited their turn. The same was the case in the joint workshop, except that institutional actors participated in the discussions on both landscapes.

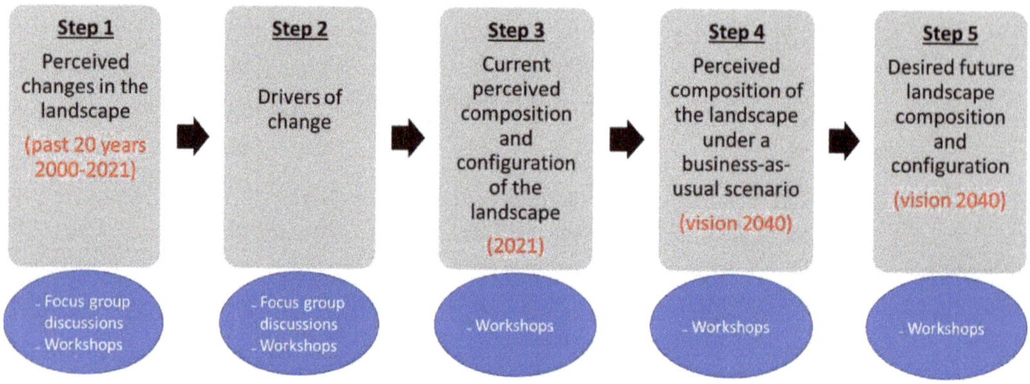

Figure 2. Participatory scenario-building methodology.

The land allocations for each land cover were displayed in a pie chart and projected for participants to visualise and generate discussion. The proportions of land-cover types in each landscape were only retained when all participants concerned with that landscape agreed. There was more debate and negotiation about future landscapes than about current landscapes. This was because participants did not easily agree on what they wanted for the future and land proportions to allot, and unsurprisingly, one could see that everyone was trying to push through a scenario that favoured their interests, with others sometimes trying to impose their own (see Section 4.5.2).

After this exercise, the participants organised themselves around the map that concerned them, following the moderator's instructions, to start the mapping. Participants had to agree on the desired configuration based on the composition of their current and future landscapes, which had been discussed in the plenary. The communities used the current state of the landscape configuration and the difficulties associated with it as a basis for deciding on the desired changes. The institutional actors, guided by national policies and laws, focused on what they considered a balanced landscape (in terms of a balance between conservation and livelihood activities). Each participant explained their point of view, trying to convince the others of the relevance of the proposed configuration. Discussions could go on for quite some time before the group agreed and started to paste the symbols of the land-use types on the map frame. This exercise was not without disagreement, which at times escalated to heated exchanges of opinion. For example, two women entered into a verbal dispute at the local community workshop because they had conflicting opinions and could not agree. The moderator had to intervene to reduce the tension (he encouraged the two women to apologise to each other and then, using small jokes, restored a good working atmosphere) and helped the participants to listen to each other to understand each other better. The joint workshop between institutional actors and local communities had moments of tension (see Section 4.5.2). The moderator's intervention always calmed the situation and allowed each participant to express themself (teasingly encouraging the more timid to offer their point of view).

2.4. Data Processing and Analysis

Some data processing commenced on-site during the workshops. Data on perceived land-cover types and their estimated proportions in current and desired future landscapes displayed as pie charts during discussions was the first layer of analysis [15]. Not all recordings could be transcribed due to the poor quality of some (poorly captured voices and incomplete sentences) and because others were in local languages (sometimes, local actors used languages not understood by the research team during mapping sessions; to ensure that we did not miss essential debates, we asked questions to receive explanations in English and noted them down). Thus, quality audible portions of discussions made in English and field notes taken by the first two authors were used for the analyses. The maps were examined separately by two members of the research team, who then discussed to harmonise their findings [15,40]. This was conducted based on a visual comparison of maps, considering the spatial distribution of land-cover types, the dominant land-cover types, the arrangement of these land-cover types, and the status of forest reserves in the landscape. The data collected during the focus group discussions and the notes taken during the workshops were compiled thematically to facilitate the analyses. Themes included: perceived changes in the landscape, drivers of change, and effects of changes on livelihoods and conservation.

3. Key Concepts: Trade-Offs, Compromises, and Consensus

Reaching consensus requires listening to and considering all viewpoints and facilitating the participation of all stakeholders in the discussion and deliberations, thus increasing the quality and legitimacy of decision-making [24]. This usually involves compromise from the actors involved, resulting in new trade-offs when a decision on an action or initiative benefits one aspect or actor at the expense of another [41]. Trade-offs can therefore be seen as the result of a disagreement when people resent having made concessions, the choice of one outcome at the expense of another, or the abandonment of the desired situation in favour of another, depending on the evolution of circumstances, contexts, and realities [42].

Related to landscape approaches, trade-offs mainly refer to those between conservation and development [25,43–45], between landscape multifunctionality and agricultural productivity and profits [32,46], and between ecosystem services [47,48]. According to Galafassi et al. [41], dealing with trade-offs requires being informed about the advantages and disadvantages of a particular choice. This can be achieved through a relational learning process involving actors with different knowledge so that the full range of visible and hidden trade-offs can be revealed [41]. However, decisions are often made without being aware of the full potential repercussions, hence the need to undertake a prior analysis of trade-offs, taking into account social and landscape dynamics in time and space to identify possible implications [41]. Therefore, landscape governance requires stakeholders to make informed choices, and participatory mapping and scenario building can help in this process [15]. Even though they have been criticised for not necessarily leading to changes in the landscapes [49,50], these processes provide opportunities for stakeholders to generate and share knowledge, discuss challenges and priorities, make future projections, and agree on shared visions for the future of the landscape [15,19].

However, as landscape stakeholders have different interests and visions, what may be seen as a loss by some may be seen as a gain by others, and what appears to be a rational decision for one stakeholder may not make any sense at all for another [8,41]. Thus, depending on interest or motivation, different people perceive trade-offs differently [42]. Given the recurrent conflicts between conservation and livelihoods, conservationists and natural resource users must understand each other by discussing all perspectives, including biodiversity and socio-economic needs [51,52]. Understanding the motivation of stakeholders is necessary to negotiate a consensus that is satisfactory to all actors to foster their commitment to the implementation of the decision taken and the sustainable governance of the landscape [53,54].

Figure 3 illustrates the iterative and adaptive nature of the sustainable landscape governance process, highlighting five major steps related to stakeholder participation, negotiation of trade-offs and consensus, and genuine stakeholder engagement in governance. Mobilising all key landscape stakeholders within a common consultation platform is a crucial step to initiate inclusive trade-off negotiations and deliberations on land use.

The aim is to reconcile various interests, which is a prerequisite for achieving consensus among stakeholders on the use of their landscape resources. When stakeholders agree on a common vision and adopt consensual decisions, their commitment to implementing actions is more meaningful. This promotes inclusive and sustainable governance of landscape resources and balances conservation and development objectives.

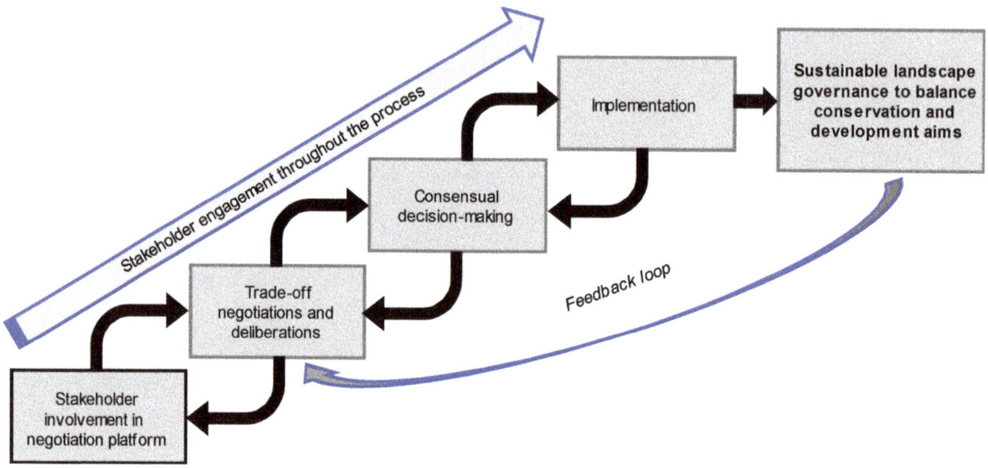

Figure 3. An iterative and adaptive conservation-development reconciliation process for sustainable landscape governance (Source: Authors).

4. Results

This section presents the perceived changes in the landscape over the last 20 years and those desired over the next 20 years. In addition, it discusses the perceptions of likely changes under a business-as-usual scenario. The study participants identified seven main land-cover types, namely, forests, waterbodies, farmlands, grasslands, tree crops, settlements, and mining, and represented them in the participatory maps. For ethical considerations, we have avoided disclosing any information that could help identify the study participants.

4.1. Perceived Changes in the Landscape, Their Drivers, and Impacts on Conservation and Livelihood

According to the actors interviewed in the WWC landscape, the significant changes include increased farmland, decreased forest, the impoverishment of soils, and the progressive siltation of water bodies. This evolution in time and space results from a growing demand for food and economic resources due to demographic growth, leading to an increase in the number and size of agricultural fields. In addition, the interviewees highlighted many other causes, namely indiscriminate bushfires, overgrazing, excessive removal of trees (for farms, charcoal, and trading), and mining. Beyond anthropogenic actions, a major cause of these changes is the amount, timing, and spatial distribution of rainfall. Most elderly groups attribute this to abandoning certain traditional rituals previously performed to attract rainfall and benefit from prosperous agricultural seasons. They also blamed it on the blatant breaches of taboos (due to the influence of new religions and modern knowledge). In contrast, young people and institutional actors attribute the decline in rainfall to climate change.

Regarding wildlife, many species are reported to have become rare in the area (Table 2). Both local communities and institutional actors view the increasing proximity of human settlements to wildlife habitats, the increasing population of hunters and poachers, and the intrusion of domestic animals and livestock into protected areas as the factors contributing to the reduction of wild animals.

Table 2. The perceived scarcity of wild animal species and declining plant species.

Common Name	Scientific Name *
Animal species	
Grasscutter	*Thryonomys swinderianus*
Crocodile	*Crocodylus niloticus*
Rabbit	*Cricetomys gambianus*
Bush fowl	*Numida meleagris*
Deer	*Cervus elaphus*
Patas monkey	*Erythrocebus patas*
Baboon	*Papio anubis*
Green monkey	*Cercopithecus aethiops*
Roan antelope	*Hippotragus equinus*
Partridge	*Alectoris rufa*
Buffalo	*Syncerus caffer*
Elephant	*Loxodonta africana*
Hyena	*Crocuta crocuta*
Bushpig	*Potamochoerus larvatus*
Plant species	
Dawadawa	*Parkia biglobosa*
Rosewood	*Pterocarpus erinaceus*
Baobab	*Adansonia digitata*
Kuka	*Kaya senegalensis*

Source: Field data, 2021. (* https://www.csir-forig.org.gh/tikfom/database (accessed on 22 December 2022); [55,56]).

Local communities perceive medicinal plants such as baobab, *kuka* (made from dried powdered baobab leaves), *dawadawa* or locust beans (*Parkia biglobosa*), and other plant-based food items (baobab fruits, *dawadawa* seeds), to have become rare due to the degradation of the forest. They also see a reduction in the population of the shea tree (*Vitellaria paradoxa*), the fruits and nuts of which are highly valued for food and trade. Women indicated that it is increasingly difficult to develop their income-generating activities based on non-timber forest products[2], especially fruits, and leaves, because of the significant decline in availability; "*Trees don't yield much anymore because they are old and few now*", they observed (Women FGD-3)[3]. In addition, many other ecosystem services have reduced significantly over time, according to the elders. These include fertile land, fish, wild fruits, clean water, protection from strong winds, and honey. The scarcity of wild animals has also negatively impacted people's livelihoods, such as hunting which is no longer able to meet market demands. Meanwhile, farmers complain that soil depletion has led to a decrease in agricultural yields, forcing them to expand their farms, whereas pastoralists lamented that the expansion of agricultural areas has reduced grazing areas.

4.2. Perceptions of Current and Future Composition of Landscapes' Future under a Business-As-Usual (BAU) Scenario

The SKGK CREMA stakeholders—representatives of different social groups, including women, youth, farmers, and CREMA leaders—perceive their present landscape to comprise 45% forest, 25% farms, 10% grassland, 10% settlements, 6% tree crops, 2% waterbodies, and 2% mining (Figure 4a). Similarly, BY-MWK stakeholders view their landscape to be dominated by forest (30%), followed by farms (21%), grassland (18%), settlements (19%), tree crops (7%), waterbodies (4%) and mining (1%) (Figure 5a). Notably, BY-MWK participants perceived approximately twice the proportions of settlement and grassland than those from SKGK. The community workshop participants unanimously agreed that their landscape would dramatically change in the next 20 years if the current rate of resource degradation continued. "*The landscape will be seriously destroyed if things continue like that. We are not happy with the current trend already*", said one participant at the workshop with communities. Additionally, according to a group of young people during the focus group discussions, "*The forest occupies too much land; we want it to reduce, but not to disappear*".[4]

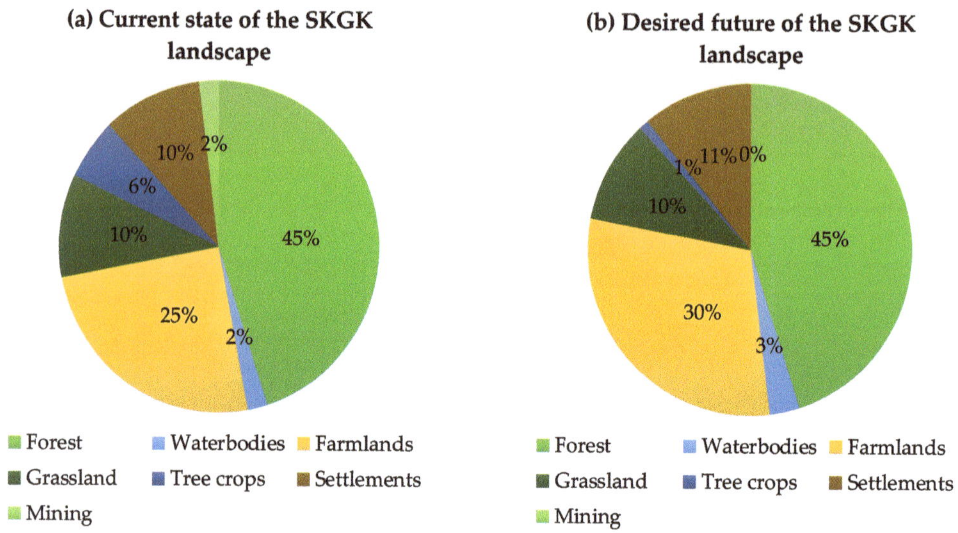

Figure 4. The current state (**a**) and desired future composition (**b**) of the Sanyiga Kasena Gavara Kara (SKGK) landscape as perceived by community actors.

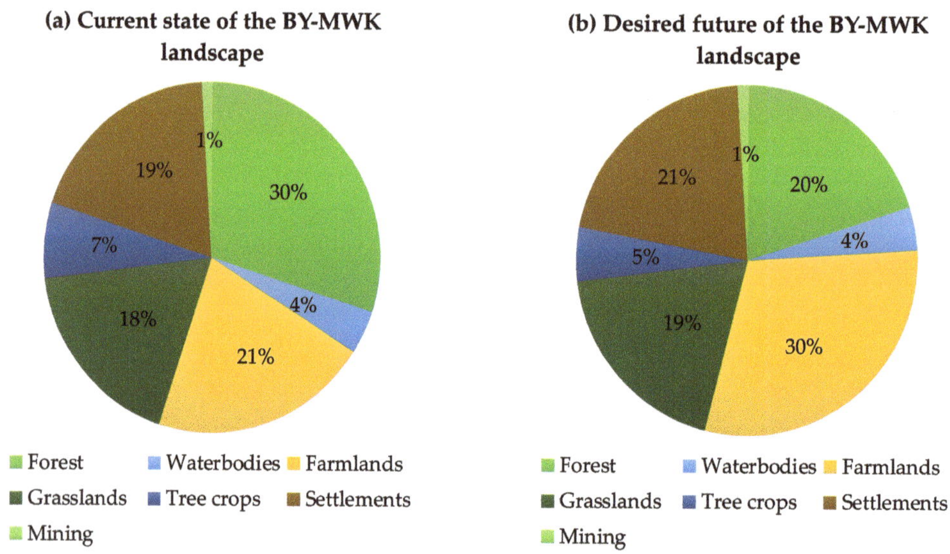

Figure 5. The current state (**a**) and desired future composition (**b**) of the Builsa Yenning (BY)-Moagduri Wuntanluri Kuwomsaasi (MWK) landscape as perceived by community actors.

Local communities envision their future landscape as heavily dominated by agricultural fields, human settlements, and tree crops under a BAU scenario. They foresee these expansions to be at the cost of forests, given that logging, farming, and mining activities are intensifying over time. Furthermore, given the projected extent of population growth, the demand for livelihoods will increase and require a larger area of agricultural land. This will contribute to deforestation through logging for timber, charcoal production, and mining that is already taking place, especially in the SKGK and MWK CREMAs. According to these communities, even the forest reserves in the area (Gia, Pudo, Chiana Hills) will be heavily destroyed, making life more difficult because there will be few ecosystem goods and services to support their survival. Settlement areas are expected to expand and densify to become more urbanised, characterised by increased road networks, buildings, and administrative services, as signs of this urbanisation process are already noticeable in Fumbisi, Nakong, Chiana, and Kayoro. In addition, there is an increasing interest in tree cash crops such as mango and cashew plantations.

Institutional stakeholders estimate the current composition of the SKGK sub-landscape to be 30% forest, 41% farmlands, 5% grassland, 15% settlements, 1% tree crops, 3% waterbodies, and 5% mining (Figure 6a), and that of the BY-MWK sub-landscape to be 40% forest, 35% farmlands, 6% grassland, 13% settlements, 2% tree crops, 3% waterbodies, and 1% mining (Figure 7a). They believe that urban development in the area will create more demand for forest products in the next 20 years. Fumbisi and Chiana already have significant markets for these products, which will probably intensify in the future. However, these actors do not see the future of the landscape under BAU as pessimistically as community actors. According to them, flora and faunal resources will decline due to agricultural, mining, and logging activities, leading to degrading natural ecosystems (forest and water bodies). However, they expect forests will always be among the dominant land-use types in the landscape, given the ongoing conservation efforts by public agencies and NGOs. In common with the local communities, they expect the area of tree crops to increase, as well as that of settlements, mines, and farms, while the area of forests, grassland, and waterbodies will decrease, but for them, all these changes are not considered to be very alarming.

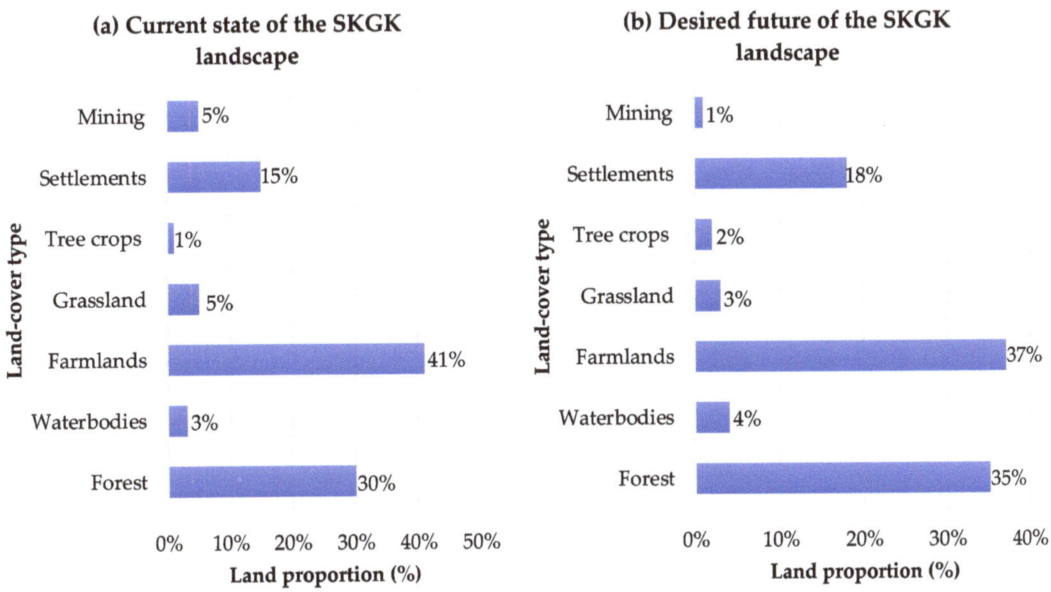

Figure 6. Perceived current state (**a**) and desired future composition (**b**) of the Sanyiga Kasena Gavara Kara (SKGK) landscape as visualised by institutional actors.

Among many other suggestions, such as promoting proper livestock-rearing practices to avoid indiscriminate grazing, good agricultural practices (sustainable agriculture, i.e., agroforestry), designing and enforcing by-laws, and greater involvement of traditional leaders, community and institutional actors supported the idea of a land-use plan that would allow for more sustainable resource use at the landscape level. The next sections, therefore, explore how stakeholders in three community management areas (CREMAs) negotiate the trade-offs and arrive at a consensual landscape scenario that allows them to achieve their objectives.

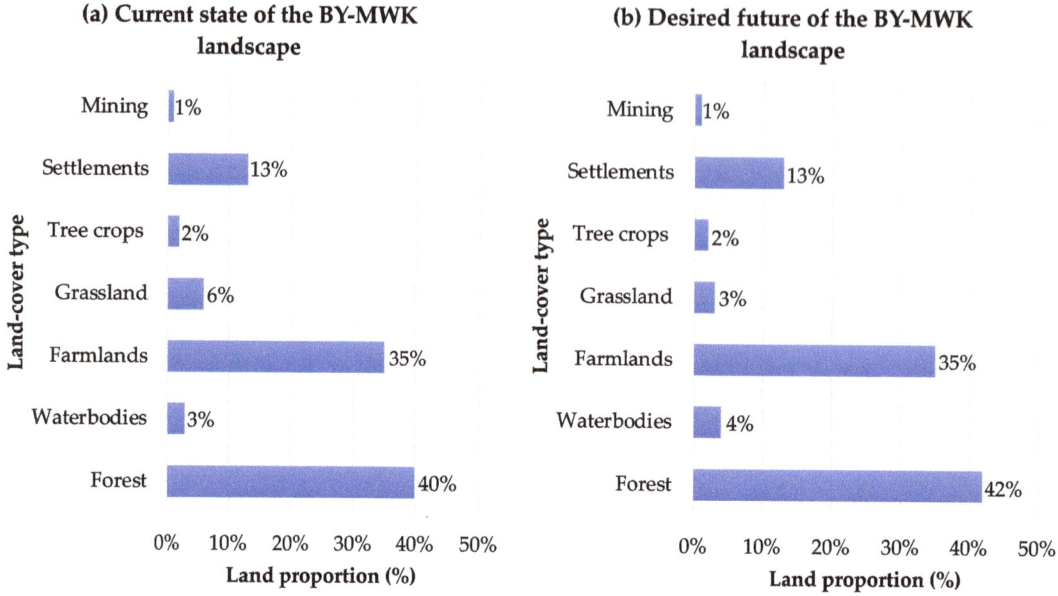

Figure 7. Perceived current state (**a**) and desired future composition (**b**) of the Builsa Yenning (BY)-Moagduri Wuntanluri Kuwomsaasi (MWK) landscape as visualised by institutional actors.

4.3. The Envisioned Composition of Desired Future Landscapes

This section analyses the composition of the future landscapes desired by local communities (Section 4.3.1), institutional stakeholders (Section 4.3.2), and the two groups together (Section 4.3.3). The desired configuration of future landscapes will be analysed in Section 4.4.

4.3.1. Desired Future Landscapes by Local Communities

The consensus among community stakeholders on the desired future SKGK landscape shows an increase in agricultural areas of 5%, and minimal increases of 1% in waterbodies and settlements, while the areas of forest and grassland remain unchanged. However, they anticipate a decline in the percentage of tree crops and total abandonment of gold mining. This is almost the same for the BY-MWK, where community stakeholders expect the projected areas of farmlands, settlements, and grasslands in the desired future landscape to increase by 9%, 2%, and 1%, respectively. In contrast, community actors hope that the forest area will decrease by 10% and that of tree crops by 2%, while waterbodies and mining will remain unchanged.

These results reflect the importance and priority local communities give to livelihoods over conservation: growth in farmland is prioritised over an increase in forest areas. It also suggests that they do not see forests as offering viable livelihood strategies now or in the future. Hence, the desire to convert portions of conservation areas (areas demarcated under the CREMA initiative) into agricultural and grazing areas. In doing so, local people hope to meet their needs for fertile land for agriculture (in the BY-MWK and SKGK) and pastoral areas (in the BY-MWK). The desire for increased water sources, particularly in the SKGK, is linked to the problems of the availability of resources in the area for off-season agricultural and pastoral activities. Communities hope this increase will be achieved by creating permanent water sources such as dams. This is likely to happen because of the current political initiative being implemented by the Government of Ghana, titled "Infrastructure for Poverty Eradication Programme", one of the projects of which is "one village, one dam" [59]. Furthermore, local communities want an increase in the area allocated for settlements, arguing that population growth in the next 20 years will require an expansion of settlement areas, necessarily encroaching on areas of lower priority in terms of livelihoods, such as forests and tree crops.

4.3.2. Desired Future Landscapes by Institutional Stakeholders

Figures 6 and 7 show the desired landscape changes of institutional actors in the WWC area in the next 20 years. Unlike local communities, the institutional actors opted for a reduction of cultivated areas in favour of an increase in conservation areas. The diagrams show a desired SKGK landscape future characterised by a decrease of 4% in the area dedicated to farmlands, 2% in that of grasslands, and 4% in that of mining activities, while the area of forest would increase by 5%, that of settlements by 3%, and that of tree crops and water resources by 1% each.

The desired composition of the BY-MWK landscape shows a 3% decrease in the area of grasslands, while the areas of farmlands, tree crops, settlements, and mines remain unchanged. On the other hand, the institutional actors wish to see a 2% increase in forest area and a 1% increase in water resources in the future BY-MWK landscape. Although the projected changes seem minimal, they suggest that institutional actors prioritise conservation aims, whereas community actors prioritise livelihood aims. This indicates a need to negotiate trade-offs and find synergies between different objectives and land uses.

4.3.3. Consensus about the Composition of Desired Future Landscapes between Institutional and Community Stakeholders

The discussions between the community and institutional stakeholders led to the elaboration of a consensual composition of the landscapes and the desired future land-use types. Figures 8 and 9 are illustrations of this.

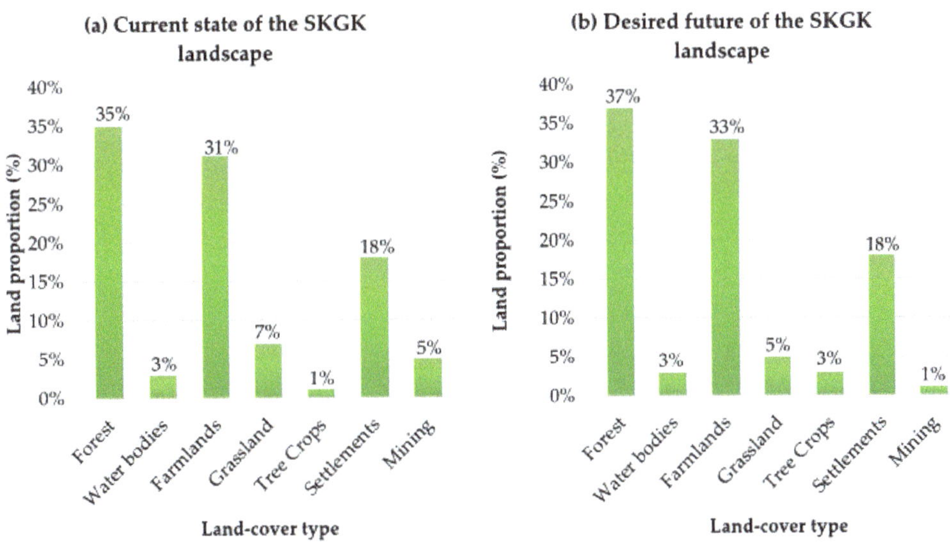

Figure 8. Consensus perception of the current state (**a**) and desired future composition (**b**) of the Sanyiga Kasena Gavara Kara (SKGK) landscape.

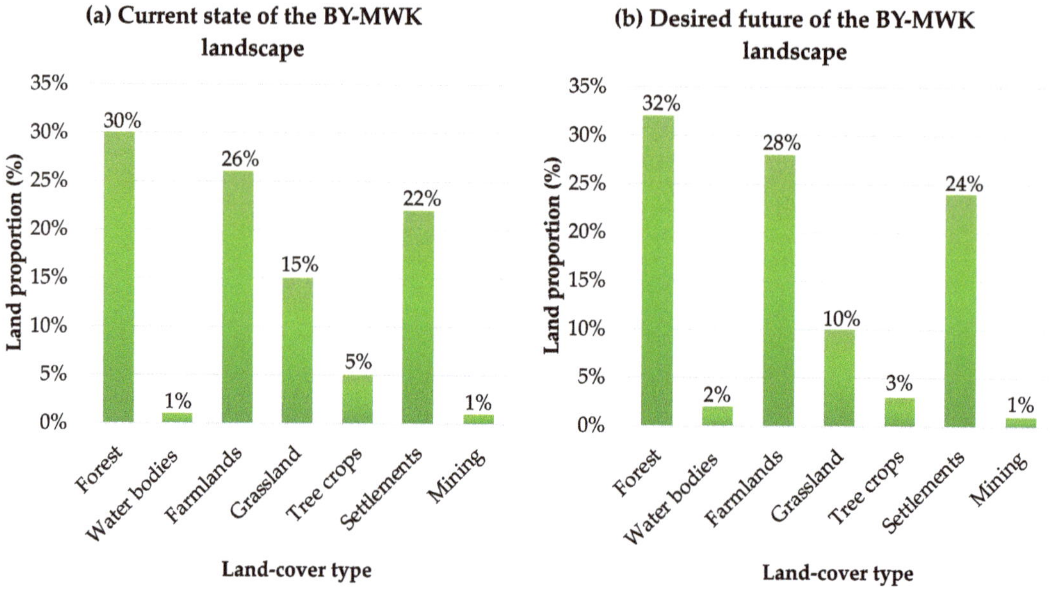

Figure 9. Consensus perception of the current state (**a**) and desired future composition (**b**) of the Builsa Yenning (BY)-Moagduri Wuntanluri Kuwomsaasi (MWK) landscape.

A comparison between the composition of the future landscape preferred by the community and institutional actors separately and the one discussed by the two actor groups together indicates that the negotiated trade-offs mainly concern agricultural and conservation areas. Initially, institutional stakeholders supported the idea of either decreasing agricultural areas (in SKGK) or keeping them unchanged (in BY-MWK) for the next 20 years to increase forest areas. This differs from the vision of the communities who prefer to reduce the extent of conservation areas or to keep them unchanged and increase the farmland. Thus, to reach a consensus, both groups of actors had to agree on reducing the areas dedicated to other land-use types, such as grasslands, mining activities, and tree crops, in favour of farmlands and forests, which are the focus of interest.

In general, and as shown in Table 3, in the SKGK, the local communities have made more concessions than the institutional actors regarding the composition of the desired future landscape. Of all the land-use types, only the consensus on the farmland corresponds to their initial preferences. For the rest, they have revised their position. This contrasts the institutional actors, who made few concessions: of the seven land-use types, they made concessions on three (water bodies, farmlands, settlements). The other four (forest, grassland, tree crops, and mining) remain consistent with their initial selections.

In the BY-MWK, on the other hand, all stakeholder groups made several concessions regarding the composition of the consensually desired future landscape. The difference lies in the types of land use on which concessions are made. For example, community actors made concessions about forests, water bodies, grassland, and mining, whereas institutional actors made concessions on farmland, tree crops, settlements, and mining (see Table 3).

Table 3. Comparative table of changes desired by institutional and community actors.

Land-Use Types	Change Wished by Communities		Change Wished by Institutional Actors		Consensus between Community and Institutional Actors	
	SKGK	BY-MWK	SKGK	BY-MWK	SKGK	BY-MWK
Forest	No change	Decrease	Increase	Increase	Increase	Increase
Water bodies	Increase	No change	Increase	Increase	No change	Increase
Farmlands	Increase	Increase	Decrease	No change	Increase	Increase
Grassland	No change	Increase	Decrease	Decrease	Decrease	Decrease
Tree crops	Decrease	Decrease	Increase	No change	Increase	Decrease
Settlements	Increase	Increase	Increase	No change	No change	Increase
Mining	Abandonment	No change	Decrease	No change	Decrease	Decrease

Source: Field data, 2021.

4.4. Configuration of Desired Future Landscapes

Concerning the future SKGK, communities prefer a more heterogeneous landscape with a large continuous block of forest surrounded by mixed areas of agriculture, settlement, and grassland, indicating a preference for a multifunctional landscape. This reflects a desire for a significant change from the current landscape they have mapped, which shows a homogeneous configuration with large strips of grassland surrounding forest reserves and small areas of farms and settlements distributed across the landscape.

In contrast, the map of the future landscape produced by institutional actors featured a relatively homogeneous area of forest and linear strips of farmlands. Grasslands and settlements are smaller, spotted in isolation next to agricultural lands for easy food supply. This indicates a slight difference from the current configuration presented by these actors, which shows a less homogeneous landscape with a mixture of small-sized farms, grassland, and settlements, next to large forest and mining areas.

The map of the future landscape jointly produced by institutional actors and communities shows a landscape configuration that draws on the two maps generated by both actor groups. The consensus map maintained existing forest reserve boundaries while adding new forests interspersed with medium settlement areas surrounded by agricultural land of almost the same size. Grasslands are rather small and dotted across the landscape (Figure 10).

It is worth noting that both the institutional and community actor groups expressed a desire for a future landscape in which forest corridors were created to connect to two forest reserves lying east and west to allow easy movement of animals. However, while the jointly conceived consensus map suggested the same, the configuration in the connecting areas is much more heterogenous with mixed forest, agriculture, and settlement areas.

As for the BY-MWK, the communities prefer a homogeneous landscape configuration for the next 20 years (Figure 11b). Indeed, the map of the desired future landscape is characterised by extensive farmlands adjacent to the settlements, which also occupy large proportions, and small forests scattered across the landscape. This configuration differs from the current landscape mapped by these communities, which presents a heterogeneous landscape marked by a mix of farms, grassland, tree crops, and settlements.

The future landscape map produced by the institutional actors (Figure 11d) shows a preference for a more homogeneous landscape with wide strips of forest interspersed with a relatively smaller size of farms, grasslands, and settlements. It also indicates a trend towards reinforcing existing forest reserves surrounded by additional forests. This contrasts somewhat with the current landscape map (Figure 11c), where farm and grassland sizes are larger, and forest blocks are fewer.

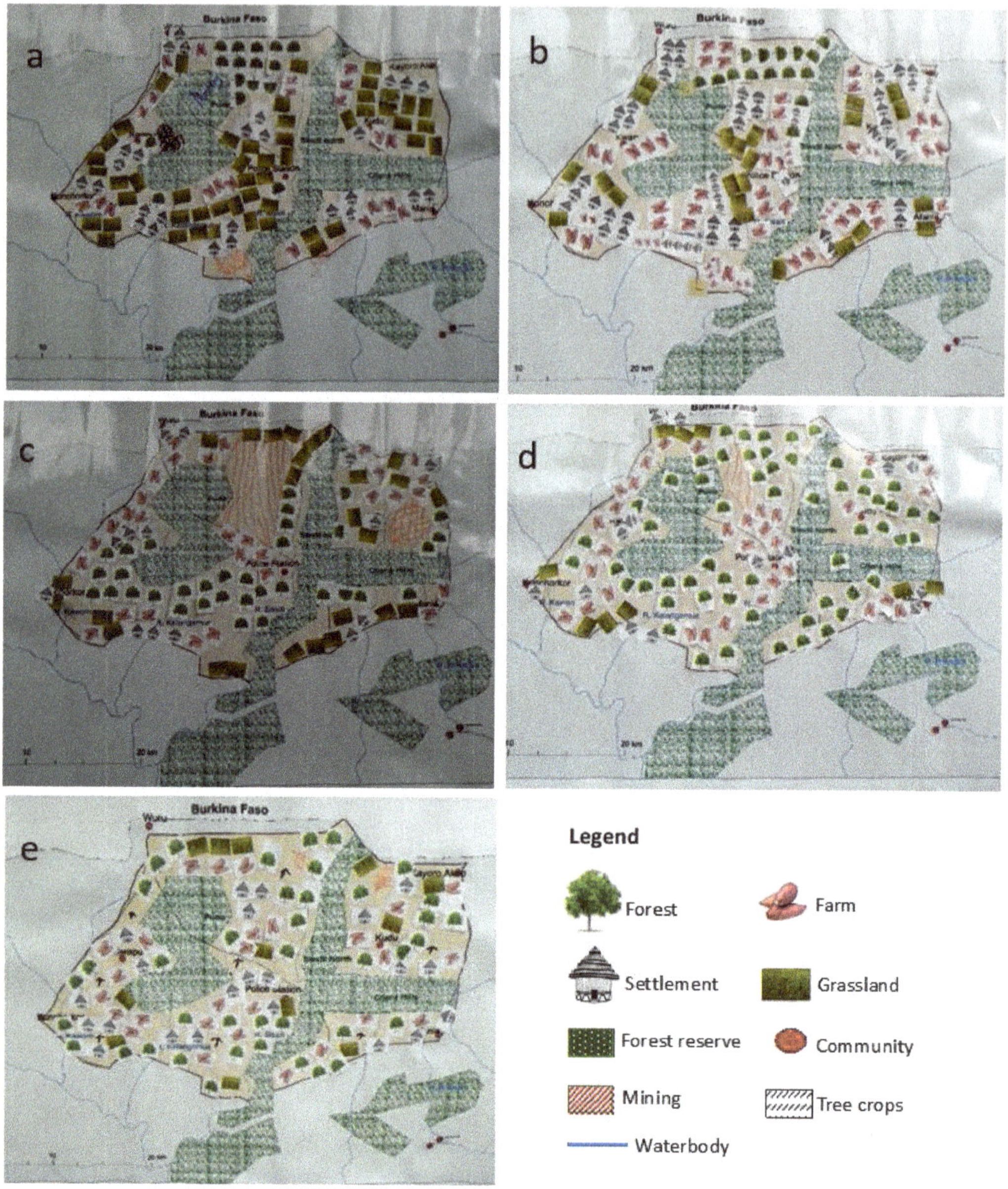

Figure 10. Participatory maps of current and desired futures of the SKGK landscape by community actors (**a** = current; **b** = future), institutional actors (**c** = current; **d** = future), and the consensus view of the future landscape (**e**).

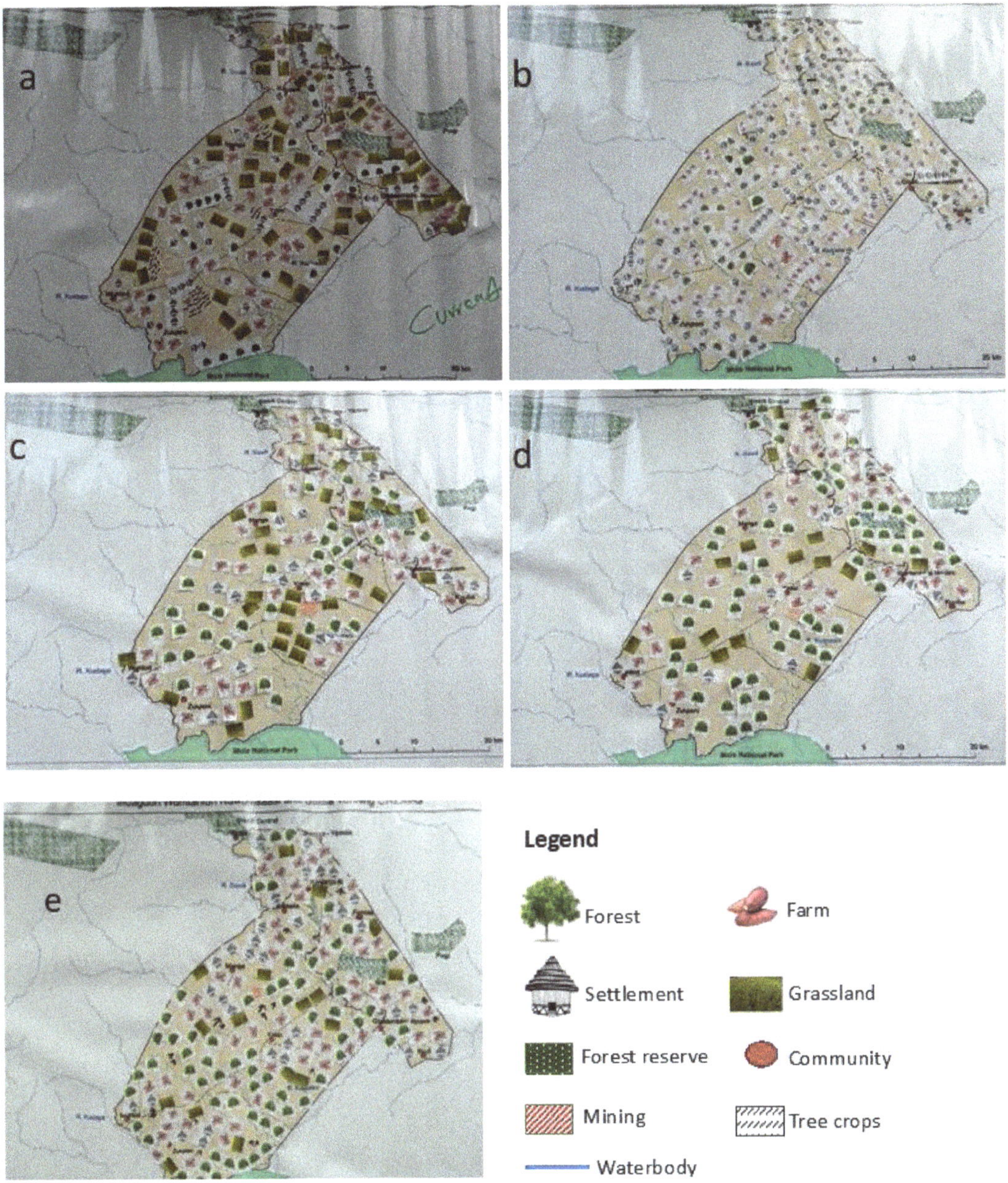

Figure 11. Participatory maps of current and desired futures of the BY-MWK landscape by community actors (**a** = current; **b** = future), institutional actors (**c** = current; **d** = future), and the consensus view of the future landscape (**e**).

The jointly conceived map representing the future desired landscape (Figure 11e) shows that the institutional and community stakeholders have agreed on a homogeneous landscape, with a wide forest strip constituting a corridor that runs from one end of the landscape to the other and a second forest block interspersed by small grasslands. The map also shows large areas on either side of the forests, with a concentration of farms and grasslands close to the settlements.

The 'consensus' was not entirely satisfactory for some participants (mainly women and elders). They felt that they were somewhat more disadvantaged with their views not sufficiently taken into account. Indeed, while women and elders pleaded for a total abandonment of gold panning because of the disastrous social and environmental consequences (insecurity, immoral behaviour, decimation of trees, and pollution), others, such as the institutional actors and the local youth, saw it as an opportunity to be exploited if the activity is well monitored and controlled. However, as everyone had an interest in the proposed scenario, concessions were made to reach a consensus.

The participants in this participatory mapping and scenario-building process expressed their satisfaction with such an initiative, as it allowed them to consult each other and harmonise their visions. Furthermore, they indicated a desire for subsequent similar experiences, noting that:

> *"This methodology of work (participatory mapping) helped us to see the picture of our area, and now, we understand better some challenges"* (A CREMA leader); *"We got a lot of insights on how realities are in the landscape on a broader way"* (NGO representative); *"Our (institutional actors') vision and community vision were different. But coming together, we were able to solve many of our differences and come out with a shared vision about the future landscape"* (public administration representative).[5]

4.5. Factors of Influence in the Trade-Off Negotiations

The study results allow us to identify three major factors influencing negotiations and consensus-building, namely culture, power, and knowledge (Table 4). We briefly elaborate on these factors below.

Table 4. Factors influencing trade-off negotiations by type of actor.

Type of Actor	Factors
Local actors (communities)	Culture; traditional norms Authority/power
Institutional actors	Knowledge
Local and institutional actors	Conflict of authority Traditional norms

Source: Field data, 2021.

4.5.1. Cultural Factors

According to tradition, women must comply with the decisions of men, who are more empowered to decide on community matters and natural resource use. These cultural factors certainly impacted their ability to effectively engage in elements of the participatory mapping process, which involves decision-making. Their contribution to the exchanges was timid during the discussions between local communities and institutional actors. However, they were much more active in defending their points of view during the discussions between community actors. The women interviewed explained this attitude by the fact that they were intimidated by the presence of institutional actors who are strangers to them, whereas among community actors, they are less intimidated because they are familiar with each other.

4.5.2. Power Relations and Knowledge

Having some authority within the community confers weight in decision-making about community resources. This was particularly noticeable in discussions between community actors. The leaders always tried to impose their views, giving little or no consideration to those of other participants, thus giving the impression that the views of others did not matter. The social position of these leaders made them dominant members in the deliberations, and until the propositions were to their liking, a decision could not be made. This seemed to be an accepted norm among the other participants.

The discussions between the institutional actors were based on intellectual, balanced, and organised exchanges, guided by forest and wildlife conservation laws, and above all, on knowledge of the terrain. The participants who were more familiar with the landscapes dominated the exchanges and succeeded in having their points of view front and centre. There were no serious conflicts of view because all were unanimous on the need to balance local development and conservation, but also on the fact that the existing agricultural areas are largely sufficient and could be reduced in some places in favour of reforestation.

However, the exchanges between the institutional and community actors were tense, marked by incomprehension and authority conflicts. Community leaders tried to impose their visions of the future, essentially based on prioritising livelihoods over conservation, while institutional actors, notably foresters, supported the opposite ideas, aiming to give more importance to conservation areas. In addition, each of these actors, feeling vested with authority based on their status of community leader or public administration officer, had difficulty accepting any opposition to their opinions. After extensive explanations by institutional actors of the benefits of biodiversity conservation and the need for sufficient forest space, stakeholders began to consider making concessions. In turn, the community actors expressed their motivations. Thus, the knowledge exchange facilitated the negotiation of compromises between the two groups of actors.

5. Discussion

5.1. From Conflicts of Interest to Trade-Offs and Consensus

This study showed that local and institutional actors have experienced significant degradation of the WWC landscape's natural ecosystems over a 20-year time spam, driven by a high demand for natural resources for people's livelihoods. Hence there is an urgent need to find a better balance between conservation and local development by negotiating trade-offs between the various stakeholders operating in this landscape; consensual decision-making is needed to ensure the sustainability of this multifunctional landscape. In this sense, the participatory mapping and the focus group discussions carried out in this study provided the stakeholders with an opportunity to express their respective views and knowledge, negotiate trade-offs, and find consensus on desired future landscapes that would be sustainable and beneficial to conservation and local development. The results of the mapping undertaken by the institutional and community stakeholder groups show minimal changes in the composition of desired future landscapes. This is certainly related to the specificity of the WWC as a biodiversity conservation area, and as such, it does not offer much potential for profound change. In terms of configuration, the maps show that institutional actors prefer more homogeneous landscapes. This is probably motivated by the will to have a more organised landscape that separates biodiversity conservation from livelihoods. The preferences of the communities depend on the current configuration of the landscape in which they are situated. Those who consider themselves to be in a heterogeneous landscape aspire to a homogeneous landscape and vice versa. This result shows their desire to benefit more from the landscape, and changing the configuration is seen as an appropriate solution. However, institutional and community stakeholders agreed on more homogeneous landscapes for the future and increasing agricultural and conservation areas at the expense of other land-use types. This reflects their desire to improve community livelihoods while enhancing biodiversity conservation. Managing and negotiating biodiversity-livelihoods trade-offs entails maximising food security benefits while minimising environmental damage [60]. This consensus indicates a willingness and ability to work with stakeholders with different statuses and interests towards a common vision and a sustainable and mutually beneficial landscape. This result implies that a governance model is needed that allows all key stakeholders to equitably participate in negotiations on how to achieve the desired changes in their landscape, including traditionally absent or marginalised actors whose interests have often been ignored or undermined.

The results further indicate that knowledge is a contributing factor to consensus building in that the decision of one stakeholder can be influenced by knowing the motivations of other stakeholders [8]. Similarly, Galafassi et al. [41] state that knowledge plays a role in consensus building because the decision to accept a trade-off requires a minimum of information as a guideline. However, even when consensus is achieved, inequities remain, as decisions at the landscape level are often influenced by the most powerful and knowledgeable actors [61].

5.2. Do Trade-Offs and Consensus Fall under the "Win More, Lose Less" Principle?

Most landscapes are affected by power differentials between stakeholders, beliefs, and traditional norms [8]. With large gaps in the capacity of stakeholders to participate and have their voices heard, decisions are not always made on an equal footing in many contexts [62]. Consensus may therefore be biased in favour of the most powerful actor. A power position implies respect and loyalty in most cultures, thus constituting a permanent source of influence [63]. Thus, no matter how well efforts are made, those with power will typically assert influence, either directly by imposing themselves or indirectly by their mere presence in decision-making [11]. This could prevent some members present around the negotiation table from expressing their views frankly and accepting the views of the influential members out of convenience, thus making the consensus skewed to the preferences of those more dominant. This attitude is usually seen in women when making decisions in the presence of men, in general, and their husbands, in particular. In Ghana, for example, traditional norms give men more power than women, particularly regarding access to natural resources and decision-making on their management [64,65], hence the low participation of women in decision-making processes. Furthermore, in many societies, decisions made by authorities with power have more weight and are more influential than those without the same power [63]. Thus, in the presence of an authority, people tend to give them priority in decision-making, as was the case in the consultation between community actors.

The results of this study also show that the expression of authority can be a constraint for the negotiation of trade-offs but also an asset for reaching a balanced consensus. Negotiations can become difficult when two (or more) authorities seek to affirm themselves, with the risk of never reaching a consensus. However, such a situation can also lead to a negotiated outcome satisfactory to both parties if the power balance is not in favour of one authority and leads to trade-offs where each party gains more and loses less. The oft-promoted win-win principle in conservation suggests that the decisions taken satisfy all landscape stakeholders and that everyone benefits from landscape governance. However, in negotiating consensus and adopting a rational decision, each stakeholder must make compromises, i.e., accept losing something to gain something else. In the debate on conservation approaches, many authors consider win-win initiatives to be elusive because rarely have they succeeded in achieving an optimal balance between conservation and development objectives [43,66–68]. In addition, because governance decisions frequently result in losers and winners, the principles of equity, equality, and win-win in conservation are often considered to be the exception [10,45,69]. This highlights the utopian nature of win-win conservation and, consequently, of fully harmonious decision-making related to landscape governance. Thus, this study claims that in landscape planning and decision-making, the principle of a negotiated "win more, lose less" [8,25,69,70] should prevail. In this respect, integrated landscape approaches offer a better chance than sectoral approaches of making equitable cross-sectoral decisions on landscapes and land uses [69]. They facilitate the creation of spaces for negotiating trade-offs between several actors to make decisions on the desired future landscapes [8,69]. The idea behind this is that the negotiation process implies compromises in terms of conservation and development. Therefore, the merit of integrated landscape approaches is that they involve a process of deliberation of trade-offs between different landscape actors with different interests and visions, thus raising awareness of multiple needs, knowledge, and beliefs, building trust among stakeholders, and promoting adaptive management in which stakeholders win more than they lose [44,69].

Nevertheless, integrated landscape approaches should not be seen as a panacea or silver bullet that can solve all landscape challenges but rather as iterative models of governance that adapt according to local, dynamic contexts and embed the necessary mechanisms to adjust as these contexts evolve [25]. Such approaches will be strengthened by good leadership and a robust governance mechanism, including sufficient financial and technical capacity and a verifiable monitoring and evaluation system [8].

This study made clear that participatory plans are not always the result of fully unanimous decisions where all stakeholders are fully satisfied. Consensus does not mean that all the expectations of the landscape stakeholders have been met. However, in the face of differing opinions, visions, cultures, personalities, and contexts, stakeholders strive to find mutually acceptable solutions to resolve conflicts and ensure the sustainable use of natural resources and the sustainability of the landscape [63]. To this end, scenario-building and participatory mapping methods in a multi-stakeholder setting offer great potential for stakeholders to engage in direct dialogue and better harmonise their respective visions, interests, and expectations.

5.3. Methodological Considerations

Participatory scenario-building and mapping were useful in harnessing place-based knowledge, creating awareness of landscape change, and making decision-making more inclusive. It embodies indigenous practices, tacit understanding and values, and knowledge held by state institutions of national laws [14,15,71,72]. However, insights into the composition of stakeholders and implementation approach are as important as outputs derived from the processes to reduce the exclusion of socially vulnerable groups [71]. Depending on the research objectives, existing studies have engaged relatively homogenous stakeholders with common interests and knowledge regarding their circumstances and future aspirations [14,73], mixed stakeholders for knowledge triangulation and confirmation [13,19], and different stakeholder groups with conflicting interests separately to identify trade-offs [15,72]. The latter studies usually do not bring together stakeholders with conflicting interests. This study has tried to fill this gap by adapting the methodology of Asubonteng et al. [15] applied in the Eastern Region of Ghana, which consisted of facilitating separate workshops with community and institutional stakeholders to discuss and map the current status and future desired landscapes. In the original methodology, there was no confrontation between the two groups of actors. The adaption consisted of conducting separate workshops first and then a joint workshop including local communities and institutional actors to allow them to discuss their visions and reach a consensus. Bringing these stakeholder groups together created a different dynamic, and contradictions became clearer. Moreover, it offered the possibility to simulate negotiations and to see what happens when people with different power positions and knowledge sit together in a negotiation process. Furthermore, we did not pay much attention to the BAU future landscape scenario, as did Asubonteng et al. [15], not only because of time constraints but also because the focus was more on the deliberation of trade-offs regarding the future desired landscapes, especially between institutional actors and local communities.

Similar to any scientific approach, the method used in this study is not without constraints. The stakeholder composition was compromised by the unwillingness of some stakeholders to participate in such processes. It was impossible to involve pastoralists in the workshops due to the ongoing tensions between farmers and pastoralists in the area (see [37,38]). Similarly, private companies that assumedly had appropriated most farmlands in the Eastern Region of Ghana excluded themselves from the participatory scenario process [15]. This shows that some actors may willingly exclude themselves from inclusive decision-making processes. Yet, in landscape governance, and especially in participatory mapping and planning, all key stakeholders should be involved because each of them has an interest in, and influence on, the landscape [2]. The presence of the pastoralist group would certainly have influenced the importance given to grasslands in future landscapes.

In addition, the current landscapes mapped by each stakeholder group display different proportions of land-use types. This makes it difficult to compare the desired future landscapes, especially with regard to the desired amounts of land. Another option would have been to use remote sensing to produce maps of the current state of each landscape (c.f. Siangulube et al., this issue), and each stakeholder group would have worked based on the same maps to project their future landscapes. This could be the subject of future studies. However, the exact proportions are not the focus of this study, but rather a means to better appreciate the evolutionary trend of land-use types, which the different stakeholders wish to have in their future landscapes. The advantage of the method is that it allows each group to visualise their landscape and represent what they believe the proportion currently occupied by each land-use unit to be in the future. In this way, participants feel more involved in the mapping process and less influenced by a pre-designed map imposed on them. To minimise inter-professional, inter-generational, and gender influences, it may be interesting to conduct participatory mapping only with each of the community groups separately (farmers, pastoralists, women, youth, elderly, and forest operators), and then a separate workshop with institutional actors before doing a joint workshop. However, this option can be very costly and time-consuming.

One of the difficulties in this study was the distance between the target communities. The communities had to come from far away to participate in the study. Therefore, participants arrived late in the morning and had to leave before it became dark for security reasons. This did not leave much time for discussion and mapping. Therefore, for the joint workshop held in Navrongo town, we had to rent hotel rooms to enable community participants to come a day before the workshop (the nearest target community was about 42 km away and the most distant was about 85 km, and the institutional actors are mainly based in Bolgatanga which is 30 km away). This is an additional cost on top of the organisational costs (materials for mapping such as the printed map frames, venue hire, meals, and fuel for participants' travel). Therefore, the financial aspect is also a factor to be considered in this and similar studies.

Four people per CREMA were chosen to avoid having a large number of participants, which makes the mapping exercise difficult. In addition, COVID restrictions had to be taken into account, which forbade groupings of more than 25 people. Thus, the workshop that brought together community and institutional actors could not have more than 25 participants, impacting stakeholder composition.

6. Conclusions

The Western Wildlife Corridor faces sustainability challenges that require diverse stakeholders to reach consensual decisions that reconcile biodiversity conservation and livelihood development. According to stakeholder perceptions, the natural ecosystems of the WWC landscape have experienced significant degradation over the past 20 years due to population pressure on natural resources. This impacted conservation through the disappearance of species and the loss of ecosystem services and livelihoods due to the loss of wildlife species, non-timber forest products, and the population of shea trees. The participatory scenario-building and mapping exercise helped participants to reflect on and visualise their different desired futures while building consensus on a future scenario of land use and land-cover change [2]. As a result of the deliberations, there was a consensus among the stakeholders on more homogenous landscapes for the future and on increasing farming and conservation areas. However, decisions taken in landscape governance typically involve trade-offs due to the difficulty of fully satisfying the interests of all stakeholders. Furthermore, during trade-off negotiations, some stakeholders are subject to the power and cultural norms, forcing them to reconsider their position. The results of this study indicate that knowledge, culture, and power inequities influence trade-off negotiations and consensus-building. Integrated landscape approaches enhance consensus-building by facilitating dialogue, negotiation, harmonising plural, and even divergent interests, and mutual understanding between stakeholders while promoting equity and power balance. A solid and inclusive platform that supports such multi-stakeholder processes is considered necessary to bring stakeholders of the same landscape to converge towards a common ideal where each stakeholder group gains more and loses less [5,7].

The lessons and recommendations that emerge from this study based on multi-stakeholder participatory mapping from the perspective of implementing integrated landscape approaches can be summarised as follows:

Firstly, consensual decision-making through a multi-stakeholder platform requires experienced facilitation of the discussions to balance the debates and enable participants to express themselves and be understood. The negotiation of trade-offs and achieving consensus in this study were largely possible due to careful moderation of the debates that enabled an open dialogue between actors and attention to and balancing of power dynamics [63]. The capacity, experience, attitude, and behaviour of the moderators are crucial to the quality of the results obtained [74].

Secondly, sufficient time must be allowed for deliberation and negotiation of trade-offs. Not all actors have the same level of understanding and may need time to discuss certain misunderstood aspects or to make themselves understood by others.

Thirdly, initiators of participatory planning through participatory mapping must ensure that they include all key stakeholders in the landscape to make the negotiation process and its outcomes as inclusive as possible. A plan may appear consensual, but if the process has not been inclusive, curtailing the interests of absent and marginalised stakeholders could perpetuate problems in the medium or long term.

Fourthly, landscape actors are often keen to engage in a dynamic of multi-stakeholder deliberation for the governance of their landscape, but multi-stakeholder processes are only facilitated and effective when there are solid multi-stakeholder platforms established to support them. Reaching a consensus on a common future landscape—through participatory mapping and other means—is a symbol of the willingness of actors to cooperate to ensure the sustainability of landscape resources and reflects the need for a multi-stakeholder framework for consultation, knowledge sharing, negotiations, and joint decision-making. The intention to provide such frameworks in community-based landscape governance systems, such as CREMAs, is expected to foster multi-stakeholder processes to address common needs and aspirations and create ecologically and socially sustainable and resilient landscapes.

Author Contributions: Conceptualisation, E.R.C.B., M.R.-T., J.R. and T.S.; Data curation, E.R.C.B.; Funding acquisition, J.R. and T.S.; Investigation, E.R.C.B. and K.O.A.; Methodology, E.R.C.B., K.O.A. and M.R.-T.; Project administration, J.R.; Resources, H.D.; Software, E.R.C.B., K.O.A. and F.S.S.; Supervision, M.R.-T., H.D., J.R. and T.S.; Validation, M.R.-T., J.R. and T.S.; Visualization, E.R.C.B., M.R.-T., H.D., J.R. and T.S.; Writing—original draft, E.R.C.B.; Writing—review & editing, K.O.A., M.R.-T., H.D., F.S.S., J.R. and T.S. All authors have read and agreed to the published version of the manuscript.

Funding: This research was funded by the German Federal Ministry for the Environment, Nature Conservation, Building and Nuclear Safety (BMUB) through a grant from the International Climate Initiative (IKI) 18_ IV_084. Additional funding for this research came from the United States Agency for International Development (USAID) Forest and Biodiversity Office and the CGIAR FTA programme.

Institutional Review Board Statement: This study was conducted in accordance with the Declaration of Helsinki and approved by the Ethics Advisory Board of the Amsterdam Institute for Social Science Research (AISSR) of the University of Amsterdam in April 2020.

Informed Consent Statement: Informed consent was obtained from all subjects involved in the study.

Data Availability Statement: The data presented in this study are available on request from the corresponding author. The data are not publicly available due to the ethical policy of the University of Amsterdam.

Acknowledgments: This study is part of the CGIAR Research Program on Forests, Trees and Agroforestry (CRP-FTA). This collaborative programme aims to enhance the management and use of forests, agroforestry and tree genetic resources across the landscape, from forests to farms. We express our sincere thanks to all the actors who participated in the interviews and all those who contributed in one way or another to the realisation of this article. We are grateful to the German Federal Ministry for the Environment, Nature Conservation, Building and Nuclear Safety (BMUB), the United States Agency for International Development (USAID) Forest and Biodiversity Office, and the CGIAR FTA programme, for funding the Collaborating to Operationalise Landscape Approaches for Nature, Development and Sustainability (COLANDS) initiative, which made this research possible. Thanks are also due to the COVID fund of the Amsterdam Institute for Social Science Research (AISSR) of the University of Amsterdam for granting an extension of the scholarship of the first author.

Conflicts of Interest: The authors declare no conflict of interest. The funders had no role in the design of the study; in the collection, analyses, or interpretation of data; in the writing of the manuscript, or in the decision to publish the results.

Appendix A

Basic map frames used for the participatory mapping.

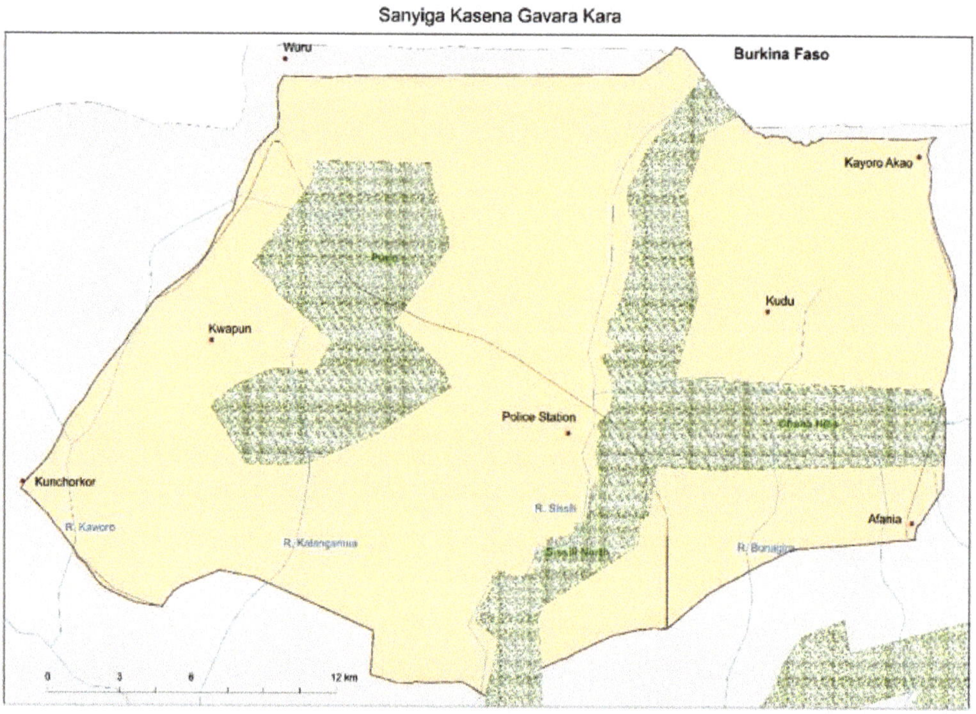

Figure A1. Map A: Sanyiga Kasena Gavara Kara (SKGK) CREMA.

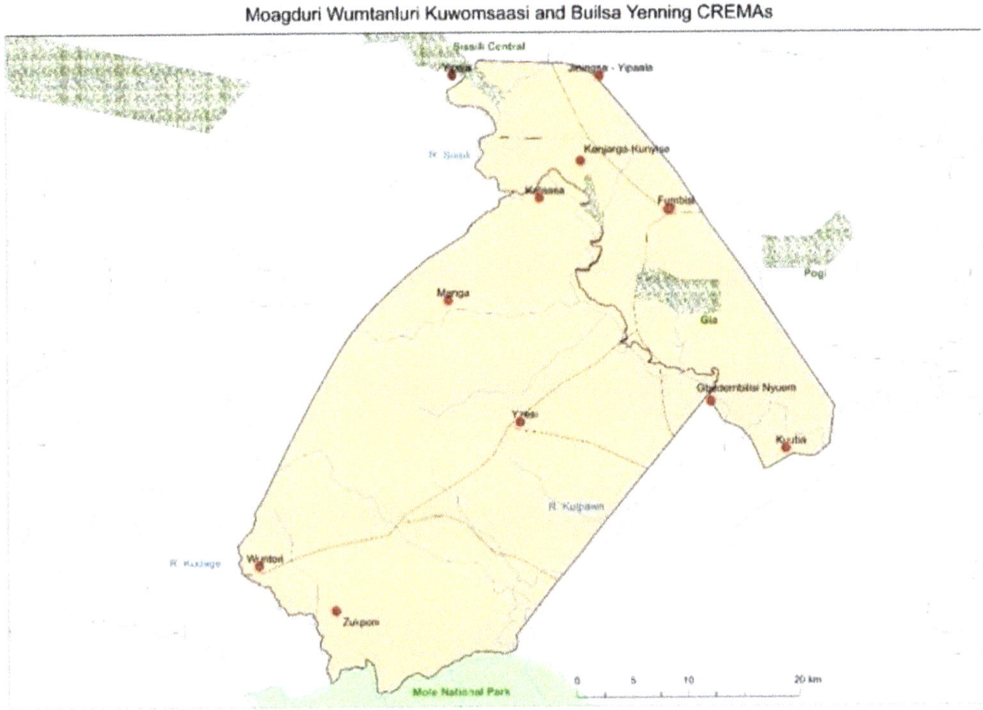

Figure A2. Map B: Builsa Yenning—Moagduri Wuntanluri Kuwomsaasi (BY-MWK) CREMAs.

Notes

1 We asked the CRMCs leaders of the target communities to identify those they considered young and those they considered old. As a result, the age ranges obtained correspond to those suggested by the communities.

2 Non-timber forest products are plant and animal species other than timber from forested ecosystems. They include a broad range of products, including food, fodder, medicinal and cosmetic plants, material for woodcarving and household tools, and bushmeat and other animal products [57,58].

3 Focus group discussion held in Fumbisi on 21 May 2021.

4 Focus group discussion held in Nakong on 19 May 2021.

5 Workshop held in Navrongo on 7 December 2021.

References

1. Kozar, R.; Buck, L.E.; Barrow, E.G.; Sunderland, T.C.; Catacutan, D.E.; Planicka, C.; Hart, A.K.; Willemen, L. Toward viable landscape governance systems: What works? *World Agroforestry Centre.* 2014. Available online: http://www.worldagroforestry.org/publication/toward-viable-landscape-governance-systems-what-works (accessed on 12 May 2021).
2. Best, L.; Fung-Loy, K.; Ilahibaks, N.; Ramirez-Gomez, S.O.I.; Speelman, E.N. Toward Inclusive Landscape Governance in Contested Landscapes: Exploring the Contribution of Participatory Tools in the Upper Suriname River Basin. *Environ. Manag.* **2021**, *68*, 683–700. [CrossRef]
3. Larson, A.M.; Sarmiento Barletti, J.P.; Heise Vigil, N. A place at the table is not enough: Accountability for Indigenous Peoples and local communities in multi-stakeholder platforms. *World Dev.* **2022**, *155*, 105907. [CrossRef]
4. Bastos Lima, M.G.; Persson, U.M. Commodity-Centric Landscape Governance as a Double-Edged Sword: The Case of Soy and the Cerrado Working Group in Brazil. *Front. For. Glob. Chang.* **2020**, *3*, 27. [CrossRef]
5. Kusters, K.; Buck, L.; de Graaf, M.; Minang, P.; van Oosten, C.; Zagt, R. Participatory Planning, Monitoring and Evaluation of Multi-Stakeholder Platforms in Integrated Landscape Initiatives. *Environ. Manag.* **2018**, *62*, 170–181. [CrossRef]
6. Pascual, U.; Phelps, J.; Garmendia, E.; Brown, K.; Corbera, E.; Martin, A.; Gomez-Baggethun, E.; Muradian, R. Social equity matters in payments for ecosystem services. *Bioscience* **2014**, *64*, 1027–1036. [CrossRef]

7. Reed, J.; Barlow, J.; Carmenta, R.; van Vianen, J.; Sunderland, T. Engaging multiple stakeholders to reconcile climate, conservation and development objectives in tropical landscapes. *Biol. Conserv.* **2019**, *238*, 108229. [CrossRef]
8. Sayer, J.; Margules, C.; Boedhihartono, A.K.; Dale, A.; Sunderland, T.; Supriatna, J.; Saryanthi, R. Landscape approaches; What are the pre-conditions for success? *Sustain. Sci.* **2015**, *10*, 345–355. [CrossRef]
9. van Oosten, C.; Runhaar, H.; Arts, B. Capable to govern landscape restoration? Exploring landscape governance capabilities, based on literature and stakeholder perceptions. *Land Use Policy* **2021**, *104*, 104020. [CrossRef]
10. Ros-Tonen, M.A.F.; Willemen, L.; McCall, M.K. Spatial Tools for Integrated and Inclusive Landscape Governance: Toward a New Research Agenda. *Environ. Manag.* **2021**, *68*, 611–618. [CrossRef]
11. Siangulube, F.S.; Ros, M.A.F.; James, T.; Djoudi, H.; Gumbo, D.; Sunderland, T. Navigating power imbalances in landscape governance: A network and influence analysis in southern Zambia. *Reg. Environ. Chang.* **2023**, *23*, 41. [CrossRef]
12. Mccall, M.K.; Dunn, C.E. Geo-information tools for participatory spatial planning: Fulfilling the criteria for "good" governance? *Geoforum* **2012**, *43*, 81–94. [CrossRef]
13. Robinson, C.J.; Maclean, K.; Hill, R.; Bock, E.; Rist, P. Participatory mapping to negotiate indigenous knowledge used to assess environmental risk. *Sustain. Sci.* **2016**, *11*, 115–126. [CrossRef]
14. Aggrey, J.J.; Asubonteng, K.O.; Ros-tonen, M.A.F. Using Participatory Spatial Tools to Unravel Community Perceptions of Land-Use Dynamics in a Mine-Expanding Landscape in Ghana. *Environ. Manag.* **2021**, *68*, 720–737. [CrossRef]
15. Asubonteng, K.O.; Ros-Tonen, M.A.F.; Baud, I.; Pfeffer, K. Envisioning the Future of Mosaic Landscapes: Actor Perceptions in a Mixed Cocoa/Oil-Palm Area in Ghana. *Environ. Manag.* **2021**, *68*, 701–719. [CrossRef]
16. Brown, G.; Kyttä, M. Key issues and priorities in participatory mapping: Toward integration or increased specialisation? *Appl Geogr.* **2018**, *95*, 1–8. [CrossRef]
17. McCall, M. Mapping Territories, Land Resources and Rights: Communities Deploying Participatory Mapping/Pgis in Latin America. *Geogr. Dep. Univ. Sao Paulo* **2014**, *94–122*, 94–122. [CrossRef]
18. Reed, M.; Kenter, J.; Bonn, A.; Broad, K.; Burt, T.; Fazey, I.; Fraser, E.; Hubacek, K.; Nainggolan, D.; Quinn, C.; et al. Participatory scenario development for environmental management: A methodological framework illustrated with experience from the UK uplands. *J. Environ. Manag.* **2013**, *128*, 345–362. [CrossRef]
19. Kabaya, K.; Hashimoto, S.; Fukuyo, N.; Uetake, T.; Takeuchi, K. Investigating future ecosystem services through participatory scenario building and spatial ecological–economic modelling. *Sustain. Sci.* **2019**, *14*, 77–88. [CrossRef]
20. Asubonteng, K.O.; Pfeffer, K.; Ros-Tonen, M.A.F.; Baud, I.; Benefoh, D.T. Integration versus segregation: Structural dynamics of a smallholder-dominated mosaic landscape under tree-crop expansion in Ghana. *Appl. Geogr.* **2020**, *118*, 102201. [CrossRef]
21. Wildlife Division. Wildlife Division Policy for Collaborative Community Based Wildlife Management; 2000; Volume 12. Available online: http://www.fcghana.org/library.php?id=19 (accessed on 23 June 2020).
22. Agyare, A. Polycentric Governance and Social-Ecological Performance of Community Resource Management Areas in Ghana. Ph.D. Thesis, University of Victoria, Victoria, BC, Canada, 2013. Available online: https://dspace.library.uvic.ca/bitstream/handle/1828/4930/Agyare_Andrew_PhD_2013.pdf?sequence=1&isAllowed=y (accessed on 23 June 2020).
23. IUCN. Upscaling Community Resource Management Areas as a Delivery Mechanism for REDD+ Implementation in Ghana. 2017, pp. 1–6. Available online: https//www.iucn.org/sites/dev/files/content/documents/upscaling_community_resource_management_-_ghana.pdf (accessed on 3 February 2020).
24. Urfalino, P. La décision par consensus apparent. Nature et propriétés. *Rev. Eur. Sci. Soc.* **2007**, *XLV-136*, 47–70. [CrossRef]
25. Reed, J.; Ickowitz, A.; Chervier, C.; Djoudi, H.; Moombe, K.; Ros-Tonen, M.; Yanou, M.; Yuliani, E.L.; Sunderland, T. Integrated landscape approaches in the tropics: A brief stock-take. *Land Use Policy* **2020**, *99*, 104822. [CrossRef]
26. Bayala, E.R.C.; Djoudi, H.; Ros-Tonen, M.; Zida, M. Context for landscape approach implementation in the Western Wildlife Corridor Landscape (Northern Ghana). In *Operationalising Iltegrated Landscape Approaches in the Tropics*; Reed, J., Ros-Tonen, M., Sunderland, T., Eds.; Centre for International Forestry Research: Jawa Barat, Indonesia, 2020; pp. 126–147. Available online: https://www.cifor.org/knowledge/publication/7808 (accessed on 22 October 2022).
27. Braimoh, A.K.; Vlek, P.L.G. Land-cover change trajectories in northern Ghana. *Environ. Manag.* **2005**, *36*, 356–373. [CrossRef]
28. Abdul-Moomin, A.; Issahaku, A.; Motin, B. The Shea Industry and Rural Livelihoods among Women in the Wa Municipality, Ghana. *J. Soc. Sci. Stud.* **2016**, *3*, 40. [CrossRef]
29. Barlow, J.; França, F.; Gardner, T.A.; Hicks, C.C.; Lennox, G.D.; Berenguer, E.; Castello, L.; Economo, E.P.; Ferreira, J.; Guénard, B.; et al. The future of hyperdiverse tropical ecosystems. *Nature* **2018**, *559*, 517–526. [CrossRef]
30. Owusu-Ansah, N. Assessing How Collaborative Resource Management Impacts Victims' Perspectives on Wildlife Crop Raids. *J. Nat. Resour. Dev.* **2018**, *8*, 115–124. [CrossRef]
31. Marchetta, F. *On the Move Livelihood Strategies in Northern Ghana*; Working Paper; CERDI: Clermont Ferrand, France, 2011.
32. Grass, I.; Kubitza, C.; Krishna, V.V.; Corre, M.D.; Mußhoff, O.; Pütz, P.; Drescher, J.; Rembold, K.; Ariyanti, E.S.; Barnes, A.D.; et al. Trade-offs between multifunctionality and profit in tropical smallholder landscapes. *Nat. Commun.* **2020**, *11*, 1186. [CrossRef]
33. Abebe, G.; Getachew, D.; Ewunetu, A. Analysing land use/land cover changes and its dynamics using remote sensing and GIS in Gubalafito district, Northeastern Ethiopia. *SN Appl. Sci.* **2022**, *4*, 30. [CrossRef]
34. Zoungrana, B.J.-B.; Conrad, C.; Amekudzi, L.K.; Thiel, M.; Da, E.D.; Forkuor, G.; Löw, F. Multi-temporal landsat images and ancillary data for land use/cover change (LULCC) detection in the Southwest of Burkina Faso, West Africa. *Remote Sens.* **2015**, *7*, 12076–12102. [CrossRef]

35. Yangouliba, G.I.; Zoungrana, B.J.-B.; Hackman, K.O.; Koch, H.; Liersch, S.; Sintondji, L.O.; Dipama, J.-M.; Kwawuvi, D.; Ouedraogo, V.; Yabré, S.; et al. Modelling past and future land use and land cover dynamics in the Nakambe River Basin, West Africa. *Model. Earth Syst. Environ.* **2022**, 1–17. [CrossRef]
36. Meissner, P.; Wulf, T. Cognitive benefits of scenario planning: Its impact on biases and decision quality. *Technol. Forecast Soc. Chang.* **2013**, *80*, 801–814. [CrossRef]
37. Tonah, S. Migration and Farmer-Herder Conflicts in Ghana's Volta Basin. *Can. J. Afr. Stud.* **2006**, *40*, 152–178. [CrossRef]
38. Bukari, K.N.; Schareika, N. Stereotypes, prejudices and exclusion of Fulani pastoralists in Ghana. *Pastoralism* **2015**, *5*, 20. [CrossRef]
39. Börjeson, L.; Höjer, M.; Dreborg, K.H.; Ekvall, T.; Finnveden, G. Scenario types and techniques: Towards a user's guide. *Futures* **2006**, *38*, 723–739. [CrossRef]
40. Bryman, A. *Social Research Methods*, 4th ed.; Oxford University Press Inc.: New York, NY, USA, 2012.
41. Galafassi, D.; Daw, T.M.; Munyi, L.; Brown, K.; Barnaud, C.; Fazey, I. Learning about social-ecological trade-offs. *Ecol. Soc.* **2017**, *22*, 2. [CrossRef]
42. Schaafsma, M.; Eigenbrod, F.; Gasparatos, A.; Gross-Camp, N.; Hutton, C.; Nunan, F.; Schreckenberg, K.; Turner, K. Trade-off decisions in ecosystem management for poverty alleviation. *Ecol. Econ.* **2021**, *187*, 107103. [CrossRef]
43. Sunderland, T.C.H.; Ehringhaus, C.; Campbell, B.M. Conservation and development in tropical forest landscapes: A time to face the trade-offs? *Environ. Conserv.* **2007**, *34*, 276–279. [CrossRef]
44. Sayer, J.; Sunderland, T.; Ghazoul, J.; Pfund, J.-L.; Sheil, D.; Meijaard, E.; Venter, M.; Boedhihartono, A.K.; Day, M.; Garcia, C.; et al. Ten principles for a landscape approach to reconciling agriculture, conservation, and other competing land uses. *Proc. Natl. Acad. Sci. USA* **2013**, *110*, 8349–8356. [CrossRef]
45. Reed, J.; Van Vianen, J.; Deakin, E.L.; Barlow, J.; Sunderland, T. Integrated landscape approaches to managing social and environmental issues in the tropics: Learning from the past to guide the future. *Glob. Chang. Biol.* **2016**, *22*, 2540–2554. [CrossRef]
46. van Noordwijk, M. Scaling trade-offs between crop productivity, carbon stocks and biodiversity in shifting cultivation landscape mosaics: The FALLOW model. *Ecol. Model.* **2002**, *149*, 113–126. Available online: http://apps.worldagroforestry.org/sea/Products/AFModels/FALLOW/FALLOW0303/EM149.pdf (accessed on 7 July 2022). [CrossRef]
47. Jafarzadeh, A.A.; Mahdavi, A.; Shamsi, S.R.F.; Yousefpour, R. Assessing synergies and trade-offs between ecosystem services in forest landscape management. *Land Use Policy* **2021**, *111*, 105741. [CrossRef]
48. Wang, J.; Cao, Y.; Fang, X.; Li, G. Identification of the trade-offs/synergies between rural landscape services in a spatially explicit way for sustainable rural development. *J. Environ. Manag.* **2021**, *300*, 113706. [CrossRef]
49. Cochrane, L.; Corbett, J. Participatory mapping. In *Handbook of Communication for Development and Social Change*; Springer: Singapore, 2018. [CrossRef]
50. Asmolov, G. *Mapping Crisis: Participation, Datafication and Humanitarianism in the Age of Digital Mapping*; University of London Press: London, UK, 2020. [CrossRef]
51. Dakwa, K.B. How does the cost of raid influence tolerance and support of local communities for a wildlife reserve? *Int. J. Biodivers. Conserv.* **2016**, *8*, 81–92. [CrossRef]
52. Bempah, G.; Dakwa, K.B.; Monney, K.A. Evaluation of the community resources management area (CREMA) programme around Ankasa conservation area, Ghana. *Cogent. Environ. Sci.* **2019**, *5*, 1592064. [CrossRef]
53. Allendorf, T.D.; Das, R.; Bose, A.; Ray, B.; Chaudhuri, K.D.; Brock, S.; Horwich, R.H. Motivations of the community forest protection forces of the Manas Biosphere Reserve in Assam, India. *Int. J. Sustain. Dev. World Ecol.* **2013**, *20*, 426–432. [CrossRef]
54. Mansourian, S.; Walters, G.; Gonzales, E. Identifying governance problems and solutions for forest landscape restoration in protected area landscapes. *Parks* **2019**, *25*, 83–96. [CrossRef]
55. Dery, G.; Imoro, A.Z.; Dzitse, S. A survey on trade in wildlife parts in the Northern Region of Ghana. *Glob. Ecol. Conserv.* **2022**, *38*, e02208. [CrossRef]
56. The West African Science Association. Wildlife Conservation in West Africa. In Proceedings of the Seventh Biennial Conference of the West African Science Association, Ibadan, Nigeria, 2 April 1970.
57. Clarck, L.E.; Sunderland, T.C.H. *The Key Non-Timber Forest Products of Central Africa: State of the Knowledge*; Technical paper No. 122; U.S. Agency for International Development: Washington, DC, USA, 2004. Available online: https://pdf.usaid.gov/pdf_docs/pnada851.pdf (accessed on 18 January 2023).
58. Ros-Tonen, M.A.F.; Wiersum, K.F. The scope for improving rural livelihoods through non-timber forest products: An evolving research agenda. *For. Trees Livelihoods* **2005**, *15*, 129–148. [CrossRef]
59. Ghansah, B.; Foster, T.; Higginbottom, T.P.; Adhikari, R.; Zwart, S.J. Monitoring spatial-temporal variations of surface areas of small reservoirs in Ghana's Upper East Region using Sentinel-2 satellite imagery and machine learning. *Phys. Chem. Earth.* **2022**, *125*, 103082. [CrossRef]
60. Sunderland, T.; Abdoulaye, R.; Ahammad, R.; Asaha, S.; Baudron, F.; Deakin, E.; Duriaux, J.-Y.; Eddy, I.; Foli, S.; Gumbo, D.; et al. A methodological approach for assessing cross-site landscape change: Understanding socio-ecological systems. *For. Policy Econ.* **2017**, *84*, 83–91. [CrossRef]
61. Kusters, K.; De Graaf, M.; Buck, L.; Galido, K.; Maindo, A.; Mendoza, H.; Nghi, T.H.; Purwanto, E.; Zagt, R. Inclusive Landscape Governance for Sustainable Development: Assessment Methodology and Lessons for Civil Society Organizations. *Land* **2020**, *9*, 128. [CrossRef]

62. Omoding, J.; Walters, G.; Andama, E.; Carvalho, S.; Colomer, J.; Cracco, M.; Eilu, G.; Kiyingi, G.; Kumar, C.; Langoya, C.D.; et al. Analysing and Applying Stakeholder Perceptions to Improve Protected Area Governance in Ugandan Conservation Landscapes. *Land* **2020**, *9*, 207. [CrossRef]
63. Sagi, S. The concept of power in decision making process: A cross cultural perspective. *Int. J. Manag. Soc. Sci. Res. Rev.* **2015**, *1*, 160–163.
64. Wrigley-Asante, C. Women Becoming Bosses: Changing Gender Roles and Decision Making in Dangme West District of Ghana. *Ghana J. Geogr.* **2011**, *3*, 60–67.
65. Britwum, A.O.; Tsikata, D.; Akorsu, A.D.; Ako, M.A. *Gender and Land Tenure in Ghana: A Synthesis of the Literature*; Institute of Statistical, Social and Economic Research (ISSER), University of Ghana: Accra, Ghana, 2014. [CrossRef]
66. McShane, T.O.; Hirsch, P.D.; Trung, T.C.; Songorwa, A.N.; Kinzig, A.; Monteferri, B.; Mutekanga, D.; Thang, H.V.; Dammert, J.L.; Pulgar-Vidal, M.; et al. Hard choices: Making trade-offs between biodiversity conservation and human well-being. *Biol. Conserv.* **2011**, *144*, 966–972. [CrossRef]
67. Muradian, R.; Arsel, M.; Pellegrini, L.; Adaman, F.; Aguilar, B.; Agarwal, B.; Corbera, E.; De Blas, D.E.; Farley, J.; Froger, G.; et al. Payments for ecosystem services and the fatal attraction of win-win solutions. *Conserv. Lett.* **2013**, *6*, 274–279. [CrossRef]
68. Chaigneau, T.; Brown, K. Challenging the win-win discourse on conservation and development: Analysing support for marine protected areas. *Ecol. Soc.* **2016**, *21*, 36. [CrossRef]
69. Ros-Tonen, M.A.F.; Reed, J.; Sunderland, T. From Synergy to Complexity: The Trend Toward Integrated Value Chain and Landscape Governance. *Environ. Manag.* **2018**, *62*, 1–14. [CrossRef]
70. Reed, J.; van Vianen, J.; Barlow, J.; Sunderland, T. Have integrated landscape approaches reconciled societal and environmental issues in the tropics? *Land Use Policy* **2017**, *63*, 481–492. [CrossRef]
71. Pfeffer, K.; Baud, I.; Denis, E.; Scott, D.; Sydenstricker-Neto, J. Participatory spatial knowledge management tools: Empowerment and upscaling or exclusion? *Inf. Commun. Soc.* **2013**, *16*, 258–285. [CrossRef]
72. Villamor, G.B.; Palomo, I.; Santiago, C.A.L.; Oteros-Rozas, E.; Hill, J. Assessing stakeholders' perceptions and values towards social-ecological systems using participatory methods. *Ecol. Process.* **2014**, *3*, 22. [CrossRef]
73. Boedhihartono, A.K.; Endamana, D.; Ruiz-Perez, M.; Sayer, J. Landscape scenarios visualised by Baka and Aka Pygmies in the Congo Basin. *Int. J. Sustain. Dev. World Ecol.* **2015**, *22*, 279–291. [CrossRef]
74. Chambers, R. Participatory Mapping and Geographic Information Systems: Whose Map? Who is Empowered and Who Disempowered? Who Gains and Who Loses? *Electron. J. Inf. Syst. Dev. Ctries.* **2006**, *25*, 1–11. [CrossRef]

land

MDPI

Article

Consistent Changes in Land-Use/Land-Cover in Semi-Arid Areas: Implications on Ecosystem Service Delivery and Adaptation in the Limpopo Basin, Botswana

Ephias Mugari [1,2,*] and Hillary Masundire [2]

1 Department of Geography & Environmental Sciences, Faculty of Science, Agriculture and Engineering, University of Venda, Private Bag X5050, Thohoyandou 0950, South Africa
2 Biological Science Department, Faculty of Science, University of Botswana, Private Bag, Gaborone 0022, Botswana
* Correspondence: mugarie@gmail.com; Tel.: +27-651-869-968

Abstract: Ecosystems in semi-arid areas remain essential to securing livelihoods and aiding climate change adaptation. However, land-use and land-cover change (LULCC) is the leading driver of biodiversity, ecosystem services, habitat, and ecosystem loss in most rural areas of developing countries. We evaluated LULCC in the Bobirwa sub-district of Botswana between 1995 and 2015. We employed the supervised classification's maximum likelihood algorithm on the 1995, 2005, and 2016 Landsat images to establish the implications of LULCC on the delivery of provisioning ecosystem services (ES) and ecosystem-based adaptation in the Limpopo Basin part of Botswana. Five major LULC classes—vegetation, cropland, bare land, built-up areas, and water bodies—were identified in the sub-district. The decline in vegetation by 50.67 km^2/year between 1995 and 2016 was characterized by an increase in croplands (34.02 km^2/year). These changes were attributed to the growing human population that induced farming households to expand croplands. Government programs also encouraged agricultural expansions by offering free inputs and compensating smallholder farmers for land preparation. Higher agricultural yields remained critically low while the loss of vegetated areas to croplands threatened biodiversity, habitats, and the sustainability of provisioning ES through impaired ecosystem functions. There is an urgent need to arrest all unnecessary agricultural expansions and enhance agricultural productivity from current land parcels. The government and other relevant stakeholders also need to strengthen the ecosystem management capacities of local communities and support them to develop and implement biodiversity-based village action plans. Engaging communities through participatory, biodiversity-based action planning promotes biodiversity conservation and the sustainable use of ecosystem resources.

Keywords: ecosystem-based adaptation; LULCC; provisioning ecosystem services; semi-arid areas; Limpopo Basin; Botswana

1. Introduction

Land-use and/or land-cover change has been identified as one of the most important drivers of change in ecosystems and their services. Changes in land use highlight the influence of human activities that have a direct impact on the delivery of ecosystem services from the landscape [1,2]. Economic growth and development often trigger such changes and result in several trade-offs with human dependence on the natural environment [3]. In recent years, humans have exerted considerable pressure on the natural environment. For instance, the growing human population has increased the expansion of human settlements [1,4,5]. This often involves the clearing of naturally vegetated areas, which fragments local ecosystems, causes loss of biodiversity, and interferes with several ecosystem functions—both underpin ecosystem service delivery—and often exceed their ability to regenerate, resulting in loss of some ecosystems and ecosystem services [6,7].

In some places, excessive human exploitation continues to threaten the sustainability of local ecosystems [2]. However, as shown elsewhere, effective policies and management regimes protect and conserve local ecosystems while also promoting sustainable utilization of their services. In Botswana, the protected areas, forest reserves, and wildlife management areas in the north, including the Chobe National Park, have not only conserved local ecosystems but have also ensured a constant delivery of ecosystem goods and services to the surrounding communities [8].

Several international agreements, such as the United Nations Sustainable Development Agenda and the Convention on Biological Diversity, have been critical in guiding and encouraging countries to strike a balance between development and environmental protection [9,10]. Moreso, the Intergovernmental Platform on Biodiversity and Ecosystem Services (IPBES) provides a platform for governments to improve the interface between science and policy on issues of biodiversity and ecosystem services [11]. Nonetheless, the diversity of landscapes, climate, ecosystem services, cultures, social values, legislation, and the state of development render different land-use and/or land-cover types to be of varying importance to local people [12].

Previous studies in the Limpopo Basin of Botswana have established the importance of provisioning ecosystem services (ES) such as fresh water, Mopane caterpillars, cultivated crops, natural pastures, firewood, medicinal plants, and wild fruits to rural livelihoods and climate change adaptation [13–15]. The Millennium Ecosystem Assessment defined and popularized provisioning ecosystem services as tangible products from ecosystems [16]. The exploitation of, and trade in, certain provisioning services form a common yet vital part of livelihood strategies the world over. Botswana's National Adaptation Plan (NAP) Framework also acknowledges the importance of ecosystem-based adaptations in the NAP process [17]. Moreover, indigenous people living and depending on the natural environment in Botswana and elsewhere further reveal the importance of local ecosystems as a key livelihood and adaptation option given the susceptibility of rain-fed agriculture. However, the incremental climate change impacts as global average temperatures are set to warm beyond the 1.5 °C threshold of the Paris Agreement are expected to further limit the potential of ecosystems in semi-arid areas [18–20]. Without studies that quantify LULCC, especially in rural areas, understanding and redressing the simultaneous threats posed by climate change and anthropogenic pressure on provisioning ES becomes limited.

The nexus between land use, ecosystem services, and climate change adaptation has become more critical for researchers in recent years. Liao et al. found that land use/cover, climate (annual average rainfall, etc.), and human disturbance are among the key factors affecting ecosystem service provision in their study of the Northeast Forest Belt in China [12]. Another study by Biratu et al. evaluated the impact of landscape management scenarios on ecosystem service values in Central Ethiopia and found that cultivated land increased by 22% under the business-as-usual (BAU) scenario, whereas forest, water body, wetland, and shrubland were reduced [3]. The study, however, showed that forest cover increased, while bare land was reduced by 21% and 25% under the Ecosystems Protection and Agricultural Development (EPAD) and the Landscape Ecosystems Restoration and Conservation (LERC) scenarios. Chen et al. examined the impacts of land use changes on five typical ecosystem services (grain production, water yield, soil conservation, habitat quality, and carbon sequestration) in Guangdong province of China and concluded that land use and its changes had a significant impact on ecosystem services [7]. Mugari et al. note that climate change and anthropogenic pressure are increasingly modifying and/or interfering with ecosystem functions. This limits the delivery of ecosystem products, threatens livelihoods, and weakens the adaptive capacity of people in developing countries where direct dependence on the natural environment remains very high [15].

Although changes in land use are mainly driven by policies and management regimes in the short term, long-term benefits from land use changes can in turn influence decision-makers and policies [7]. However, information on the consequences of land use change, for instance on ecosystem services and climate change adaptation, is largely lacking at local scales [1]. Thus, examining the consequences of changes in land use on the delivery of ecosystem services and adaptation can provide the necessary evidence to inform policies and decisions that contribute to sustainable adaptation and development.

Various techniques have been used to detect LULCC, including land cover classification for multiple periods, change vector analysis, and normalized index differencing techniques [21–23]. Freely available remotely sensed data allows similar analyses to be conducted inexpensively in data-poor areas to aid decision-making. Landsat data provides high spatial resolution images where variations can be observed at a scale of 30 m [24]. Supervised land classification techniques have been used extensively to classify land and detect changes in LULC over time [23,25,26].

A serious challenge to the sustainable management of local ecosystems, adaptation, and development in data-scarce regions, is how the required evidence is derived and utilized [1]. However, in order to proffer solutions to societal challenges, the process needs to engage local communities to co-produce the required knowledge. Cognisant of this, we have attempted to plug this knowledge gap, including how evidence for supporting decisions can be derived in data-scarce areas, by quantifying land-use and land-cover (LULC) change at a local scale, and engaging surrounding communities to identify bundles of ecosystem services derived from different LULC types. Similar to the studies mentioned above, we also apply the supervised classification technique to quantify LULC changes. We then discuss the consequences of recent changes in different LULC types on the delivery of provisioning ecosystem services and adaptation to climate change. The analyses in this study thus aim to provide the evidence needed to support local-level decisions and planning processes that can also contribute to the sustainable management of local ecosystems, adaptation, and development in Botswana.

The analyses in this study were guided by the following research question: What are the implications of recent changes in LULC on the delivery of key provisioning ES and ecosystem-based adaptations in the Limpopo Basin of Botswana? This work was part of the Adaptation at Scale in Semi-Arid Regions (ASSAR) and the South Africa/Flanders Climate Adaptation Research and Training Partnership (SAF-ADAPT) projects. Bobirwa sub-district was selected for this study since it exhibited desired attributes of rapid urbanization, increasing demand for agricultural land, high human dependence on the natural environment, proximity to protected areas, and sharing national boundaries with Zimbabwe and South Africa.

2. Materials and Methods

2.1. Study Area

The Limpopo River Basin part of Botswana is ecologically and economically significant to indigenous people and surrounding communities. Our case study, the Bobirwa sub-district is situated between 28°09′10″ E to 29°21′42″ E and 22°35′17″ S to 21°35′56″ S and lies entirely within the Limpopo River Basin part of Botswana. The sub-district has an altitude ranging between 590 and 886 m, making it the lowest part of Botswana, hence it has a network of channels that drain into the Limpopo River [14]. The sub-district boundary forms the national boundary with Zimbabwe to the northeast and South Africa to the southeast where the Limpopo River marks the boundary (Figure 1). According to the 2011 National Population and Housing Census report, the population of the Bobirwa sub-district was 71,936, comprising 34,247 males and 37,689 females from 19,213 households with an average household size of 3.74 and a population density of 5.05 people/km^2 [27].

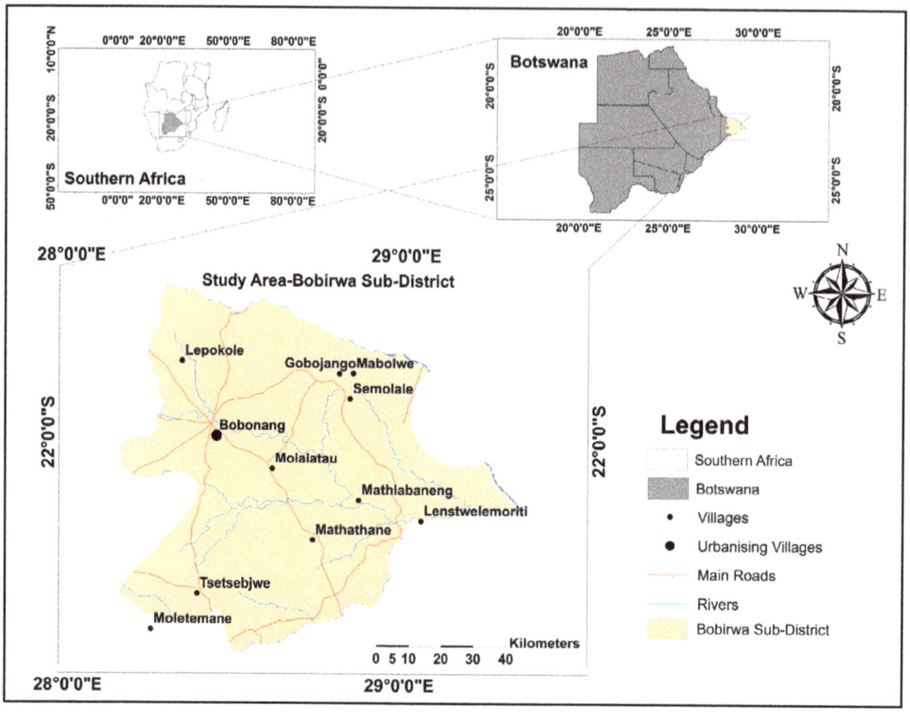

Figure 1. Map of Bobirwa sub-district showing study villages.

The study area is highly susceptible to droughts and erratic rainfall which fluctuates well below 400 mm/year in most years [18]. The recent (2010–2016) average winter and summer temperatures were 15.8 °C and 24.5 °C, respectively. However, summer temperatures have often exceeded 38 °C in the last 5 years with the occurrence of heat waves [18]. Summers are hot and dry between August and November (pre-rainy season) and become hot and wet between December and February/March (rainy season). Thus, the period March–April is considered the post-rainy season. Winters are cold and dry and occur from late April to early August. Nonetheless, the study area experiences a less severe climate than other parts of the country, allowing crop and livestock production and supporting considerable biodiversity that underpins the delivery of several timber and non-timber forest products [14,28].

Natural vegetation in the study area is mostly represented by desert-like vegetation and a variety of species, such as grasses and grass-like plants, shrubs, and trees with patches of bare land [29]. *Colophospermum mopane*, commonly known as mopane, dominates the tree, shrubs, and other desert-like vegetation. The Limpopo Basin is richly endowed with resources such as wood fuel, medicinal plants, and Mopane caterpillars (larvae of *Imbrasia belina* moth), which thrive on Mopane trees (*Colophospermum mopane*) abundant in the area. The sub-district is well-known for Mopane caterpillars, which are an important source of protein and income, thus attracting harvesters and buyers from elsewhere in Botswana, Zimbabwe, and South Africa [14,30]. Village settlements are separated from "cattle posts" and crop fields, resulting in a 'three-home' system for many households whereby there is a temporary and seasonal movement across the three locations, i.e., village—crop fields—cattle posts.

The main livelihood activities and economy revolve around rain-fed agriculture and exploitation of timber, firewood, and non-timber forest products such as Mopane caterpillars (*Imbrasia belina*), wild fruits, thatch, and palm leaves (*Hyphaene petersiana*) for basketry [14,30,31]. The main crops grown under rain-fed conditions are maize, sorghum, millet, cowpeas, groundnuts, round nuts, and watermelons. Subsistence livestock and poultry production, a mainstay of the local economy, is mainly characterized by rearing cattle, goats, and chickens under free-range [14]. Other land uses include game reserves and privately owned farms with abundant wildlife. In recent years, tourism development in the sub-district, for instance in the Tuli Block area, has been strengthened in part due to government support of local tourism and to a larger extent by the abundance of wildlife in the area.

2.2. Data Sources

Three sets of Landsat images with less than 0.1% cloud contamination were downloaded from the USGS GloVis for the post-rainy period of the years 1995, 2006, and 2016 (each year with 4 scenes) through the URL: http://glovis.usgs.gov/ (accessed on 16 August 2019) using 170-1/075-6 as path/row, respectively, as detailed in Table 1. Landsat offers the longest global record of the earth's surface and is operated by USGS/NASA. The dates when the Landsat images were acquired (Table 1) are for the period after the rainy season; these were specifically chosen as vegetation, croplands and bare lands are more distinct after the rainy season and thus reduced chances of confusion. Since the study was limited by the unavailability of cloud-free imagery on desired dates, we appreciate the slightly lower classification accuracy when using Landsat imagery acquired on different dates for land-cover classification. Nonetheless, the multi-temporal analysis, use of imagery before the leaf senescence (greenness transition) period, and extensive ground-truthing minimized these errors [32].

Table 1. Landsat images used for this study.

Month	Year	Path/Row							
		170/075		170/076		171/075		171/076	
		Sensor	Date Acquired	Sensor	Date Acquired	Sensor	Date Acquired	Sensor	Date Acquired
March/April	1995	TM	17/03/95	TM	17/03/95	TM	09/04/95	TM	09/04/95
	2006	ETM	08/04/06	ETM	10/05/06	ETM	07/04/06	ETM	07/04/06
	2016	OLI	02/03/16	OLI	02/03/16	OLI	02/04/16	OLI	02/04/16

As shown in Table 1, the satellite images used in this study were imagery from Landsat-5 Thematic Mapper (TM) Sensor (1995), Landsat-7 Enhanced Thematic Mapper (ETM) Sensor (2006), and Landsat-8 Operational Land Imager (OLI) Sensor (2016) sensors with a resolution of 30 m.

2.3. Image Pre-Processing

The pre-processing of the acquired Landsat images was executed in ENVI (v5.3) and included radiometric and atmospheric correction using the Fast Line-of-sight Atmospheric Analysis of Spectral Hypercubes (FLAASH). This is a first-principles atmospheric correction modeling tool for retrieving spectral reflectance from hyperspectral radiance images. FLAASH can accurately compensate for atmospheric effects and correct wavelengths in the visible through near-infrared and short-wave infrared range [33]. The Landsat images were corrected to remove atmospheric effects and then re-sampled to 30 m pixel size for all bands using the nearest neighbor method. The nearest neighbor algorithm is applied where knowledge of the underlying probabilities is limited.

2.4. Image Processing

2.4.1. Supervised Image Classification

The pre-processed satellite images for the post-rainy period (March/April) were classified using the maximum likelihood algorithm (MLA) using ENVI v5.3 software.

Initially, regions of interest (ROIs) were carefully selected on each satellite image to create a training sample for supervised classification. Each ROI had between 80 and 150 polygons as training sites, and these were chosen randomly depending on the area covered by the class. Five ROIs were selected for this study, as described in Table 2. The MLA uses the ROIs to classify a data point, i.e., classifies a data point based on how its neighbors are classified [34]. Despite a wide choice of algorithms for supervised classification, MLA was most suitable for classifying at least two classes with overlapping spectral characteristics, simple point clouds, and some correlation between the brightness in different ranges of the spectrum [34]. Several comparable studies elsewhere also used MLA [26–28].

The LULC classes in this study (Table 2) were distinguished clearly from each other with the aid of a feature space image, which plots a scatter of pixel values of two bands of the imagery and different color reflectance. The intensity of each portion of the feature space image shows all the pixels with the same reflectance values, which reduces the chances of mixing of any two classes Therefore, classification was performed after there was minimal mixing.

Table 2. Land cover classes used in the classification of Landsat images.

Land Cover Class	Description
Built-up areas	Village settlements, rock surfaces, paved roads, etc.
Croplands	Cultivated areas under rain-fed and irrigation, plantations
Water bodies	Dams, ponds, rivers, streams, etc.
Vegetation	Predominantly trees, shrubs, or grasslands
Bare land	Bare lands with exposed soil surfaces, with no vegetation all year round, or with very sparse vegetation

2.4.2. Accuracy Assessment

Accuracy assessments were performed to establish the relationship between the classified LULC images and the control reference data similar to previous studies [5,26,35]. This was based on ground data collected from the respective land-use and land-cover types in the study area. For the 2016 image, a minimum of 50 ground truth points was generated per class in 8 villages except for water bodies, which had about 25 ground control points as there were fewer open water bodies. For the 1995 and 2006 images, historical Google Earth was used for validation of the image accuracy assessments. Similarly, each class, except for water bodies, had a minimum of 80 ground control points taken for accuracy assessment.

The user's accuracy was calculated as the number of correctly classified pixels as a proportion of the total number of pixels assigned to a specific category and accounts for errors of commission. The producer's accuracy was computed to show the number of pixels correctly classified in a specific category as a proportion of the total number of pixels actually belonging to that category and accounts for errors of omission. The overall classification accuracy was also computed to show the total of all the correctly classified sample pixels as a proportion of the total number of reference pixels.

2.4.3. Error Matrix

An error matrix showing errors of omission (producer's accuracy) and commission (user's accuracy), overall classification accuracy, and Kappa coefficient was used to determine the accuracy of the LULC maps [26]. Error matrices are commonly used to report classification accuracies, as they compare the relationship between known reference data (ground truth) and the corresponding results of an automated classification on a category-by-category basis [36]. The error matrices used in this study to compare the reference class values with the classified class values are a 5 $\times$ 5 matrix, where 5 is the total number of classes used for this study. The overall accuracy and the Kappa coefficient values were calculated using the producer's and user's accuracy for all the LULC classes similar to several studies elsewhere [22,24,37]. The Kappa coefficient measures the deviation of chance agreement from the actual agreement. A minimum overall classification accuracy of 80% is considered efficient.

2.5. Household Survey

The study applied the sampling error formula [1] for a target population exceeding 10,000 to determine the sample size. The sampling unit for this study was the household, and the target population was the 10,896 households in the study area [2]. Stratified simple random sampling was then employed to select 317 study participants in the eight villages using numbered village lists from the local Sub-Land Board and random numbers generated in R studio software. Commands for random number generation required the first and last positions of households on each village list, including the total number required from that range without repeating. All participants were informed prior to the interview with the assistance of Village Development Committees in each village, and written consent was sought before each interview.

A semi-structured questionnaire prepared in collaboration and with input from members of the Adaptation at Scale in Semi-Arid Regions (ASSAR) project Southern Africa regional team was administered to household heads or the most senior member available. This study only focused on questions that sought to establish the provisioning of ecosystem services in the Limpopo Basin, their relative importance, land-use and/or land-cover types providing them, and their recent trends. However, only 310 questionnaires were considered complete and usable for analyses.

2.6. Data Analysis

Maps were used to show the spatial distribution of the different LULC types, and band math were then computed on the classified LULC images. The absolute area (in km^2) and the proportion covered (in %) by different LULC class categories were calculated for 1995, 2006, and 2016 using the number of pixels covered by each LULC class. The overall classification accuracy and the Kappa coefficient were used to analyze the efficiency of the classified LULC classes similar to other studies [22,24]

Post-classification change detection was performed by computing the changes in the area covered by each LULC class between 1995 and 2006, 2006 and 2016, and 1995 and 2016. Land cover classification images for the different years were compared on a class-by-class basis, and the differences in terms of the number of pixels covered by each category in the different time periods were used to compute the area statistics for these changes. The absolute changes (in km^2 and %) and average rate of changes (in km^2/year and %/year) in the area covered were computed to detect changes in each of the three time periods. Splitting the analyses into decadal periods was key to showing the period with significant changes and unmasking the most recent changes. The sections below present the results of the change matrix for the periods 1995–2006, 2006–2016, and 1995–2016. LULCC change matrices were used to examine how a decline in one LULC was distributed among other LULC classes. Similarly, they also revealed how an increase in one LULC occurred.

Data from the household survey were analyzed using STATA software (version 14.2). Ranking of provisioning ecosystem services by households considered their relative importance, i.e., the contribution of each provisioning ecosystem service towards household food security, income, employment, and provision of raw materials. It also considered the possible impacts of a decline in the ecosystem service on the same. Ranking was achieved by averaging all the individual responses for each of the provisioning ecosystem services identified into a single metric ranging between 1 and 100 points. Thus, those provisioning ecosystem services within the 66–100 points range were highly ranked, while those in the range of 33–66 were ranked medium and those lowly ranked (0–33) provisioning ecosystem services. Highly ranked provisioning ecosystem services contributed more to livelihoods, while the lowly ranked contributed the least.

3. Results

3.1. Land-Use and/or Land-Cover Change in Bobirwa Sub-District

The spatial distribution of LULC types in Bobirwa for 1995, 2006, and 2016 is shown in Figure 2 below.

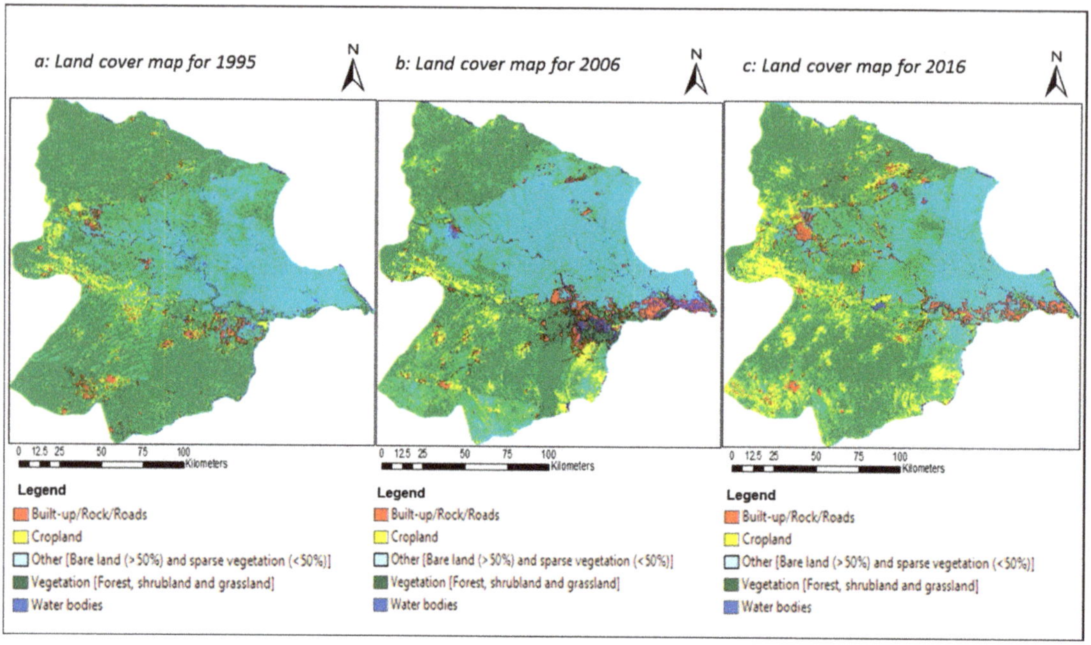

Figure 2. Land-use and land-cover maps for the years 1995, 2006, and 2016.

Figure 3 summarizes the absolute areas (km^2) and proportions (%) covered by the different LULC types in 1995 and 2016. In 1995, approximately 5130 km^2 (68%) of the total surface area in the Bobirwa sub-district was covered by natural vegetation such as woodlands, shrublands, and grasslands. Croplands covered only 395 km^2 (5.24%). The overall classification accuracy and the Kappa coefficient for the 1995 LULC were 79.82% and 0.7035, respectively. In 2016, the area covered by natural vegetation was about 4066 km^2 (53%) of the total surface area of the Bobirwa sub-district. Croplands spanned nearly 1110 km^2 (14.71%), while bare lands were distributed over 1798.33 km^2 (23.84%). The overall accuracy and the Kappa coefficient of the 2016 classified image were 78.51% and 0.7249, respectively. The area covered by natural vegetation was significantly different from the area covered by built-up areas (p = 0.0080) and water bodies (p = 0.0068). The area covered by bare lands was also significantly different from the area covered by built-up areas (p = 0.0040) and water bodies (p = 0.0006).

Table 3 indicates the dynamics of LULCC in the Bobirwa sub-district during study periods. Between 1995 and 2006, croplands increased by approximately 169 km^2 (42.65%), which translates to an increase of 15.33 km^2/year. Natural vegetation declined by about 1048 km^2 (20.42%), an average decline of 95.25 km^2/year during the same decade. Between 2006 and 2016, croplands increased by approximately 546 km^2 (96.74%), at a rate of 54.57 km^2/year. Natural vegetation and bare land decreased by about 16.37 km^2 (0.40%) and 612.83 km^2 (25.42%) at an average of 1.64 km^2/year and 61.28 km^2/year, respectively. Overall, between 1995 and 2016, croplands and bare land increased by 714.39 km^2 (180.65%) and 115.11 km^2 (6.84%), respectively (Table 3). This was an average increase of 34.02 km^2/year and 5.48 km^2/year in croplands and bare lands. Natural vegetation declined by 1064.15 km^2 (20.74%) at an average rate of 50.67 km^2/year during the 21 years. None of these changes were significant (p = 0.1906).

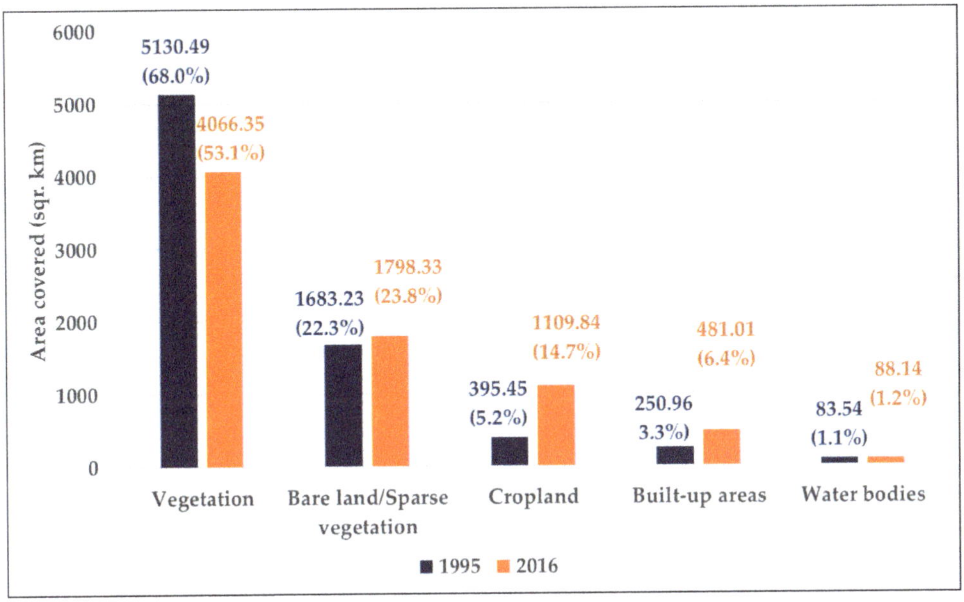

Figure 3. Comparison of LULC classes in Bobirwa sub-district between 1995 and 2016.

Table 3. Comparison of LULCCs in Bobirwa sub-district between 1995 and 2016.

LULC Class	Changes from 1995–2006		Changes from 2006–2016		Changes from 1995–2016	
	km^2 (km^2/Year)	% (%/Year)	km^2 (km^2/Year)	% (%/Year)	km^2 (km^2/Year)	% (%/Year)
Bare land	727.94 (66.18)	43.25 (3.93)	−612.83 (−61.28)	−25.42 (−2.54)	115.11 (5.48)	6.84 (0.33)
Built-up areas	165.18 (15.02)	65.82 (5.98)	64.87 (6.49)	15.59 (1.56)	230.05 (10.95)	91.67 (4.37)
Croplands	168.66 (15.33)	42.65 (3.88)	545.72 (54.57)	96.74 (9.67)	714.39 (34.02)	180.65 (8.60)
Water bodies	−14.01 (−1.27)	−16.77 (−1.52)	18.61 (1.86)	26.77 (2.68)	4.60 (0.22)	5.51 (0.26)
Vegetation	−1047.78 (−95.25)	−20.42 (−1.86)	−16.37 (−1.64)	−0.40 (−0.04)	−1064.15 (−50.67)	−20.74 (−0.99)

Figure 4 shows how gains or losses in each of the five LULC classes came about during the period under study. Figure 4a shows that from 1995 to 2006, 404.31 km^2 (7.88%) of vegetation was converted into croplands, and a further 1123.23 km^2 (21.89%) was converted into bare lands. A total of 3457.67 km^2 (67.39%) of natural vegetation remained unchanged. During the same decade, 262.71 km^2 (15.61%) of bare land and 238.27 km^2 (60.25%) of croplands were converted into naturally vegetated areas. A total of 108.6 km^2 (27.46%) remained as croplands, while the rest was converted to other LULC types.

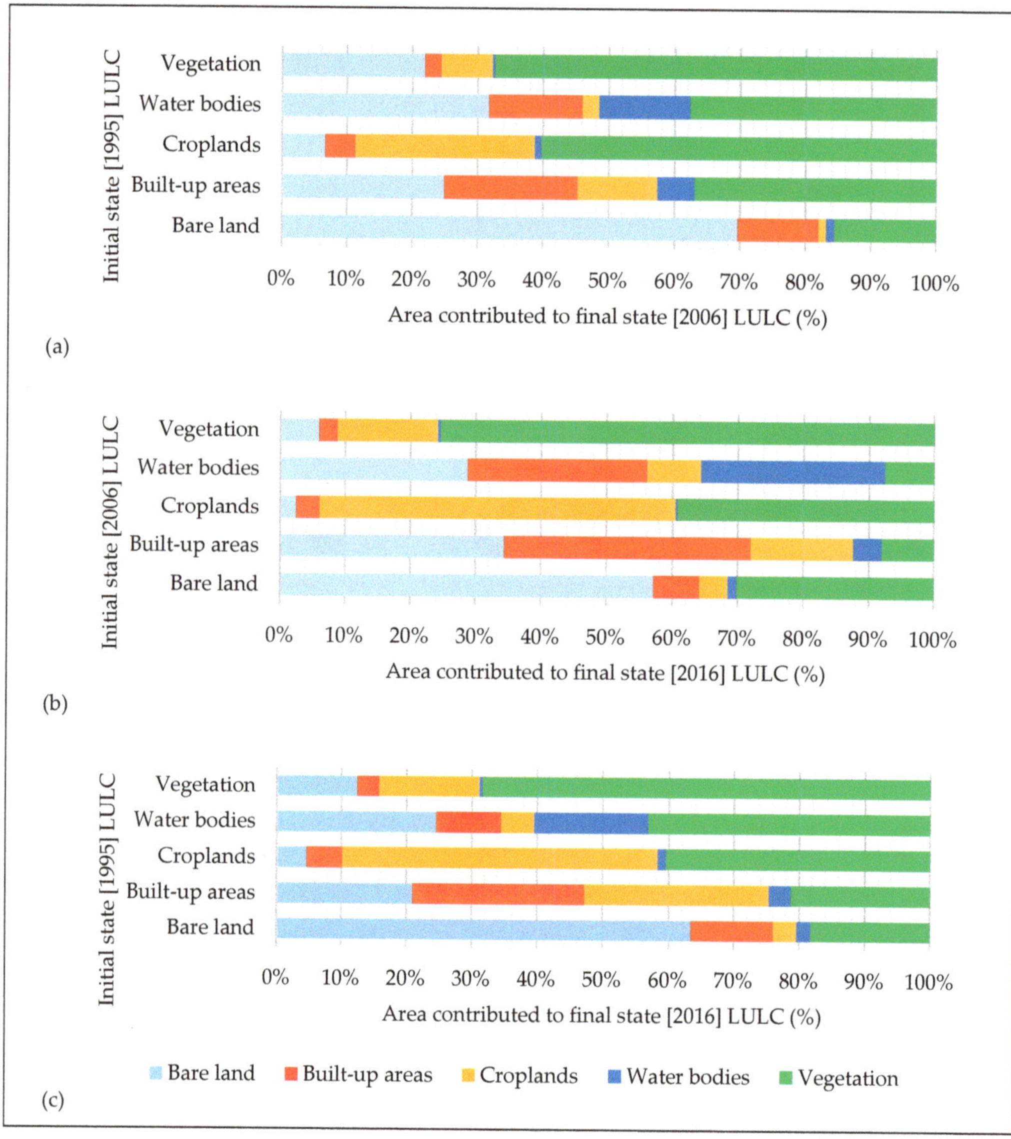

Figure 4. Proportion of area contributed initial state LULC class to the final LULC class for the period: (**a**) 1995–2006; (**b**) 2006–2016; (**c**) 1995–2016.

From 2006 to 2016, Figure 4b shows that 627.83 km^2 (15.38%) of vegetation was converted into croplands, while 3079.28 km^2 (75.42%) remained unchanged. Only 243.40 km^2 (5.96%) of vegetation was converted into bare land. A total of 221.42 km^2 (39.25%) of croplands and 726.81 km^2 (30.14%) of bare land were revegetated. Approximately 306.44 km^2 (54.32%) of croplands remained unchanged between 2006 and 2016. Overall, considering the entire 21 years from 1995 to 2016, Figure 4c shows that 785.62 km^2 (15.31%) of vegetation was converted into croplands, and 639.20 km^2 (12.46%) was converted into bare land. About 3509.01 km^2 (68.40%) remained as natural vegetation between 1995 and 2016. Only 172.24 km^2 (3.36%) of vegetation was converted into built-up areas. A total of

159.93 km^2 (40.44%) of croplands and 308.17 km^2 (18.31%) of bare land were revegetated during this period. Approximately 190.58 km^2 (48.19%) of croplands in 1995 remained as croplands in 2016.

3.2. Household Socio-Economic Characteristics

Table 4 summarizes the socio-economic characteristics of survey participants and presents their frequencies (or means) and proportions (%), including Chi-square tests (and p-values) of differences among participants. There were 79.7% females and 20.3% males with an average age of 56.3 years. Only 27.1% of the participants had not gone past primary school, while 32.3% reported being full-time farmers and 67.7% as part-time farmers. The average size of farmland owned by a household was 8.85 ha, but the cultivated area was 2.32 ha. Average household size was 5.9 (3.3 females and 2.6 males). A proportion of 37.4%, 26.1%, 55.5%, and 10.3% of the households owned cattle, donkeys, goats, and sheep, respectively.

Table 4. Socio-economic characteristics of survey participants.

Socio-Economic Characteristics		**N/Mean**	**% of Cases**	χ^2	***p*-Value**
Age of household head	Young (20–40)	57	18.4	9.332	0.009 **
	Adult (40–60)	128	41.3		
	Elderly (>60)	125	40.3		
Gender of household head	Female	247	79.7	7.688	0.006 **
	Male	63	20.3		
Education level of household head	None	84	27.1		0.014 *
	Primary	122	39.4		
	Secondary	71	22.9		
	Tertiary	33	10.7		
Employment status of household head	Full-time farmer	100	32.3	0.160	0.689
	Part-time farmer	210	67.7		
Household size		5.88	-		
Arable farm size (ha)	Total land owned	8.85	-		
	Cultivated area	2.32	26.2		
Annual income of household (BWP) (USD 1 ≈ BWP 13.4)	<5000	193	62.3	4.160	0.385
	5000–10,000	39	12.6		
	10,001–15,000	19	6.1		
	15,001–20,000	15	4.8		
	>20,000	44	14.2		
Household sources of income	Crop production	164	52.9		
	Casual labor (*Ipelegeng*)	115	37.1		
	Mopane caterpillars	91	29.4		
	Formal employment	86	27.7		
	Small livestock	76	24.5		
	Poultry products	76	24.5		
	Informal trade	64	20.6		
	Cattle	48	15.5		
	Timber/firewood	48	15.5		
	NTFPs	10	3.2		

Table 4. *Cont.*

Socio-Economic Characteristics		N/Mean	% of Cases	χ^2	*p*-Value
Livestock owned	Cattle	6.27	37.4	0.201	0.654
	Donkeys	1.42	26.1	5.117	0.077
	Goats	7.77	55.5	9.633	0.008 **
	Sheep	1.02	10.3	0.649	0.421
Government (ISPADD) input support received	Full subsidy	145	46.8		
	Partial subsidy	17	5.5	1.145	0.887
	None	148	47.7		
Major farm implements used	Hand-held	204	65.8	3.514	0.037 *
	Animal-drawn	137	44.2	9.257	0.026 *
	Tractor-drawn	11	3.5	0.004	0.950
Access to climate and agricultural information	Yes	282	91.0	0.071	0.790
	No	28	9.0		
Average grain yield (kg/ha) [last 3 seasons]	Sorghum	168.9			
	Millet	167.5			
	Maize	164.3			
	Cowpeas	122.3			

Significance level; * $p < 0.05$, ** $p < 0.01$. Source: Household Survey Data, 2017.

More than 62% of the households had income below BWP 5000 (USD 374). Sources of income for households included crop production (52.9%), casual labor (37.1%), selling Mopane caterpillars (29.4%), formal employment (27.7%), selling small livestock and poultry (24.5%), selling timber and firewood (15.5%), and selling other ecosystem products such as palm wine and wild fruits (3.2%). At least half of the participants had received free inputs from the government through the Integrated Support Programme for Arable Agricultural Development (ISPAAD). Households used hand-held (65.8%), animal-drawn (44.2%), and tractor-drawn (3.5) farm implements. The average yield for sorghum, millet, maize, and cowpeas under rain-fed conditions for the past 3 seasons ranged between 122 and 170 kg/ha.

3.3. *Key Provisioning Ecosystem Services in Bobirwa Sub-District*

Figure 5 shows that a total of 15 provisioning ecosystem services were identified as critical to livelihoods from the survey (see Supplementary Table S1 for a more detailed list); seven contributed directly towards household food requirements, nine contributed towards household income and four (4) provided raw materials, e.g., palm leaves for weaving. Reserved areas delivered almost all the identified provisioning ES (14), followed by woodlands (13), croplands (9), areas within village settlements (9), water bodies (3), and bare lands (1).

Figure 5 also shows that the highly ranked provisioning ES in terms of relative importance and contribution to household food requirements, income, raw materials, and employment opportunities were natural pastures, livestock production (cattle, donkeys, goats, sheep, and poultry), freshwater, Mopane caterpillars, and firewood. The medium ranked were timber and poles, palm plants, wild fruits, and thatch. The lowly ranked included natural medicines and medicinal plants, natural dyes (for decorating palm leaves used for weaving), wild animals, fisheries and sand, and precious stones. These rankings also considered the effects of marginal changes in the delivery of each ES on household food security, income, provision of raw materials, and employment opportunities.

Provisioning Ecosystem Service	Rank	Wood-lands	Crop-lands	Water Bodies	Bare lands	Village/ Settlement	Reserved Areas
Cultivated crop production	High	✓	✓			✓	✓
Natural pastures	High	✓	✓				✓
Livestock production	High	✓				✓	✓
Fresh water [human & livestock consumption]	High	✓		✓		✓	✓
Mopane caterpillars [*Imbrasia belina spp.*]	High	✓	✓			✓	✓
Firewood [*esp. Colophospermum mopane*]	High	✓					✓
Timber/poles	Medium	✓	✓				✓
Palm leaves/plants [*Hyphaene petersiana*]	Medium	✓	✓			✓	✓
Wild fruits	Medium	✓	✓			✓	✓
Thatching grass	Medium	✓					✓
Natural medicines/medicinal plants	Low	✓	✓			✓	✓
Natural dyes [*for decorating palm leaves for weaving*]	Low	✓	✓			✓	✓
Game meat	Low	✓	✓			✓	✓
Fish	Low			✓			✓
Sand/stone for construction	Low			✓	✓		
Total number of provisioning ES		13	9	3	1	9	14

Figure 5. Key provisioning ecosystem services provided by different land-use and land-cover types in the Limpopo Basin of Botswana (n = 310). Rank in terms of relative importance to household requirements: Green = high; Amber = medium; Red = low.

A general decline in accessibility to several timber and non-timber forest products was reported by at least 78% of participants in the study area (Figure 6). Access to palm plants and natural dyes was reported to be largely unchanged though concerns were raised over the destructive harvesting and/or extraction of palm wine. Access to water for livestock consumption at the communal grazing areas (cattle posts) was also reported to have declined by 34% of the participants. Almost 23% of the participants reported difficulties in accessing fertile land for cultivating crops, while about 60% had the same access. Only 7% reported more access to fertile land for cultivation.

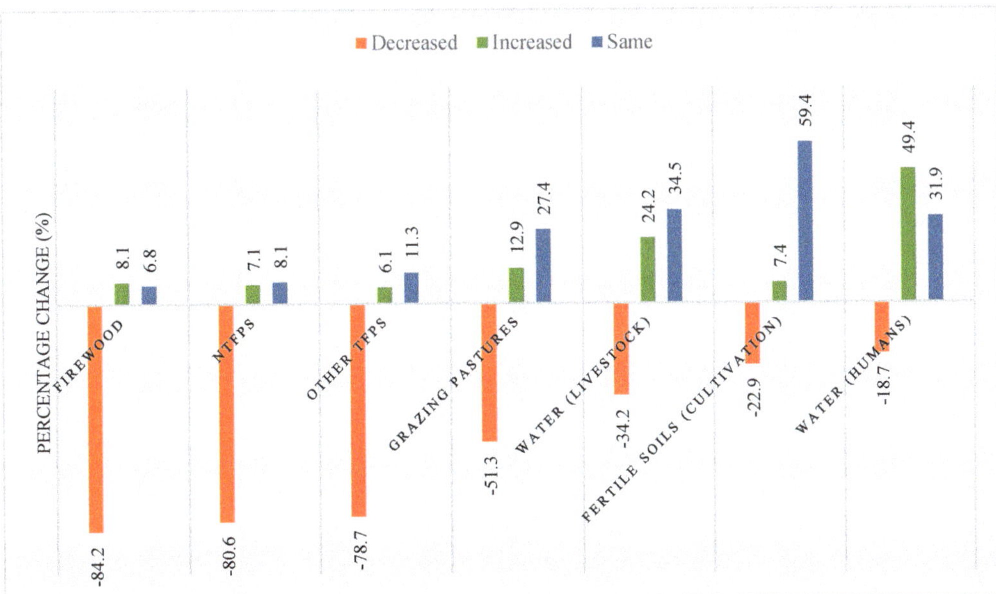

Figure 6. Changes in accessibility to key provisioning ecosystem services in Bobirwa (2006–2017) (n = 310).

3.4. Perceived Drivers of Change in Provisioning Ecosystem Services

Figure 7 shows the perceived drivers of change in (a) total area (size) of ES sites and (b) walking distance to sites providing ES. The perceived drivers of change included primary (e.g., adverse climate) or secondary, i.e., those driven by one or more of the stated (primary) drivers (e.g., land degradation).

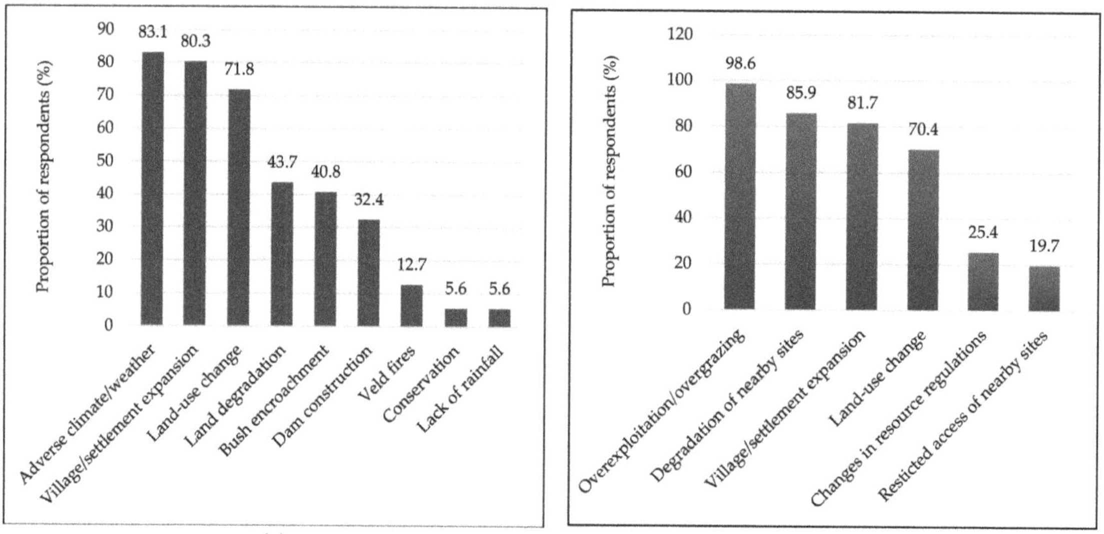

Figure 7. Drivers of change influencing delivery of provisioning ecosystem services in Bobirwa (n = 310): (**a**) Drivers of changes in total area/size of ES sites; (**b**) Drivers of changes in distance to ES sites.

Fluctuations in the total area or size of historical sites delivering provisioning ES (Figure 7a) were perceived to be driven by adverse climate and/or weather (83.1%, $p = 0.000$), expansion of human villages (80.3%, $p = 0.000$) and changes in land-use (71.8%, $p = 0.002$). Land degradation (43.7%, $p = 0.132$), bush encroachment (40.8%, $p = 0.217$), dam construction (32.4, $p = 0.000$), and veld fires (12.7%, $p = 0.000$) were perceived by fewer participants. Similarly, causes of changes in distance to sites delivering provisioning ES were perceived as overexploitation and/or overgrazing (98.6%, $p = 0.000$), degradation of nearby sites (85.9%, $p = 0.000$), village and/or settlement expansion (81.7%, $p = 0.000$) and land-use change (70.4%, $p = 0.001$). Fewer participants perceived changes in regulations (25.4%, $p = 0.000$) and restricted access to privately owned farms (19.7%, $p = 0.000$) as limiting access to nearby sites.

4. Discussion

4.1. Dynamics of Change in Bobirwa Sub-District

The majority of provisioning ES in the Limpopo Basin are derived from, or depend on the extent and richness of, natural vegetation. Agriculture is also an important livelihood, thus our discussion mainly focuses on changes in natural vegetation and croplands. Although this study did not quantify the changes in individual provisioning ES, our analyses provide a suitable quantitative measure to understand the implications on bundles of provisioning ES delivered by natural vegetation and croplands.

The dominant land-cover type in Bobirwa during the study period was natural vegetation followed by bare lands (with sparse vegetation) and croplands. The decline in natural vegetation between 1995 and 2016 at a rate of 50.67 km^2/year, was characterized by a significant increase in croplands of 34.02 km^2/year during the same period. This indicates an increased demand for agricultural land, particularly for crop production similar to studies in Central Ethiopia [3].

Besides the expansion of croplands, the decline in natural vegetation is also attributed to the expansion of village settlements caused by a growing human population. According to the 2011 Botswana National Census, the total human population of the sub-district grew by 8.2% between 2001 and 2011, which translates to an annual growth rate of 0.8% during that decade [27]. For instance, the expansion of villages such as Bobonong, Tsetsebjwe, Mathathane, Motlhabaneng, and Molalatau to establish built-up areas involved clearing vegetated areas, particularly the *Colophospermum Mopane* woodlands that are prevalent in the area as noted by earlier studies [30,38]. With such growth and urbanization of some villages, pressure on surrounding land for settlement and croplands could further reduce natural vegetation in the Bobirwa sub-district.

The three-tier household system among many of the rural households in Botswana, i.e., spread across the village settlement, communal grazing areas, and crop fields also has implications on LULCC in the study area. The increase in built-up areas and bare lands around the villages in Bobirwa is consistent with the effects of a growing human population [35]. Expansion of settlements that involve the clearing of woodlands, shrublands, and grasslands has also been reported in the Likangala catchment in Malawi [39] and Mukuvisi and Marimba catchments in the outskirts of Harare [40]. The Mopane woodlands are also sources of firewood, some of which are harvested illegally for commercial purposes. Thus, an increase in human population in the Bobirwa sub-district, coupled with limited livelihood opportunities, and a unique dependency on the natural environment, particularly agriculture, exerts pressure on land [5].

The increase in croplands in Bobirwa during the assessment period is consistent with most rural communities in Africa where subsistence, rain-fed crop, and livestock production are critical sources of livelihoods, as shown elsewhere in Botswana [2]. Given that only 25% of the arable farmland owned by households was under cultivation, expansions in croplands are likely to perpetuate. However, increases in croplands often fragment natural ecosystems and could significantly impair some ecosystem functions [41]. For instance, some households in the study area resorted to establishing croplands at the cattle posts, areas only reserved for communal grazing, thereby reducing natural pastures among other natural products obtained therefrom. Such actions were mostly driven by delays in obtaining land officially. In addition, more frequent droughts experienced in recent years were cited by some farmers as forcing them to clear more land for growing crops. The critically low grain crop yields (sorghum, millet, maize, and cowpeas), i.e., <170 kg/ha, obtained by farmers in the study area highlight the precarity and susceptibility of rain-fed agriculture. An earlier study in the area shows a decline in rainfall events and increasing deviations below average rainfall [38]. The same study suggests an increased potency of droughts in the Limpopo Basin of Botswana and highlights the adverse effects on rain-fed crop production. Thus, some farmers viewed the expansion of area under cultivation as a coping strategy and insurance against drought-induced total crop failure. Kosonei et al. [42] also found increased demand for agricultural land in the Marigat sub-county of Kenya driven by the need to salvage yields in case of severe droughts. This is in contrast to studies elsewhere in Botswana that found increased abandonment of croplands due to the increasing severity of droughts [2].

Another driver of expansions of croplands in Botswana is the Integrated Support Programme for Arable Agricultural Development (ISPAAD) by the Ministry of Agriculture. The program supports smallholder farmers with free inputs such as fertilizer, seeds, herbicides, tillage, and access to credit [43]. While the study shows that at least half of the surveyed households were beneficiaries of the ISPAAD program, the effect of such programs is that they attract new farmers and influence current farmers to increase the area under cropland. Considering that the average area under cultivation by the surveyed households was about 2 ha, a quarter of total arable landholdings, and that the ISPAAD program gives free inputs for up to 5 ha, farmers were likely to expand their croplands. Our findings, similar to others before, show that the program has not enhanced crop productivity over the years [44]. Although current efforts to deviate from the traditional system where grazing and croplands are separate, to integrated farms of up to 15 ha where farmers can simultaneously practice crop and livestock production, more efforts need to go towards enhancing agricultural productivity. This could arrest unnecessary agricultural expansions since land fragmentation interferes with ecosystem functions and threatens biodiversity and the delivery of essential ES [2].

4.2. Impacts of LULCC on Ecosystem Service Delivery

The findings show that the increase in croplands, built-up areas, bare lands, and water bodies was mainly at the expense of natural vegetation. While increasing area under croplands could increase food production, the average grain yield obtained in the study area (Table 4) paints a gloomy picture. Instead, the observed decline in natural vegetation threatens biodiversity and reduces primary production [45]. Consequently, the delivery of key provisioning services such as timber, firewood, Mopane caterpillars, natural medicines, and natural dyes by vegetated areas becomes threatened. This further threatens household sources of income since some of these services feature among the reported household sources of income, e.g., Mopane caterpillars and firewood. With at least 60% of the households earning below BWP5000 (USD 373) per year, the loss of these sources of livelihoods could push them further into poverty. There is widespread evidence of the effectiveness of ecosystem services in safeguarding livelihoods, as well as providing a sustainable adaptation pathway, as also highlighted in Botswana's National Adaptation Framework and National Biodiversity Strategy and Action Plan [17,46]. Ecosystem services from natural systems are particularly important in the study area given the precarity of rain-fed agriculture. Thus, a further loss of vegetated areas in the Limpopo Basin of Botswana could have cascading negative effects, especially on low-income households, which are the majority of the surveyed households.

Globally, the impairment of the delivery of provisioning ES such as timber, medicinal plants, wild fruits, resins, and natural dyes due to a decline in biodiversity is well-documented [47–49]. LULCC has been reported as the most severe direct driver of biodiversity loss by several studies [50–52]. The loss of vegetated areas in the study area, particularly at the communal grazing areas (cattle posts), also threatens livestock production, which, for low-income households, largely depends on natural pastures [53]. Elsewhere, land fragmentations have also caused habitat changes and significant biodiversity losses often through sudden disruptions of the balance in the ecosystem [41]. As natural habitats decline further in the Limpopo Basin of Botswana, conflicts between humans and wildlife could become rife, particularly for villages such as Mabolwe, Tsetsebjwe, Gobojango, and Motlhabaneng, which are close to protected areas. All these interactions and trade-offs negatively affect the well-being of communities, more so the poor.

The increasing severity of droughts in the Bobirwa sub-district, coupled with a lack of irrigation, presents immediate challenges to smallholder agriculture in the study area. Despite government support of smallholder farmers with free productive inputs and access to cheap credit under ISPAAD, without addressing the drought-induced moisture deficits during the growing season, cultivated crop yields are likely to remain low. Studies elsewhere in Botswana have highlighted the precarity of rain-fed crop production [2]. Thus, without addressing productivity issues that limit smallholder agriculture, trade-offs between forest ES and agricultural ES will leave local communities worse off [54]. There is an urgent need to ensure that the ISPAAD program is climate-proofed, for instance by promoting climate-smart agricultural practices and crop varieties and types that suit local agroecological conditions.

Besides provisioning ES, the loss of natural vegetation also has negative implications on several supporting and regulating ES. For instance, the important functions of regulating the micro-climate [55], water regulation [56], and water purification [57] performed by vegetation are also lost. Without vegetation to regulate the micro-climate, human and animal health will be highly exposed, e.g., to heat stress [58]. The loss of vegetated areas could be reducing infiltration, hydrological regulation, and accelerating run-off, soil erosion, and siltation of water bodies such as the Thune dam. The consequences include limited recharging of the groundwater system. Studies elsewhere have associated declining woodlands with land degradation, soil erosion, and the formation of gullies [59].

The expansion of croplands and the increased use of agrochemicals such as fertilizers, pesticides, and herbicides issued under the ISPAAD program may be contaminating surface water and groundwater sources. An increase in nitrates (from nitrogenous fertilizers) in water bodies such as the Thune Dam would cause algal blooms and increased biological oxygen demand. These threaten fish and other aquatic life if the dissolved oxygen in water bodies becomes insufficient to sustain these organisms. This would be a threat to an emerging fish industry in the study area. The contamination of water bodies and loss of aquatic organisms such as fish due to increased pollution from agricultural lands has been a huge problem elsewhere [60].

5. Conclusions and Recommendations

LULCC in the Bobirwa sub-district was characterized by declining vegetation and expansion of croplands. These LULCC regimes, which also fragment land, threaten the sustainability of many provisioning ES through impairing ecosystem functions and loss of biodiversity that underpins ES delivery. If unchecked, LULCC in the study area could irreversibly destroy biodiversity and natural habitats. Despite the support with free inputs from ISPAAD, the incessantly low crop yields among smallholder farmers highlight the ineffectiveness of the program and/or the precarity of rain-fed agriculture. The observed patterns of LULCC in the study area are expected to perpetuate, driven by the growing human population and more intense climate change impacts, and more efforts are needed to enhance crop productivity. However, the current conversion of natural vegetation to croplands needs a more balanced approach to reduce the fragmentation of ecosystems. This is required to ensure the sustainability of agricultural and natural systems and is necessary for reducing pressure on biodiversity, which underpins ecosystem service delivery.

A growing population and increasing demand for food from agriculture is a unique challenge to ecosystem management. Such actions as providing drought-tolerant, high-yielding crop varieties and fostering climate-smart agricultural practices could increase crop productivity, reduce agricultural expansions, and relieve pressure on natural ecosystems. Currently, natural vegetation delivers a wide range of provisioning ecosystem services that are critical for adaptation. Therefore, any unnecessary agricultural expansions, which severely threaten ecosystem-based livelihoods and adaptations need to be curtailed. In addition, there is a need to strengthen and capacitate traditional systems to take a leading and active role in managing local resources. Engaging communities through participatory, biodiversity-based action planning may play a significant role in promoting biodiversity conservation and the implementation of associated measures. Thus, the government and other relevant stakeholders need to enhance the ecosystem management capacities of local communities and support them in developing and implementing village action plans. However, village action plans need to mainstream biodiversity actions and be smoothly integrated with development and adaptation planning processes at the district and national levels. Moreover, such community- and ecosystem-based initiatives are emphasized in the country's National Adaptation Plan (NAP) Framework and also contribute to achieving international biodiversity targets enshrined in the National Biodiversity Strategy and Action Plan [17,46].

Supplementary Materials: The following supporting information can be downloaded at: https://www.mdpi.com/article/10.3390/land11112057/s1, Table S1: Extended list of provisioning ecosystem services in Bobirwa sub-district.

Author Contributions: Conceptualization, methodology, validation, formal analysis, investigation, writing—original draft preparation, E.M.; writing—review and editing, supervision, project administration, funding acquisition, H.M. All authors have read and agreed to the published version of the manuscript.

Funding: This research was supported through the Adaptation at Scale in Semi-Arid Regions (ASSAR) and South Africa/Flanders Climate Adaptation Research and Training Partnership (SAF-ADAPT) projects. ASSAR was funded by IDRC and the UK FCDO as part of the Collaborative Adaptation Research Initiative in Africa and Asia (CARIAA) research programme. SAF-ADAPT is funded by the Government of Flanders and is a 4.5-year collaborative project between University of Cape Town, University of Fort Hare, and University of Venda, KLIMOS Interuniversity Platform, and the South African Adaptation Network. All opinions, interpretations and conclusions expressed are entirely those of the authors and do not reflect the views of the funders, IDRC, UK FCDO and the Government of Flanders.

Institutional Review Board Statement: The study was conducted in accordance with the Declaration of Helsinki, and approved by the Institutional Review Board of University of Botswana (UBR/RES/IRB/1590 approved on 28 June 2015).

Informed Consent Statement: Informed consent was obtained from all participants involved in the study.

Data Availability Statement: Satellite data used for this study are publicly available from the USGS/NASA website (URL: http://glovis.usgs.gov/) (accessed on 16 August 2019). Data for the household survey is readily available from the authors upon request.

Acknowledgments: We appreciate the Southern Africa ASSAR team who developed the household survey questionnaire, particularly those at the University of Botswana who localized the questionnaire to the Botswana context. We are also grateful to the local communities, local leadership, and the study participants for participating in the survey.

Conflicts of Interest: The authors declare no conflict of interest. The funders had no role in the design of the study; in the collection, analyses, or interpretation of data; in the writing of the manuscript, or in the decision to publish the results.

References

1. Najmuddin, O.; Li, Z.; Khan, R.; Zhuang, W. Valuation of Land—Use/Land—Cover—Based Ecosystem Services in Afghanistan—An Assessment of the Past and Future. *Land* **2022**, *11*, 1906. [CrossRef]
2. Matlhodi, B.; Kenabatho, P.K.; Parida, B.P.; Maphanyane, J.G. Evaluating Land Use and Land Cover Change in the Gaborone Dam Catchment, Botswana, from 1984–2015 Using GIS and Remote Sensing. *Sustainability* **2019**, *11*, 5174. [CrossRef]
3. Biratu, A.A.; Bedadi, B.; Gebrehiwot, S.G.; Melesse, A.M.; Nebi, T.H.; Abera, W.; Tamene, L.; Egeru, A. Impact of Landscape Management Scenarios on Ecosystem Service Values in Central Ethiopia. *Land* **2022**, *11*, 1266. [CrossRef]
4. Amini Parsa, V.; Salehi, E. Spatio-Temporal Analysis and Simulation Pattern of Land Use/Cover Changes, Case Study: Naghadeh, Iran. *J. Urban Manag.* **2016**, *5*, 43–51. [CrossRef]
5. Othow, O.O.; Gebre, S.L.; Gemeda, D.O. Analyzing the Rate of Land Use and Land Cover Change and Determining the Causes of Forest Cover Change in Gog District, Gambella Regional State, Ethiopia. *J. Remote Sens. GIS* **2017**, *6*, 218. [CrossRef]
6. Lagrosa, J.J.; Zipperer, W.C.; Andreu, M.G. An Ecosystem Services-Centric Land Use and Land Cover Classification for a Subbasin of the Tampa Bay Watershed. *Forests* **2022**, *13*, 745. [CrossRef]
7. Chen, H.; Chen, Y.; Chen, X.; Zhang, X.; Wu, H.; Li, Z. Impacts of Historical Land Use Changes on Ecosystem Services in Guangdong Province, China. *Land* **2022**, *11*, 809. [CrossRef]
8. Kolawole, O.I.; Mbaiwa, J.E.; Mmopelwa, G. The Environmental Impacts of Tourism on Community People's Quality of Life in Maun, Botswana. *Afr. J. Hosp. Tour. Leis.* **2016**, *5*, 1–14.
9. *Secretariat of the Convention on Biological Diversity Voluntary Guidelines for the Design and Effective Implementation of Ecosystem-Based Approaches to Climate Change Adaptation and Disaster Risk Reduction and Supplementary Information;* Secretariat of the Convention on Biological Diversity: Montreal, QC, Canada, 2019.
10. UNDP Sustainable Development Goals—United Nations. Available online: http://www.un.org/sustainabledevelopment/sustainable-development-goals/ (accessed on 20 March 2017).
11. Díaz, S.; Demissew, S.; Carabias, J.; Joly, C.; Lonsdale, M.; Ash, N.; Larigauderie, A.; Adhikari, J.R.; Arico, S.; Báldi, A.; et al. IPBES The IPBES Conceptual Framework—Connecting Nature and People. *Curr. Opin. Environ. Sustain.* **2015**, *14*, 1–16. [CrossRef]
12. Liao, Z.; Su, K.; Jiang, X.; Zhou, X.; Yu, Z.; Chen, Z.; Wei, C.; Zhang, Y.; Wang, L. Ecosystem and Driving Force Evaluation of Northeast Forest Belt. *Land* **2022**, *11*, 1306. [CrossRef]

13. Motlhanka, D.M.T.; Motlhanka, P.; Selebatso, T. Edible Indigenous Wild Fruit Plants of Eastern Botswana. *Int. J. Poult. Sci.* **2008**, *7*, 457–460. [CrossRef]
14. Dube, O.P.; Sekhwela, M.B.M. *Community Coping Strategies in Semiarid Limpopo Basin Part of Botswana: Enhancing Adaptation Capacity to Climate Change*; AIACC: Washington, DC, USA, 2007; Volume 47.
15. Mugari, E.; Masundire, H.; Bolaane, M. Adapting to Climate Change in Semi-Arid Rural Areas: A Case of the Limpopo Basin Part of Botswana. *Sustainability* **2020**, *12*, 8292. [CrossRef]
16. Millenium Ecosystem Assessment. Ecosystems and Their Services. In *Ecosystems and Human Well-Being: A Framework for Assessment*; Island Press: Washington, DC, USA, 2003; pp. 49–70. ISBN 9781559634038.
17. GoB. *National Adaptation Plan Framework for Botswana*; Ministry of Environment, Natural Resources Conservation and Tourism: Gaborone, Botswana, 2020.
18. Nkemelang, T.; New, M.; Zaroug, M. Temperature and Precipitation Extremes under Current, 1.5 °C and 2.0 °C Global Warming above Pre-Industrial Levels over Botswana, and Implications for Climate Change Vulnerability. *Environ. Res. Lett.* **2018**, *13*, 065016. [CrossRef]
19. Maúre, G.; Pinto, I.; Ndebele-Murisa, M.; Muthige, M.; Lennard, C.; Nikulin, G.; Dosio, A.; Meque, A. The Southern African Climate under 1.5 °C and 2 °C of Global Warming as Simulated by CORDEX Regional Climate Models. *Environ. Res. Lett.* **2018**, *13*, 065002. [CrossRef]
20. Daron, J.D. *Regional Climate Messages for Southern Africa. Scientific Report from the CARIAA Adaptation at Scale in Semi-Arid Regions (ASSAR) Project*; ASSAR: Cape Town, South Africa, 2014.
21. Abd, H.A.A.; Alnajjar, H.A. Maximum Likelihood for Land-Use/Land-Cover Mapping and Change Detection Using Landsat Satellite Images: A Case Study "South Of Johor. " *Int. J. Comput. Electron. Res.* **2013**, *3*, 26–33.
22. Ganasri, B.P.; Dwarakish, G.S. Study of Land Use/Land Cover Dynamics through Classification Algorithms for Harangi Catchment Area, Karnataka State, INDIA. *Aquat. Procedia* **2015**, *4*, 1413–1420. [CrossRef]
23. Khan, S.; Qasim, S.; Ambreen, R.; Syed, Z.-U.-H. Spatio-Temporal Analysis of Landuse/Landcover Change of District Pishin Using Satellite Imagery and GIS. *J. Geogr. Inf. Syst.* **2016**, *08*, 361–368. [CrossRef]
24. Kamusoko, C.; Gamba, J.; Murakami, H. Monitoring Urban Spatial Growth in Harare Metropolitan Province, Zimbabwe. *Adv. Remote Sens.* **2013**, *2*, 322–331. [CrossRef]
25. Haque, M.I.; Basak, R. Land Cover Change Detection Using GIS and Remote Sensing Techniques: A Spatio-Temporal Study on Tanguar Haor, Sunamganj, Bangladesh. *Egypt. J. Remote Sens. Space Sci.* **2017**, *20*, 251–263. [CrossRef]
26. Bektas, B.F.; Karakacan, K.A. Determination of Land Cover/Land Use Using Spot 7 Data With Supervised Classification Methods. *Int. Arch. Photogramm. Remote Sens. Spat. Inf. Sci.* **2016**, *42*, 143–146. [CrossRef]
27. Botswana, S. *Central Bobonong Sub-District Population and Housing Census 2011 Selected Indicators*; Ministry of Finance and Development Planning: Gaborone, Botswana, 2015.
28. Dube, O.P.; Sekhwela, M.B.M. Indigenous Knowledge, Institutions and Practices for Coping with Variable Climate in the Limpopo Basin of Botswana. In *Climate Change and Adaptation*; Leary, N., Adejuwon, J., Barros, V., Burton, I., Kulkarni, J., Lasco, R., Eds.; Earthscan: Gaborone, Botswana, 2012; pp. 71–89, ISBN 9781849770750.
29. Mogotsi, K.; Nyangito, M.M.; Nyariki, D.M. The Role of Drought among Agro-Pastoral Communities in a Semi-Arid Environment: The Case of Botswana. *J. Arid Environ.* **2013**, *91*, 38–44. [CrossRef]
30. Mugari, E.; Masundire, H.; Bolaane, M.; New, M. Perceptions of Ecosystem Services Provision Performance in the Face of Climate Change among Communities in Bobirwa Sub-District, Botswana. *Int. J. Clim. Chang. Strateg. Manag.* **2018**, *11*, 265–288. [CrossRef]
31. Masundire, H.; Morchain, D.; Raditloaneng, N.; Hegga, S.; Ziervogel, G.; Molefe, C.; Angula, M.; Omari, K. *About ASSAR Reports Vulnerability and Risk Assessment in Botswana's Bobirwa Sub-District: Fostering People-Centred Adaptation to Climate Change*; ASSAR: Gaborone, Botswana, 2016.
32. Khatami, R.; Mountrakis, G. The Interacting Effects of Image Acquisition Date, Number of Images, Classifier, and Number of Training Samples on Accuracy of Binary Classification of Impervious Cover. *Remote Sens. Lett.* **2018**, *9*, 189–198. [CrossRef]
33. ENVI. *FLAASH Module User's Guide*; Research Systems, Inc.: Washington, DC, USA, 2005.
34. Keller, J.M.; Gray, M.R. A Fuzzy K-Nearest Neighbor Algorithm. *IEEE Trans. Syst. Man Cybern.* **1985**, *15*, 580–585. [CrossRef]
35. Coulter, L.L.; Stow, D.A.; Tsai, Y.-H.; Ibanez, N.; Shih, H.; Kerr, A.; Benza, M.; Weeks, J.R.; Mensah, F. Classification and Assessment of Land Cover and Land Use Change in Southern Ghana Using Dense Stacks of Landsat 7 ETM+ Imagery. *Remote Sens. Environ.* **2016**, *184*, 396–409. [CrossRef]
36. Rawat, V.; Puri, M. Land Use/Land Cover Change Study of District Dehradun, Uttarakhand Using Remote Sensing and GIS Technologies. *Int. J. Adv. Remote Sens. GIS* **2017**, *6*, 2223–2233. [CrossRef]
37. Mosammam, H.M.; Nia, J.T.; Khani, H.; Teymouri, A.; Kazemi, M. Monitoring Land Use Change and Measuring Urban Sprawl Based on Its Spatial Forms. The Case of Qom City. *Egypt. J. Remote Sens. Space Sci.* **2016**, *20*, 103–116. [CrossRef]
38. Mugari, E.; Masundire, H.; Bolaane, M. Effects of Droughts on Vegetation Condition and Ecosystem Service Delivery in Data-Poor Areas: A Case of Bobirwa Sub-District, Limpopo Basin and Botswana. *Sustainability* **2020**, *12*, 8185. [CrossRef]
39. Pullanikkatil, D.; Palamuleni, L.; Ruhiiga, T. Assessment of Land Use Change in Likangala River Catchment, Malawi: A Remote Sensing and DPSIR Approach. *Appl. Geogr.* **2016**, *71*, 9–23. [CrossRef]

40. Gumindoga, W.; Rientjes, T.; Shekede, M.D.; Rwasoka, D.T.; Nhapi, I.; Haile, A.T.; Gumindoga, W.; Rientjes, T.; Shekede, M.D.; Rwasoka, D.T.; et al. Hydrological Impacts of Urbanization of Two Catchments in Harare, Zimbabwe. *Remote Sens.* **2014**, *6*, 12544–12574. [CrossRef]
41. Liu, J.; Wilson, M.; Hu, G.; Liu, J.; Wu, J.; Yu, M. How Does Habitat Fragmentation Affect Biodiversity and Ecosystem Functioning Relationship? *Landsc. Ecol.* **2018**, *33*, 341–352. [CrossRef]
42. Kosonei, R.C.; Abuom, P.O.; Bosire, E.; Huho, J.M. Effects of Drought Dynamics on Vegetation Cover in Marigat Sub-County, Baringo County, Kenya. *Int. J. Sci. Publ.* **2017**, *7*, 89–98.
43. Ministry of Agriculture (MOA) Integrated Support Programme for Arable Agriculture Development (ISPAAD). Available online: http://www.gov.bw/en/Ministries--Authorities/Ministries/MinistryofAgriculture-MOA/Tools--Services/Support-Schemes-and-Initiatives/ISPAAD/ (accessed on 10 July 2018).
44. UNDP-UNEP PEI. *Support to Smallholder Arable Farmers in Botswana: Agricultural Development or Social Protection? Results and Policy Implications from a Poverty and Social Impact Analysis*; UNDP: Gaborone, Botswana, 2013.
45. Owusu, K.; Obour, P.B.; Nkansah, M.A. Downstream Effects of Dams on Livelihoods of River-Dependent Communities: The Case of Ghana's Kpong Dam. *Geogr. Tidsskr. J. Geogr.* **2017**, *117*, 1–10. [CrossRef]
46. Department of Environmental Affairs Botswana's National Biodiversity Strategy and Action Plan. Available online: https://www.cbd.int/doc/world/bw/bw-nbsap-v3-en.pdf (accessed on 11 October 2022).
47. Bennett, E.M.; Cramer, W.; Begossi, A.; Cundill, G.; Diaz, S.; Egoh, B.N.; Geijzendorffer, I.R.; Krug, C.B.; Lavorel, S. Linking Biodiversity, Ecosystem Services, and Human Well-Being: Three Challenges for Designing Research for Sustainability. *Curr. Opin. Environ. Sustain.* **2015**, *14*, 76–85. [CrossRef]
48. Polasky, S.; Nelson, E.; Pennington, D.; Johnson, K.A.; Polasky, S.; Pennington, D.; Johnson, K.A.; Nelson, E. The Impact of Land-Use Change on Ecosystem Services, Biodiversity and Returns to Landowners: A Case Study in the State of Minnesota. *Env. Resour. Econ* **2011**, *48*, 219–242. [CrossRef]
49. Abate, T.; Angassa, A. Conversion of Savanna Rangelands to Bush Dominated Landscape in Borana, Southern Ethiopia. *Ecol. Process.* **2016**, *5*, 1–18. [CrossRef]
50. Laurance, W.F.; Sayer, J.; Cassman, K.G. Agricultural Expansion and Its Impacts on Tropical Nature. *Trends Ecol. Evol.* **2014**, *29*, 107–116. [CrossRef] [PubMed]
51. Perrings, C. Biodiversity, Ecosystem Services and Climate Change. *Econ. Probl. Environ. Dep. Pap.* **2010**, *120*, 39.
52. Sandifer, P.A.; Sutton-Grier, A.E.; Ward, B.P. Exploring Connections among Nature, Biodiversity, Ecosystem Services, and Human Health and Well-Being: Opportunities to Enhance Health and Biodiversity Conservation. *Ecosyst. Serv.* **2015**, *12*, 1–15. [CrossRef]
53. Fenetahun, Y.; Yong-dong, W.; You, Y.; Xinwen, X. Dynamics of Forage and Land Cover Changes in Teltele District of Borana Rangelands, Southern Ethiopia: Using Geospatial and Field Survey Data. *BMC Ecol.* **2020**, *20*, 55. [CrossRef]
54. Akinyemi, F.O.; Mashame, G. Analysis of Land Change in the Dryland Agricultural Landscapes of Eastern Botswana. *Land Use Policy* **2018**, *76*, 798–811. [CrossRef]
55. Garcia-Nieto, A.P.; Quintas-Soriano, C.; Garcia-Llorente, M.; Palomo, I.; Montes, C.; Martin-Lopez, B. Collaborative Mapping of Ecosystem Services: The Role of Stakeholders' Profiles. *Ecosyst. Serv.* **2015**, *13*, 141–152. [CrossRef]
56. Quintas-Soriano, C.; Castro, A.J.; Castro, H.; García-Llorente, M. Impacts of Land Use Change on Ecosystem Services and Implications for Human Well-Being in Spanish Drylands. *Land Use Policy* **2016**, *54*, 534–548. [CrossRef]
57. Habib, T.J.; Heckbert, S.; Wilson, J.J.; Vandenbroeck, A.J.K.; Cranston, J.; Farr, D.R. Impacts of Land-Use Management on Ecosystem Services and Biodiversity: An Agent-Based Modelling Approach. *PeerJ* **2016**, *4*, e2814. [CrossRef]
58. Moses, O. Heat Wave Characteristics in the Context of Climate Change over the Past 50 Years in Botswana | Botswana Notes and Records. *Botsw. Notes Rec.* **2017**, *49*, 13–25.
59. Wang, Z.; Wang, Z.; Zhang, B.; Lu, C.; Ren, C. Impact of Land Use/Land Cover Changes on Ecosystem Services in the Nenjiang River Basin, Northeast China. *Ecol. Process.* **2015**, *4*, 11. [CrossRef]
60. Sunitha, V.; Reddy, B.M.; Reddy, M.R. Groundwater Contamination from Agro-Chemicals in Irrigated Environment: Field Trials. *Adv. Appl. Sci. Res.* **2012**, *3*, 3382–3386.

MDPI AG
Grosspeteranlage 5
4052 Basel
Switzerland
Tel.: +41 61 683 77 34

Land Editorial Office
E-mail: land@mdpi.com
www.mdpi.com/journal/land

www.ingramcontent.com/pod-product-compliance
Lightning Source LLC
Chambersburg PA
CBHW061448210926
51850CB00008B/1646
9783725864041